Rapid Analysis Techniques in Food Microbiology

Rapid Analysis Techniques in Food Microbiology

Edited by

P.D. PATEL
Leatherhead Food Research Association
Surrey

BLACKIE ACADEMIC & PROFESSIONAL
An Imprint of Chapman & Hall

London · Glasgow · Weinheim · New York · Tokyo · Melbourne · Madras

Published by
Blackie Academic and Professional, an imprint of Chapman & Hall,
Wester Cleddens Road, Bishopbriggs, Glasgow G64 2NZ

Chapman & Hall, 2-6 Boundary Row, London SE1 8HN, UK

Blackie Academic & Professional, Wester Cleddens Road, Bishopbriggs, Glasgow G64 2NZ, UK

Chapman & Hall GmbH, Pappelallee 3, 69469 Weinheim, Germany

Chapman & Hall Inc., One Penn Plaza, 41st Floor, New York, NY 10119, USA

Chapman & Hall Japan, Thomson Publishing Japan, Hirakawacho Nemoto Building, 6F, 1-7-11 Hirakawa-cho, Chiyoda-ku, Tokyo 102, Japan

DA Book (Aust.) Pty Ltd, 648 Whitehorse Road, Mitcham 3132, Victoria, Australia

Chapman & Hall India, R. Seshadri, 32 Second Main Road, CIT East, Madras 600 035, India

First edition 1994

© 1994 Chapman & Hall

Typeset in 10/12pt Times by Greenshires Icon, Exeter, Devon
Printed in Great Britain by St Edmunsbury Press, Bury St Edmunds, Suffolk

ISBN 0 7514 0030 0

A catalogue record for this book is available from the British Library
Library of Congress Catalog Card Number: 94-70712

∞ Printed on acid-free text paper, manufactured in accordance with ANSI/NISO Z39.48-1992 (Permanence of Paper)

Preface

The food industry, with its diverse range of products (e.g. short shelf-life foods, modified atmosphere packaged products and minimally processed products) is governed by strict food legislation, and microbiological safety has become a key issue. Legally required to demonstrate 'due diligence', food manufacturers are demanding analytical techniques that are simple to use, cost effective, robust, reliable and can provide results in 'real time'.

The majority of current microbiological techniques (classical or rapid), particularly for the analysis of foodborne pathogens, give results that are only of retrospective value and do not allow proactive or reactive measures to be implemented during modern food production. Rapid methods for microbial analysis need to be considered in the context of modern Quality Assurance (QA) systems.

This book addresses microbiologists, biochemists and immunologists in the food industry, the public health sector, academic and research institutes, and manufacturers of kits and instruments. This volume is an up-to-date account of recent developments in rapid food microbiological analysis, current approaches and problems, rapid methods in relation to QA systems, and future perspectives in an intensely active field.

P.D.P.

Contributors

F.J. Bolton   Public Health Laboratory, Royal Preston Hospital, PO Box 202, Sharoe Green Lane North, Preston PR2 4HG, UK.

D. M. Gibson   Ministry of Agriculture, Fisheries and Food, Torry Research Station, 135 Abbey Road, Aberdeen AB9 8DG, Scotland.

P.A. Hall   Microbiology and Food Safety, Kraft General Foods, 801 Waukegan Road, Glenview, Illinois 60025, USA.

W.E. Hill   Seafood Products Research Center, Seattle District Office, Food and Drug Administration, Bothell, Washington 98041-3012, USA.

A.D. Hocking   CSIRO Food Research Laboratory Division of Food Science and Technology, PO Box 52, North Ryde, New South Wales 2113, Australia.

A.L. Kyriakides   J. Sainsbury plc, Scientific Services Division, Stamford House, Stamford Street, London SE1 9LL, UK.

C.M.L. Marengo   Joint Research Centre of the European Community, Ispra, Italy.

D.A.A. Mossel   Eijkman Foundation, Utrecht University, PO Box 6024, 3503 PA Utrecht, The Netherlands.

S.H. Myint   Department of Microbiology, University of Leicester, Leicester LE1 9HN, UK.

Ø. Olsvik   Norwegian College of Veterinary Medicine, Oslo, Norway.

P.D. Patel   Rapid Methods Section, Leatherhead Food Research Association, Randalls Road, Leatherhead, Surrey KT22 7RY, UK.

J.I. Pitt   CSIRO Food Research Laboratory, Division of Food Science and Technology, PO Box 52, North Ryde, New South Wales 2113, Australia.

A.N. Sharpe   Bureau of Microbial Hazards, Food Directorate, Health Protection Branch, Health and Welfare Canada, Ottawa, Ontario K1A 0LZ, Canada.

C.B. Struijk   Eijkman Foundation, Utrecht University, PO Box 6024, 3503 PA Utrecht, The Netherlands.

D. W. Williams   Rapid Methods Section, Leatherhead Food Research Association, Randalls Road, Leatherhead, Surrey KT22 7RY, UK.

69—

Contents

1 History of and prospects for rapid and instrumental methodology for the microbiological examination of foods

D.A.A. MOSSEL, C.M.L. MARENGO and C.B. STRUIJK

1.1 Introduction

The modest start, made in the 1920s, with microbiological safety assurance – initially of milk and dairy products, later, to a lesser extent, of other foods – was modelled after the well-established matrix of ensuring the chemical integrity of food products. This originated from a branch of science termed 'bromatology'. It attempted to attain two main purposes: (i) to avoid the ingestion of foods contaminated with toxic elements including arsenic, mercury and lead; and (ii) to control the nutritive value of staple foods by detecting elevated water content or increasing the weight by adulteration, i.e. the addition of non-nutritive materials. The public was protected against such frauds by monitoring the food supply at points of sale. If a contaminant or untoward stretching was observed, the food was eliminated from the trade. This strategy was quite successful in assuring the chemical safety and quality of foods for two reasons: (i) the unwanted constituents were rather homogeneously distributed in the food, so that any sample of sufficient size drawn from a consignment for analysis represented the lot; and (ii) the concentration of the analytes sought was fairly constant in time, further contributing to the reliability of data obtained on samples (Mossel *et al.*, 1994).

It could have been anticipated, right from the beginning, that this scenario could not even be *expected* to be applicable to ensuring microbiological food safety. First and foremost none of the circumstances identified above as contributory factors to the efficacy of the retrospective approach apply in microbiology. In the vast majority of foods, microorganisms are erratically distributed, depriving negative results of tests of *any* significance (Mossel and Drion, 1954; Habraken *et al.*, 1986). Moreover, except for endospores of bacteria and ascospores of moulds and yeasts, microbial populations of foods bear an outspoken dynamic character; as a rule they either decrease or increase in numbers of viable cells during storage and distribution (Mossel and Struijk, 1992), making the prediction of the microbiological condition of foods at the moment of ingestion quite insecure. This awkward situation was compounded by a tremendous shortage of readily available reliable analytical techniques (Mossel, 1987). Whereas bromatological examination of foods dated back to early in the 19th century, selective-diagnostic methods required in the microbiological monitoring of foods had to be borrowed from clinical microbiology until about 1960 – and still partly have to. Finally, while in bromatology, as well as clinical microbiology, one discipline, i.e chemistry and medicine respectively, was responsible for scientific progress and strategic decisions, food microbiology was practised by six different professional groups. These include food science, veterinary medicine, pharmacy, biology, agricultural sciences and, to a lesser extent, medicine, which markedly hampered progress and above all the elaboration of effective management policies (Mossel, 1991a).

It was therefore not at all surprising that the transmission of foodborne diseases with a microbial aetiology was far from being brought under control (Mossel, 1989; Bean and Griffin, 1990; Skjerve and Johnson, 1991; Bautista *et al.*, 1992; Du Pont, 1992). On the contrary, intoxinations provoked by *Staphylococcus aureus, Bacillus cereus* and a few allied bacilli and a scala of pressor-amine-producing bacteria (Mossel *et al.*, 1994), but particularly the incidence of the most prevalent food-transmitted infectious enteric disease – salmonellosis – increased rather than decreased (Hedberg *et al.*, 1991; Tauxe, 1991; Luby *et al.*, 1993). Meanwhile, the aggressive serotype *Salmonella enteritidis* came to the fore (Hedberg *et al.*, 1991; Barnes and Edwards, 1992; Van der Giessen *et al.*, 1992; Vugia *et al.*, 1993). It was joined by a multitude of enteropathogenic agents whose aetiological role was identified, or that re-emerged, since the early 1960s. A selection of the most striking examples is collected in Table 1.1. This dim picture was aggravated by the identification of a broad spectrum of systemic complications, often very serious, occurring as a sequel to

Table 1.1 A few enteropathogenic agents transmitted by foods, identified after about 1960, when salmonellae, *Staphylococcus aureus*, *Clostridium perfringens* and *Clostridium botulinum* had been well established as foodborne pathogens

Pathogen	Main source of transmission	Reference
Adenovirus 40, 41	Faecal contamination	Jarecki-Khan *et al.* (1993)
Aeromonas hydrophila	Waterborne contamination	Thomas *et al.* (1990)
Astrovirus	Faecal contamination	Lew *et al.* (1991)
Campylobacter spp.	Chicken and pork	Nachamkin *et al.* (1992)
Citrobacter spp.	Faecal contamination	Schmidt *et al.* (1992)
Cryptosporidium parvum	Calf, lamb, poultry, pig; waterborne contamination	Gatti *et al.* (1993)
Cyclospora cayetanensis	Not yet clearly established	Bean and Griffin (1990); Long *et al.* (1991)
E. coli, enterohaemorrhagic pathotype(s)	Beef	Le Saux *et al.* (1993)
Hafnia alvei	Not yet clearly established	Westblom and Milligan (1992); Albert *et al.* (1992a); Reina *et al.* (1993)
Listeria monocytogenes, serotype 4b	Ubiquitous in the farm environment	Amgar (1991); Goulet *et al.* (1993)
Norwalk virus group	Faecal contamination	Kapikian (1993)
Providencia spp.	Faecal contamination	Albert *et al.* (1992b)
Shigella spp.	Faecal contamination	Hedberg *et al.* (1992b)
Toxocara canis	Animal environment	Salem and Schantz (1992)
Vibrio vulnificus	Waterborne contamination	Wachsmuth *et al.* (1993)

Table 1.2 Recommendations for microbiological safety assurance of foods, relying on intervention, made since about 1930

Country and era	Reference	Pathogen to be brought under control	Food vehicle
USA 1920–1935	Meyer (1931)	*Clostridium botulinum*	Canned vegetables and cured meats
UK 1930–1935	Prescott (1920)	General	Dried foods
	Wilson (1955, 1964)	Enterobacteriaceae, *Mycobacterium bovis*, group A streptococci, *Corynebacterium diphtheriae* (*Listeria monocytogenes*)	Milk, ice cream, cheese, egg products
France 1953–1956	Buttiaux *et al.* (1956)	*Clostridium botulinum*, *Staphylococcus aureus*, Enterobacteriaceae, *Clostridium perfringens*	Canned, large size hams, infant food
	Cheftel (1955)		
USA 1960	Dack (1956)	*Salmonella* spp. and allied enteropathogens	General
Netherlands 1975–1980	Kampelmacher (1983)	*Salmonella* spp.	Fresh and frozen poultry
USA	Roberts (1985)	*Campylobacter* spp.	
Netherlands 1978–1985	Van Netten *et al.* (1984)	*Salmonella*, *Campylobacter*, *Yersinia* and enterovirulent *Escherichia coli*	Fresh meat and poultry
	Smulders *et al.* (1986)		

a primary spell of gastroenteritis, in itself of a relatively mild nature (Mossel, 1989; Mossel *et al.*, 1994).

There was, nonetheless, no shortage, in academic circles, of recognition of the futility of simply mimicking, in attempts to assure microbiological safety, what had ensured chemically sound food. As shown in Table 1.2, since about 1920 professorial ranks in the USA, the UK and France alike have emphasised that the retrospective approach had to be replaced by a prospective one (Mossel, 1989). Their messages were not heeded, however, until the 1970s. At that time Dr H. Bauman, chief microbiologist for a leading American food manufacturing company, suggested a complete change in course with respect to ensuring microbiological food safety (Bauman, 1974). Instead of relying on *post mortem* inspections of doubtful significance of samples of uncontrolled history, he advocated the introduction of a forward control strategy. Microbiological hazards had to be identified and faulty practices and procedures to be rectified before any monitoring would make sense. Bauman introduced the term 'hazard analysis and control of critical points', abbreviated to HACCP (Bauman, 1990). This strategy became extremely popular and is now, some 20 years after its introduction, generally accepted by professional circles (Amgar, 1992; Bryan, 1992; Mossel *et al.*, 1992; Pierson and Corlett, 1992; Shakespeare *et al.*, 1992; Macler and Regli, 1993). In Europe, Lord Hugo Plumb of Coleshill, a leading politician with an agricultural background, strongly recommended to *extend* HACCP from raw material to, and including, serving – 'from farm to fork' (Mossel, 1991b; Mossel and Struijk, 1992; Altekruse *et al.*, 1993). The term 'longitudinally integrated safety assurance', or LISA, had earlier been suggested for this most reasonable and effective strategy (Mossel, 1983; Jakobsen and Lillie, 1992).

1.2 The contemporary role and the character of microbiological examination of food samples

1.2.1 Principles

Substitution of the forward control approach for the ineffective retrospective scenario also completely changed the role of microbiological monitoring of foods. Had this previously and unsuccessfully been used to *attain* a safe food supply, it would henceforth serve to *assess* whether good manufacturing and distribution practices had been strictly followed. It would *inter alia* be utterly unwise to refrain from such validation steps within the HACCP framework. First and foremost, unfortunately, food manufacturers and caterers far too frequently fail to allow well established practice guidelines to guide practice *at all*. In many instances the LISA-maxim has indeed been adopted, but incidental breakdown of effective control may nonetheless sometimes occur, due to instrumental or human failure. Such hiatuses as a rule bring about only minor adverse effects, but may sometimes entail dramatic consequences and most expensive recalls of distributed merchandise. The earlier reliable data, confirming or refuting adherence to safe practices, are obtained, the more rapidly rectification can be applied and, consequently, the more consistently will the public be protected against products that have lost their microbiological integrity.

This calls for the introduction of a few essential elements into microbiological inspection of food samples. First of all, examination of line specimens including the food production environment (Slade, 1992) has an absolute priority over analysing finished products. Moreover, data should become available as fast as possible, because it allows earliest corrective action to be taken against hiatuses. It is therefore not at all surprising that food microbiology, ever since the 1970s, has been challenged to achieve the same speed, reliability and facility that chemical examination of foods has displayed since the introduction of the first 'auto-analyser'.

In addition, it is worth noting that acceptance of, and adherence to, the HACCP/LISA strategy will ensure that the majority of the samples reaching the laboratory are of good microbiological quality. Consequently, a very minor fraction will be found contaminated or colonised at a high level: in popular laboratory jargon most specimens will give 'negative results'. Laboratory procedures have to be geared to this situation, which is essentially different from that prevailing in the pre-proactive scenario era, when many trade samples would, unfortunately, contain high levels of organisms of concern.

1.2.2 The part to be played by 'rapid' methods – semantics

This new situation entails two effects of a most important nature. On the one hand, it markedly facilitates routine monitoring. However, it calls for a substantially increased sensitivity of methodology, which, not infrequently, conflicts with the desired rapidity, as is elucidated in detail below. In view of the main subject of this presentation it seems therefore most desirable to define precisely what the popular, customary term 'rapid' methods really wishes to convey.

In fact, the analyst seeks at least five attributes in methods aiming at validating microbiological integrity of end-product samples or compliance with hygiene standards in line specimens. These include: (i) facility, (ii) rapidity, (iii) consistency, (iv) intrinsic guarantees for avoidance of errors, e.g. through the exclusive use of reagents or ingredients certified by the supplier, and (v) mechanisation, if not automation. The often-used term *instrumental* methods covers these requirements fairly well in that laborious and subjective elements of analytical methods have been eliminated; the designation does not, however, explicitly include rapidity. In this chapter, the term 'rapid' will be used to describe methods which have most of the advantages listed, though not necessarily all; and even not consistently extreme rapidity, i.e. having data available within an hour or so, if not instantaneously.

1.3 Pitfalls in introducing 'rapid' methods

Unfortunately cardinal differences between the mechanisms of loss of chemical integrity and microbial deterioration of foods interfere in the pursuit of elaboration of rapid methods. First, as emphasised previously, the pertinent levels of detection in foods processed for safety are often extremely low, e.g. 1 cfu kg^{-1}; but worse, these low concentrations have sometimes to be isolated amongst innocuous populations exceeding the target organism by a factor of up to 10^6. The combination of the required sensitivity and the necessary selectivity is of an order of magnitude of 10^{-9}, calling for extremely selective procedures. These include the following steps: (i) concentration of primary food macerates by centrifugation (Mossel and Visser, 1960; Hawa et al., 1984; Van Netten et al., 1987; Fleet et al., 1991; Mossel et al., 1991) or filtration (section 1.5.1); or else by the advanced technique of immunoabsorption onto magnetic beads (Skjerve et al., 1990; Cudjoe et al., 1991; Lund et al., 1991; Vermunt et al., 1992; Mansfield and Forsythe, 1993); and (ii) highly selective enrichment and isolation procedures which are not yet as perfect as one would wish or suppose.

Problems surrounding the latter methods are compounded by the observation, made, for the first time, by Eijkman (1908) that the majority of microorganisms of significance in foods have incurred sublethal lesions as a result of having been exposed to adverse external conditions. These are either directly injurious, like heating, or indirectly so, e.g. lowered food pH or a$_w$, and sometimes even both (Mossel and Van Netten, 1984; Ray, 1989; Turpin et al., 1993). If highly selective procedures, including the use of particular antimicrobial agents or increased incubation temperature, are applied to such debilitated populations, the combined stress will result in cell death, causing erroneously low results (Sallam and Donnelly, 1992; Morinigo et al., 1993). This would lead to failure to take corrective measures where these were required. Consequently, meticulously elaborated *resuscitation* procedures (Figures 1.1–1.3) are required to restore the viability and unlimited culturability (Roszak et al., 1984; Jones et al., 1991; Nilsson et al., 1991; Saha et al., 1991) of debilitated populations, ensuring their inclusion in colony counts or most probable number (MPN) determinations.

A third factor accounting in part for the slow progress made in introducing more substantial modernisation in analytical food microbiology and particularly with respect to the use of molecular microbiological methodology is related to the nature of foods themselves. Methods that work remarkably well with pure cultures of target organisms, like the polymerase chain reaction (PCR)-approach (Section 1.5.2) failed initially when applied to 'real world' specimens, e.g. chicken carcasses. This results from the presence in many foods of contaminating inhibitory material (De Leon et al., 1992; Abbaszadigan, et al., 1993; Payne et al., 1993; Bej et al., 1994). Such interferences were overcome by previous concentration and purification of target organisms, obviously at the expense of simplicity and rapidity. A remaining difficulty arises from the failure

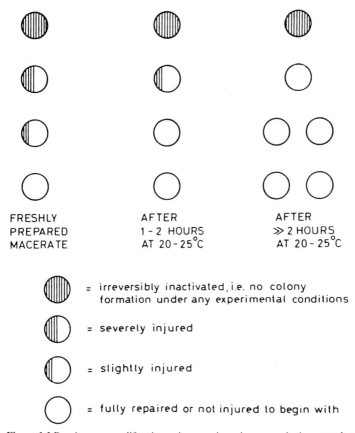

Figure 1.1 Repair versus proliferation as it occurs in various resuscitation procedures.

of PCR techniques to allow determination of viability of bacteria whose presence they visualize (Bej *et al.*, 1994).

A fourth hurdle is raised by rather successful novel rapid methods measuring parameters distinct from the classically accepted ones. This compounds the already, in general, most difficult problem of interpreting the results of micro-biological examination of foods and particularly gauging analytical data against reference ranges, the much disputed microbiological specifications for foods (Mossel and Van Netten, 1991).

In essence, methods yielding non-conventional data may be very useful and should not, therefore, be rejected lightheartedly. They may provide most serviceable information of the *semaphora* ('traffic light') type. This indicates that a specimen belongs to one of the following three broad categories: pass ('green'), doubtful ('orange') or reject ('red').

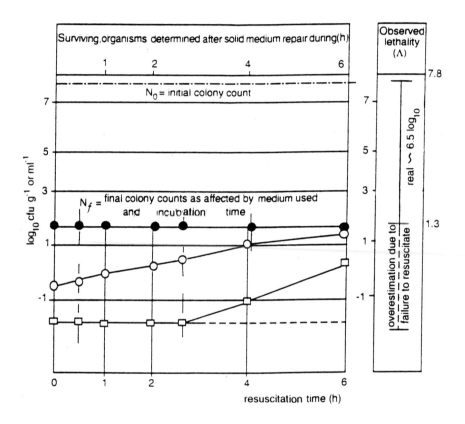

Figure 1.2   Destruction–repair (DR) curves. ●, colony counts obtained on non-selective, optimal recovery medium, with or without resuscitation; ○, *ditto*, obtained with optimal selective medium, affected by resuscitation time; □, *ditto*, obtained with suboptimal selective medium.
In some cases, no better selective enumeration medium than the suboptimal one (□) is yet available. A procedure to assess true lethality (Λ) has then yet to be elaborated.

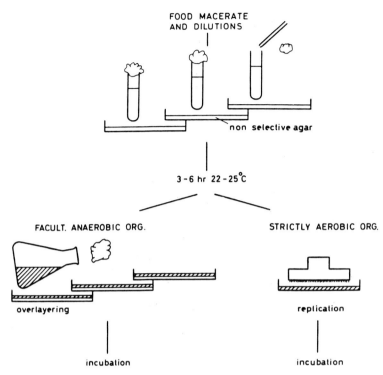

Figure 1.3 Repair of stressed organisms on a solid recovery medium, followed by overlaying or replication onto a selective-diagnostic enumeration medium.

1.4 An overall appraisal of analytical principles suggested

Even before the present scramble for resources occurred, for example in times such as during the listeriosis scare, immense workloads compelled the search for methods which would relieve the pressure on laboratory facilities – quite distinct from the tendency to sophistication indicated above. This often lead to severe disillusionment – in fact, so frequently that a special term was coined: the excitement → disappointment syndrome (Persing, 1991; Wolcott, 1991; Yolken, 1991).

To allow a critical evaluation of available methodology in general and also to enable appreciation of progress made as well as identification of needs for future development, a distinction will be made between two essentially different analytical purposes. These are (a) *identification* of well-defined food isolates; and (b) *enumeration* of organisms belonging to groups of significance from the point of view of consumer protection or microbiological quality, including sensory properties and time to spoilage under specified conditions. An inventory of this sort is presented in Table 1.3.

Table 1.3 1994 Review of realised and unrealised potentials of simplification, mechanisation, automation and acceleration in the microbiological monitoring of foods

---

(1) *Identification methodology*
  (i) Simplification of in essence conventional approaches
    — Galleries of microsize tubes
      • Customary reading times
      • Accelerated tests: reading time approx 2 h
    — Microtitre plates ('trays') with prepared diagnostic media
      • Classical I: 'fermentation' (dissimilation) tests
      • Classical II: assimilation tests
      • Enzyme profile assays
    — Immunological principles: ELISA, latex agglutination, etc.
  (ii) Novel techniques, ready for application
    — Turbidometry
    — Radiometric methods
    — Flow cytometry
    — DNA/DNA hybridisation
      • First generation radioisotope and colorimetric labels
      • PCR approach
      • Restriction fragment length polymorphism probes
    — Pyrolysis derivatisation GC–MS
  (iii) Innovative advances not yet endowed to affect working practices
    — Fourier transform infrared spectroscopy
    — Ultraviolet resonance Raman spectroscopy
    — Polarisation light scattering

(2) *Enumeration of cfu or derived attributes*
  (i) Simplifications of essentially customary measuring principles
    — Agar droplet method
    — Spiral plating methodology
    — Loop plating instruments
    — HGMF technique
    — Automatic microcolony (Frost's 'little plate') counting
  (ii) Novel principles, ready for application
    — Conductimetric and impedimetric methods
    — Turbidometry
    — Fluorometric immunological visualised flow cytometry

(3) *Presence-or-absence (semi-quantitative) methodology*
  (i) Luminometric ATP assays
  (ii) Biosensor principle
  (iii) Semi-solid motility related detection methods
  (iv) Sub-enrichment, immunomagnetic concentration, PCR method

---

Source: Mossel *et al.* (1994)

When assessing the suitability of any technique offered or considered for application, a rationale, taking into account all aspects of the procedure, has to be followed. A key which may be useful in a first-step appreciation is suggested in Table 1.4. Obviously, the classical analytical principles, summarised in Table 1.5, will have to be considered with respect to section 1 of Table 1.4. It is of utmost importance, in addition, to divorce facts from fables with respect to the attribute 'rapidity'. In Table 1.6 an attempt has been made to put this important

Table 1.4 Attributes to be considered in the primary selection, and with respect to the ultimate adoption of instrumental methods, aiming at markedly enhanced simplification and acceleration, linked to mechanisation or even automation in analytical food microbiology[a]

(1) *Strictly microbiological criteria*
   Sensitivity
   Productivity
   Selectivity
   Electivity
   Correlation of data with those obtained in reference method
(2) *Logistic aspects*
   Sample preparation
   Operation
   Attributes of reagents: preparation, stability, availability, consistency
   Required training and whether possible on site or mandatorily elsewhere
   Speed
   Capacity: number of samples/day
   Space requirements
   Reputation of company marketing hardware/software
   Technical service: availability, speed, quality
(3) *Economic considerations*
   Acquisition
   Reagents
   Maintenance
(4) *Strategic facets — acceptability of results by*
   Suppliers
   Buyers
   Regulatory agencies: local, national, regional, international

[a] Data taken from Perales and Mossel (1989).

element into proper perspective. It is quite obvious from this review that when the term rapid methods is used, in fact rapidity, facility, mechanisation and automation are often to some extent confused, as noted earlier.

Once again, with respect to the crucial parameter rapidity, it should be emphasised that there is a limit to what is achievable within the microbiological framework. Three traits of microorganisms impede speeding-up of methods to the extent that they become genuine 'real-time' techniques. First and foremost, as indicated before, virtually all microorganisms occurring in foods originating from production/distribution chains meticulously adhering to LISA practices are *ipso facto* sublethally stressed, requiring resuscitation procedures lasting between 0.5 and > 8 h (Figure 1.2). In addition, the very low colony levels to be expected in such foods, correctly processed for safety and not recontaminated or recolonised subsequently (Figure 1.4), forcibly call for a short period of enrichment, if necessary completed by concentration, to attain detectable ('threshold') levels of the order 10^5–10^6 cfu ml^{-1}. Finally, when using more or less rapid methods, not intrinsically relying on isolation of target organisms, in many instances isolation will nonetheless have to be attempted, because, for example, in legal or commercial disputes the organism itself rather than a signal indicating its presence is demanded as proof.

Table 1.5 Elements to be taken into account in the development and application of analytical methods in food microbiology[a]

---

(1) *Sampling plan*
  Randomisation
  Numbers to be drawn per predefined lot
(2) *Handling before examination*
  Transportation — time/temperature integral
  Challenge
    Incubation
    Inoculation in case of survival studies or assessment of resistance against colonisation
(3) *Preparation for examination*
  Defrosting — time/temperature integral; squeezing to release fluid
  Cleaning/disinfection of containers
(4) *Drawing of sub-samples ('aliquots')*
  Randomisation
  Size of sub-sample
  Diminution and homogenisation
(5) *Preparation of macerate and dilutions*
  Size of second sub-sample, i.e. aliquot to be examined
  Preparation of macerate
    Composition of maceration fluid
    Procedure, including time/temperature regimen
    Mode of dispersion
  Preparation of serial dilutions
    Composition of diluent
    Procedure, including time/temperature regimen
(6) *Monitoring of culture media*
  Choice of test strains
  Selection of inoculation procedure
  Incubation, time, and temperature
  Reading
  Reference values to be used
(7) *Resuscitation procedure*
  Composition of resuscitation medium
  Time/temperature programme, including tolerance
  Mode of processing of resuscitated system
(8) *Enumeration procedure*
  Composition of medium
  Preparation of medium
  Decontamination of medium
  Tempering of medium
  Holding or drying time/temperature
  Aseptic precautions during inoculation
  Inoculation
    Procedure, qualitative
    Procedure, quantitative
  Incubation
    Procedure, qualitative
    Circulation, ventilation, and temperature tolerances
    Duration, including tolerances
  Reading
    Definition of target colonies
    Accuracy of colony counting
    'Emergency' handling of plates showing too few or too many colonies
    Reference ranges
  Confirmation/identification
    Extent of picking per given type of colony
    Subculture for examination
    Preliminary taxonomic grouping
    Presumptive grouping identification
    Complementary testing — intramural/extramural
    Expression of results

Essential annexes
  Justification and documentation of techniques
  Laboratory precautions to be observed ('GLP')
  Reference centres available for consultation
  Recommended reporting form(s); storage and retrieval of data

---

[a]Data taken from Mossel (1991a).

Table 1.6 Benefits of 'rapid' methods categorised according to contributions made to the ultimate goal

---

Simplification = labour saving
  — Industrially prepared, conventional galleries, saving 'drudgery'
  — Hydrophobic grid membrane filtration, allowing to save the effort of preparing dilutions
  — Use of quadrant plates, also allowing to avoid errors during crowding of the laboratory upon receipt of many samples at the same time

Acceleration = genuine rapidity
  — Adoption of industrially manufactured galleries relying on biochemical traits of target organisms, requiring no more than 2–4 h of incubation, instead of 18–48 h
  — Centrifugation of food macerates, markedly reducing the time to attain reliable estimates of colonisation
  — Microscopical enumeration reducing reading times to about 0.5 h (viable + dead cells classically) and selective counts by modern cytochemical methods to 4 h when relying on Frost's 'little plate' or 'microcolony' approach
  — ATP estimation by bioluminometry: almost instantaneously, but, so far, lacking selectivity

Mechanisation and automation = facilitating the work load and personnel requirements
  — Camera-operated colony counting
  — Conductimetry/impedometry
  — Turbidometry
  — Automated 'reductase' techniques
  — Glucose/nitrate dissimilation methodology
  — Automated DEFT techniques

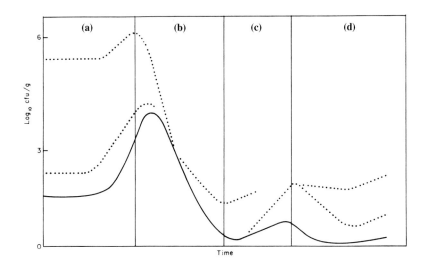

Figure 1.4 Critical points and stages. The extent to which various practices affect contamination and colonisation of perishable foods. (a) Raw material. (b) Processing for safety. (c) Distribution. (d) Culinary preparation.————, Longitudinally integrated safety assurance through meticulous adherence to good manufacturing and distribution practices; ......, more or less deficient practices followed.

1.5 Suitability of the most promising, available principles

1.5.1 Enumeration

One of the most versatile, simple and selective more-or-less conventional approaches is constituted by the hydrophobic grid membrane filtration (HGMF) principle (Entis, 1990; Szabo *et al.*, 1990; Peterkin *et al.*, 1991, 1992). It has the following advantages: (i) relying on sound classical quantitative microbiological foundations; (ii) using thoroughly tested, well-established selective media; (iii) allowing tentative taxonomic, and recently also molecular, grouping of primary, well-isolated colonies; and (iv) enabling pertinent organisms to be obtained in pure culture without much effort.

It hence satisfies all requirements of quantitative-differential examinations. Clearly, HGMF has the disadvantage common to all colony-count techniques – intrinsic slowness. However, this can be overcome by linking HGM filtration to Frost's principle (Frost, 1921) of microscopic identification of microcolonies within 4–6 h incubation (Nickerson, 1943; Winter *et al.*, 1971; Rodrigues and Kroll, 1989). Automatic pattern recognition by video-image analysis is within reach in this area (Brodsky *et al.*, 1982; Pettipher *et al.*, 1992).

A fascinating novel principle for the quantitative-differential microbiological examination of biological specimens can be found in flow cytometry. This relies on the forcing of cell suspensions out of a small orifice and their subsequent illumination, e.g. by a laser source (Hutter and Eipel, 1978; Donnelly and Baigent, 1986; Donnelly *et al.*, 1988; Nir *et al.*, 1990a, b; Goldschmidt, 1991; Patchett *et al.*, 1991). The intensity of the light that is scattered at different angles provides information about the cell that can be translated by suitable software into data allowing the microbiologist to count and identify microbial populations to a certain level. A second generation of flow cytometry uses various fluorochromes to label surface antigens and DNA in an attempt to increase the selectivity of the procedure (Bauman and Bentvelzen, 1988; Edwards *et al.*, 1992). Flow cytometry allows the enumeration of at least 10^6 cells min^{-1} and, hence, constitutes a most attractive method, once problems with cross-reactions are solved. User-friendly automation is consequently eagerly anticipated by the food microbiology profession.

Sometimes conventional quantitative differential results are not the prime objective of the analysis. Examples are (i) studies aiming at predicting shelf-life; (ii) estimation of levels of various toxins; (iii) determination of metabolites with marker value, e.g. compounds released by bacteria or moulds before their viable cells were eliminated by heat processing (Mossel *et al.*, 1994). In such instances direct tests for such factors rather than viable cell counts are in order.

Customary colony counting for the estimation of 'total' colonisation or of specific groups of organisms has also been replaced by less laborious procedures. These include spiral plating of food macerates (Gilchrist *et al.*,

1973; Manninen, 1991), automatic colony counting (Sharpe *et al.*, 1986), conductimetric or impedimetric (Stewart, 1899; Cady *et al.*, 1978; Martins and Selby, 1980; Sorrells, 1981; Firstenberg-Eden, 1986; Mirhabibollahi *et al.*, 1991; Mossel, 1991b; Parmar *et al.*, 1992) methods, turbidometry (Mattila, 1987; Mattila and Alivehmas, 1987; Wirtanen *et al.*, 1991), use of radioactive-labelled substrates (Mafart *et al.*, 1981), reductometric assessments (Alexander and Rothwell, 1950; Bartlett and Tetro, 1992; Casemore *et al.*, 1992), ELISA-techniques (Cousin *et al.*, 1990; Notermans and Kamphuis, 1990; Kamphuis *et al.*, 1992) and the estimation of the production of catalase (Dodds *et al.*, 1983; Wang and Fung, 1986; Kroll *et al.*, 1989; Bendall *et al.*, 1993) or cellular ATP (Kennedy and Oblinger, 1985; Tebbutt and Midwood, 1990; Boismenu *et al.*, 1991) – though the latter procedure requires thorough precautions to avoid inclusion of masquerading eukaryotic ATP in determinations; cf. section 1.6.2. Another elegant example is constituted by the selective glucose dissimi-lation/nitrate reduction test (SGDNR)-principle, providing reasonably reliable data on colonisation levels of foods or catered meals with Enterobacteriaceae, *Staphylococcus aureus* and yeasts, within 2–6 h (Van Netten *et al.*, 1987).

1.5.2 Presence-or-absence tests

An important advance, to be discussed in section 1.5.3 for application to pure isolates and of great importance in that area, is presented by capitilisation on molecular attributes of target organisms. Previous attempts to apply probes to biological specimens were hampered, as discussed before, by (i) lack of sensi-tivity (Donnelly *et al.*, 1988; Ninet *et al.*, 1992; Ogden and Strachan, 1993); and (ii) interference by food constituents (De Leon *et al.*, 1992; Payne *et al.*, 1993). This was not at all an exceptional situation: similar difficulties were encountered when approaches to some extent failed when applied to the complex biological materials to be examined in clinical or environmental microbiology (Stewart, 1899; Alexander and Morris, 1991; Miller and Rhoden, 1991; Pfaller *et al.*, 1991; Stager and Davis, 1992). Upon combining meticulously designed enrichment steps with immunomagnetic separation (Lund *et al.*, 1991; Skjerve *et al.*, 1990; Cudjoe *et al.*, 1991; Fratamico *et al.*, 1992; Vermunt *et al.*, 1992; Mansfield and Forsythe, 1993) and concentration, followed by PCR-identifi-cation (Lampel *et al.*, 1990; Gannon *et al.*, 1992; Giesendorf *et al.*, 1992; Islam and Lindberg, 1992; Starbuch *et al.*, 1992; Grant *et al.*, 1993; Islam *et al.*, 1993; Kopecka *et al.*, 1993; Muir *et al.*, 1993) both shortcomings indicated above can apparently be overcome. Most promising results have already been scored with enteric bacterial and viral pathogens (Islam and Lindberg 1992; Widjojoatmodjo *et al.*, 1992; Muir *et al.*, 1993) and an important agent of spoilage of meat products (Grant *et al.*, 1993).

What is called for next in this sector are concerted efforts to avoid a repetition of the excitement → disappointment syndrome discussed before. For this purpose every candidate method – like any other innovative approach – should be evaluated along the lines of the fully objective *referee procedure*, substantiated by *proficiency testing* of laboratories using it (In't Veld and Notermans, 1992; Donnison *et al.*, 1993; Peterz and Steneryd, 1993; Scotter *et al.*, 1993), that, in the past, has resulted in the emergence of excellent recommended standard operating procedures, now in general use.

1.5.3 Identification

One of the main successes in pursuing facilitation and also acceleration of micro-biological methods is, beyond a shadow of doubt, the introduction of galleries relying on sets of small plastic tubes or microtitre plate wells containing suitable substrates and pH-indicators, or else a selection of antibiotics aiding in identification. Many of such sets produce results of high reliability within 2–4 h (Heard and Fleet, 1990; Freney *et al.*, 1992; Gruner *et al.*, 1992; O'Hara *et al.*, 1992; Shakespeare *et al.*, 1992; Visser *et al.*, 1992; York *et al.*, 1992; Kerr *et al.*, 1993).

Prerequisites for success in these instances are (i) starting from entirely pure colonies; (ii) having sufficiently large quantities of biomass available. This invariably calls for experience in purification of primary isolates and moreover requires the customary amount of time for producing adequately sized colonies.

A major contribution was made by the molecular microbiological techniques referred to in section 1.5.2. These introduced an entirely novel approach, although neither necessarily real time or even very rapid, nor open to minimally trained line staff or even laboratory personnel as we demonstrated in section 1.3.1. Molecular methods include the now already well established DNA/DNA hybridisation methods (Bauman and Bentvelzen, 1988; Van Damme-Jongsten *et al.*, 1990; Nesbakken *et al.*, 1991; Peterkin *et al.*, 1991; Wolcott, 1991; Kwaga *et al.*, 1990; Peterkin *et al.*, 1991, 1992; Goverde *et al.*, 1993) and serological approaches (Szabo *et al.*, 1990; van Poucke, 1990; Cudjoe *et al.*, 1991; Nørrung *et al.*, 1991; Wieneke, 1991) including highly specific latex agglutination methodology (Perales and Audicana, 1989; Bouvet and Jeanjean, 1992; Hazeleger *et al.*, 1992; Hansen and Freney, 1992).

1.6 Achievements and prospects 1994

1.6.1 Overview

Much has been achieved since, in 1985, Jarvis emphatically called for immediately available and reliably performing instrumentation (Jarvis, 1985). The era of all promises and no performance has certainly passed. Nonetheless much remains to be done before the innovations through which chemistry laboratories have gone since the second World War will also prevail in laboratories for applied microbiology.

On the other hand, a word of caution should be said against exercising too much orthodoxy with respect to accuracy and precision of novel rapid or instrumental methods. First and foremost it cannot be emphasised too frequently that the intrinsic variability of enumerations of viable (Barnard and Glass, 1975; Mossel and Van Netten, 1991), but particularly debilitated (Rodrigues and Kroll, 1989, 1990) and even more so of in no way culturable (Bouwer-Hertzberger and Mossel, 1982) microbial populations in foods is immense and quite outside the range accepted by chemical analysts (Fleisher, 1990). This is compounded by the far from perfect functioning of almost all selective-diagnostic methods in current use for the enumeration of pathogenic organisms transmitted by foods. These include *Salmonella* (Perales and Audicana, 1989; Rodrigues and Kroll, 1990; De Zutter *et al.*, 1991; Mossel and Van Netten, 1991; D'Aoust *et al.*, 1992; Eckner *et al.*, 1992; Feng, 1992; In't Veld and Notermans, 1992; Rohner *et al.*, 1992; Tate *et al.*, 1992), *Campylobacter jejuni* (Endtz *et al.*, 1991; Peterz, 1991; Giesendorf *et al.*, 1992; Wang *et al.*, 1992; Scotter *et al.*, 1993), *Yersinia enterocolitica* (Kwaga *et al.*, 1990, 1992; Nesbakken *et al.*, 1991; Goverde *et al.*, 1993), *Listeria monocytogenes* (Cassiday *et al.*, 1990; Lund *et al.*, 1988; Slade, 1992; Van Netten *et al.*, 1991; Jensen, 1993), *Clostridium* species (Ingram, 1963; De Mesquita *et al.*, 1991; Weenk *et al.*, 1991, 1994) and shigellae (Gannon *et al.*, 1992; Islam *et al.*, 1993). Rapid methods should obviously approach the best conventional equivalent in accuracy and precision, but not necessarily exceed it (Nemes and Altwegg, 1992; Imperatrice and Nachamkin, 1993).

This introductory overview cannot pursue a detailed appreciation of those methods that are more or less ready for immediate, or short-term adoption in laboratories engaged in the microbiological monitoring of foods. Such efforts will be made in Chapters 2–8 by specialists in the various approaches. The presently available results of one of our own investigations in this area will be summarised. Finally, an attempt will be made to arrive at a broad general evaluation of the prospects for instrumental and real time methodology for laboratory managers – in trivial terms: what are, in the middle of 1994, the benefits for the laboratory chief and staff?

1.6.2 A truly real time check on adequate 'sanitation'

As indicated before in section 1.2.1, inspection of the food environment is one of the monitoring priorities. It is only too obvious from the data collected in Table 1.1 that contamination or recontamination of initially microbiologically safe commodities is amongst the most frequent causes of food-borne infections and intoxinations of microbial aetiology. Consequently, within the HACCP/LISA concepts, verification of adequate cleaning and disinfection – termed sanitation in the US literature – of all food contact surfaces has to constitute one of the major assignments of those employed to assure food safety and quality (Mossel, 1983).

As also emphasized in section 1.2.1, particularly in this instance a real time method that can be applied on-site would be most useful in an attempt to mend the pace of rectification of incidental hiatuses in overall excellent hygienic practices. We have adopted bioluminometry to this end (Poulis *et al.*, 1993). This was achieved by linking the use of one of the many, commercially available, portable ATP-monitoring instruments to examination of sufficiently large surfaces by test paper methodology currently used in bedside and consultation room diagnostics. The latter included paper strips allowing the detection of minute traces of peptides, glucose, myoglobin and nitrate reducing bacteria as used for the detection of bacteriuria. In order to be able to also detect significant taxa failing to reduce nitrate, including a most important class of so-called marker organisms (Mossel, 1982), *viz. Enterococcus* spp., a strip is being developed relying on rapid dissimilation of glucose by these – and most other – bacteria (Van Netten *et al.*, 1987). The combined use of bioluminometry and, ideally, one e.g. 10 × 10 cm strip containing all five reagents, will enable immediate differentiation between ATP of respectively eukaryotic, prokaryotic and inanimate origin. This far from constitutes an *l'art-pour-l'art* exercise, because it allows identification of the source of inadequately eliminated contamination, e.g. meat tissue and juice, fruit juice, skin exudate, bacteria, and hence enables to apply improved cleaning and/or disinfection within minutes.

In instances where a more differentiated diagnosis is required to design systematic improvement of 'sanitizing' regimes, recourse to classical microbiology is unavoidable. This is not a marked drawback, since immediate corrective action has already been implemented. Current methods, like swabbing or impression release of target organisms will require, where an expert choice of media and incubation conditions has been made, no more than some 14–18 h, an acceptable delay. All elements of this new approach require a minimum of microbiological training and expertise. This constitutes an additional asset, as it allows checking to be done and corrective action to be taken by minimally skilled personnel when the professional laboratory staff are fully occupied elsewhere, or not readily available, e.g during night or weekend shifts.

1.6.3 Theorising on a Quo vadis – a presumptuous exercise

The experience with the promising enrichment-immunomagnetic concentration-PCR triad (Fluit *et al.*, 1993a,b) indicates that progress will, probably, result in molecular traits *enhancing* rather than *supplanting* conventional methodology (Starbuch *et al.*, 1992). This is likely to be the case for yet another reason which deserves reiteration. In delicate matters like the rejection of consignments of foods because of the 'presence' of an aggressive pathogen, like *Listeria monocytogenes*, serotype 4b, observed in a sample drawn from the lot, the case will often – for scientific as well as legal reasons – call for the bacterium to be isolated, allowing third parties to pass a verdict. This will anyway require some element of classical methodology. This does not detract, however, from the *essential* advantage of an early indication of the *absence* of the pathogen.

In summary, when reason is allowed to substitute ritual it seems that flasks and Petri dishes will remain in use, also in the future when probes and other analytical tools derived from molecular microbiological advances will become part and parcel of the food microbiologist's analytical armamentarium.

1.7 Impact for education at undergraduate and postgraduate level

From the previous considerations it follows, that, in the not too distant future, simplified and accelerated methods, appropriately associated with the required classical scenarios will become the current test repertoire in analytical food microbiology. This prospect calls for timely incorporation of a 'rapid methods' module into curricula of food microbiology at all the customary educational levels (Park, 1990). Where this need would be ignored, lack of familiarity with promising novel key approaches, secondary to infrequent and superficial confrontation with the new methodology, would not allow graduates to arrive at balanced verdicts in cardinal instrumentation matters.

A curriculum aimed at such teaching goals should definitely include thorough familiarisation with the essential elements of the microbiological monitoring of biological specimens in general, summarised in Table 1.5. As in surgery, lectures and demonstrations cannot possibly suffice to acquire the necessary proficiency. Well designed, elaborated, taught, supervised and examined practical exercises are imperative; and these should include, as expounded above, training in statistically sound procedures for the comparison of data obtained by novel methods with those acquired with standard operating procedures generally accepted by trade, industry and national and international legislation.

Such training programmes at undergraduate as well as postgraduate levels will not only be essential for the future of the profession, and hence for industrial and regulatory employers of graduates in the food and catering industries. Manufacturers of laboratory equipment and instruments constitute an additional group of beneficiaries. This educational strategy will ensure a constant flow of qualified staff at research, development and sales level – the latter being allowed to arrive at a thorough consensus with potential customers. The patterns presented in Tables 1.3 and 1.4 may facilitate such an unbiased evaluation of new developments in simplified, facilitated, accelerated and automated methods.

As illustrated in the previous sections, the field of instrumentation is a rapidly changing and developing one. Food microbiologists cannot possibly be expected to keep abreast of new achievements by following the literature or attending occasional meetings in this area. Enrolling in refresher courses taught by academically employed leaders in this field (Huis in't Veld *et al.*, 1988; Adams and Hope, 1989; Stannard *et al.*, 1990; Fung, 1991, 1992) constitutes an essential part in the general framework of recredentialling of graduate scientists in more general terms (Skovgaard, 1990; Gellhorn, 1991; Davis *et al.*, 1992; Manning and DeBakey, 1992).

References

Abbaszadigan, M., Huber, M.S., Gerba, C.P. and Pepper, I.L. (1993) Detection of enteroviruses in groundwater with the polymerase chain reaction. *Appl. Environ. Microbiol.*, **59**, 1318–1324.

Adams, M.R. and Hope, C.F.A. (eds) (1989) *Rapid Methods in Food Microbiology*, Elsevier, Amsterdam.

Albert, M.J., Faruque, M., Ansaruzzaman, M.M. *et al.* (1992a) Sharing of virulence-associated properties at the phenotypic and genetic levels between enteropathogenic *Escherichia coli* and *Hafnia alvei. J. Med. Microbiol.*, **37**, 310–314.

Albert, M.J., Alam, K., Ansaruzzaman, M. *et al.* (1992b) Pathogenesis of *Providencia alcalifaciens*-induced diarrhoea. *Infect. Immunol.*, **60**, 5017–5024.

Alexander, J. and Rothwell, J. (1970) A study of some factors affecting methylene blue test and the effect of freezing on bacterial content of ice-cream. *J. Food Technol.*, **5**, 387–402.

Alexander, L.M. and Morris, R. (1991) PCR and environmental monitoring – the way forward? *Water Sci. Technol.*, **24**, 291–294.

Altekruse, S.F., Tollefson, L.K. and Bögel, K. (1993) Control strategies for *Salmonella enteritidis* in five countries. *Food Control*, **4**, 10–16.

Amgar, A. (1991) *Listeria and Food Safety*. Asept, Laval, France.

Amgar, A. (1992) *Predictive Microbiology and HACCP*. Asept, Laval, France.

Barnard, S.E. and Glass, E.D. (1975) Collecting and handling milk samples. *J. Milk Food Technol.*, **38**, 108–110.

Barnes, G.H. and Edwards, A.T. (1992) An investigation into an outbreak of *Salmonella enteritidis* phage-type 4 infection and the consumption of custard slices and trifles. *Epidemiol. Infect.*, **109**, 397–403.

Bartlett, F.M. and Tetro, J. (1992) Rapid method for assessing microbiological quality of egg washwater using resazurin. *J. Food Protect.*, **55**, 837.

Bauman, H.E. (1974) The HACCP concept and microbiological hazard categories. *Food Technol.*, **28**, (nr. 9) 30–34, 74.

Bauman, H. (1990) HACCP: Concept, development and application. *Food Technol.*, **44**, (nr. 5) 156–158.

Bauman, J.G.J. and Bentvelzen, P. (1988) Flow cytometric detection of ribosomal RNA in suspended cells by fluorescent *in situ* hybridization. *Cytometry*, **9**, 517–524.

Bautista, D.A., McIntyre, L., Laleye, L. *et al.* (1992) The application of ATP bioluminescence for the assessment of milk quality and factory hygiene. *J. Rapid Methods Automat. Microbiol.*, **1**, 179–193.

Bean, N.H. and Griffin, P.M. (1990) Foodborne disease outbreaks in the United States, 1973–1987: pathogens, vehicles, and trends. *J. Food Protect.*, **53**, 804–817.

Bej, A.K., Mahbubani, M.H., Boyce, M.J. and Atlas, R.M. (1994) Detection of *Salmonella* spp. in oysters by PCR. *Appl. Environ. Microbiol.*, **60**, 368–373.

Bendall, R.P., Lucas, S., Moody, A. *et al.* (1993) Diarrhoea associated with cyanobacterium-like bodies: a new coccidian enteritis of man. *Lancet*, **341**, 590–592.

Boismenu, D., Lépine, F., Thibault, C. *et al.* (1991) Estimation of bacterial quality of cod fillets with the disc flotation method. *J. Food Sci.*, **56**, 958–961.

Bouvet, P.J.M. and Jeanjean, S. (1992) Evaluation of two colored latex kits, the Wellcolex colour *Salmonella* test and the Wellcolex colour *Shigella* test, for serological grouping of *Salmonella* and *Shigella* species. *J. Clin. Microbiol.*, **30**, 2184–2186.

Bouwer-Hertzberger, S.A. and Mossel, D.A.A. (1982) Bacterioscopic examination of specimens possibly involved in diseases of bacterial etiology transmitted by foods. In *Isolation and Identification Methods for Food Poisoning Organisms* (eds Corry, J.E.L., Roberts, D. and Skinner, F.A.). Academic Press, London, pp. 25–33.

Brodsky, M.H., Entis, P., Sharpe, A.N. and Jarvis, G.A. (1982) Enumeration of indicator organisms in foods using the automated hydrophobic grid-membrane filter technique. *J. Food Protect.*, **45**, 292–296.

Bryan, F. (1992) *Hazard Analysis Critical Control Point Evaluations.* World Health Organization, Geneva.

Buttiaux, R., Le Breton, A. and Gaudier, B. (1956) Sur la qualité bactériologique des laits sterilisés. *Archives Françaises Pédiatrie*, **13**, 543–550.

Cady, P., Hardy, D., Martins, S. *et al.* (1978) Automated impedance measurements for rapid screening of milk microbial content. *J. Food Protect.*, **41**, 277–283.

Casemore, D.P., Richardson, K., Sands, R.L. and Stevens, G. (1992) A re-assessment of the modified methylene blue and viable count methods for the bacteriological grading of ice-creams. *PHLS Microbiol. Digest*, **9**, 166–171.

Cassiday, P.K., Graves, L.M. and Swaminathan, B. (1990) Replica plating of colonies from *Listeria*–selective agars to blood agar to improve the isolation of *Listeria monocytogenes* from foods. *Appl. Environ. Microbiol.*, **56**, 2274–2275.

Cheftel, H. (1955) Remarques à propos du controle bactériologique des jambons consérves en boites. *Annales Institut Pasteur Lille*, **7**, 256–262.

Cousin, M.A., Dufrenne, J., Rombouts, F.M. and Notermans, S. (1990) Immunological detection of *Botrytis* and *Monascus* species in food. *Food Microbiol.*, **7**, 227–235.

Cudjoe, K.S., Thorsen, L.I., Sorensen, T. *et al.* (1991) Detection of *Clostridium perfringens* type A enterotoxin in faecal and food samples using immunomagnetic separation (IMS)-ELISA. *Int. J. Food Microbiol.*, **12**, 313–321.

Dack, G.M. (1956) Evaluation of microbiological standards for foods. *Food Technol.*, **10**, 507–509.

D'Aoust, J.Y., Sewell, A.M. and Warburton, D.W. (1992) A comparison of standard cultural methods for the detection of foodborne *Salmonella*. *Int. J. Food Microbiol.*, **16**, 41–50.

Davis, D.A., Thomson, M.A., Oxman, A.D. and Haynes, R.B. (1992) Evidence of the effectiveness of [continuing medical education] CME. *J. Am. Med. Assoc.*, **268**, 1111–1117.

De Leon, R., Matsui, S.M., Baric, R.S. *et al.* (1992) Detection of Norwalk virus in stool specimens by reverse transcriptase-polymerase chain reaction and nonradioactive oligoprobes. *J. Clin. Microbiol.*, **30**, 3151–3157.

De Mesquita, M.M.F., Evison, L.M. and West, P.A. (1991) Removal of faecal indicator bacteria and bacteriophages from the common mussel (*Mytilus edulis*) under artificial depuration conditions. *J. Appl. Bacteriol.*, **70**, 495–501.

De Zutter, L., de Smedt, J.M., Abrams, R. *et al.* (1991) Collaborative study on the use of motility enrichment on modified semisolid Rappaport-Vassiliadis medium for the detection of *Salmonella* from foods. *Int. J. Food Microbiol.*, **13**, 11–20.

Dodds, K.L., Holley, R.A. and Kempton, A.G. (1983) Evaluation of the catalase and Limulus amoebocyte lysate tests for rapid determination of the microbial quality of vacuum-packed cooked turkey. *Can. Inst. Food Sci. Technol. J.*, **16**, 167–172.

Donnelly, C.W. and Baigent, G.J. (1986) Method for flow cytometric detection of *Listeria monocytogenes* in milk. *Appl. Environ. Microbiol.*, **52**, 689–695.

Donnelly, C.W., Baigent, G.J. and Briggs, E.H. (1988) Flow cytometry for automated analysis of milk containing *Listeria monocytogenes*. *J. A. O. A. C.*, **71**, 655–658.

Donnison, A.M., Ross, C.M. and Russell, J.M. (1993) Quality control of bacterial enumeration. *Appl. Environ. Microbiol.*, **59**, 922–923.

Du Pont, H.L. (1992) How safe is the food we eat? *J. Am. Med. Assoc.*, **268**, 3240.

Eckner, K.F., McIver, D., Lepper, W.A. *et al.* (1992) Use of an elevated temperature and novobiocin in a modified enzyme-linked immunosorbent assay for the improved recovery of *Salmonella* from foods. *J. Food Protect.*, **55**, 758–762.

Edwards, C., Porter, J., Saunders, J.R. *et al.* (1992) Flow cytometry and microbiology. *Soc. Gen. Microbiol. Quart.*, **19**, 105–108.

Eijkman, C. (1908) Die Ueberlebungskurve bei Abtötung von Bakterien durch Hitze. *Biochem. Zeitschr.*, **11**, 12–20

Endtz, H.P., Ruijs, G.J.H.M. and Zwinderman, A.H. (1991) Comparison of six media, including a semisolid agar, for the isolation of various *Campylobacter* species from stool specimens. *J. Clin. Microbiol.*, **29**, 1007–1010.

Entis, P. (1990) Improved hydrophobic grid membrane filter method, using EF-18 agar, for detection of *Salmonella* in foods: collaborative study. *J. A. O. A. C.*, **73**, 734–742.

Feng, P. (1992) Commercial assay systems for detecting foodborne *Salmonella*: a review. *J. Food Protect.*, **55**, 927–934.

Firstenberg-Eden, R. (1986) Electrical impedance for determining microbial quality of foods. In

Foodborne Microorganisms and their Toxins: Developing Methodology (eds Pierson, M.D. and Stern, N.J.). Marcel Dekker, New York, pp. 129–144.

Fleet, G.H., Karalis, T., Hawa, A. and Lukondeh, T. (1991) A rapid method for enumerating *Salmonella* in milk powders. *Lett. Appl. Microbiol.*, **13**, 255–259.

Fleisher, J.M. (1990) The effects of measurement error on previously reported mathematical relationships between indicator organism density and swimming-associated illness: a quantitative estimate of the resulting bias. *Int. J. Epidemiol.*, **19**, 1100–1106.

Fluit, A.C., Torensma, R. and Visser, M.J.C. (1993a) Detection of *Listeria monocytogenes* in cheese with the magnetic immuno–polymerase chain reaction assay. *Appl. Environ. Microbiol.*, **59**, 1289–1293.

Fluit, A.C., Widjojoatmodjo, M.N., Box, A.T.A. *et al.* (1993b) Rapid detection of salmonellae in poultry with magnetic immuno-polymerase chain reaction assay. *Appl. Environ. Microbiol.*, **59**, 1342–1346.

Fratamico, P.M., Schultz, F.J. and Buchanan, R.L. (1992) Rapid isolation of *Escherichia coli* O157:H7 from enrichment cultures of foods using an immunomagnetic separation method. *Food Microbiol.*, **9**, 105–113.

Freney, J., Bland, S., Desmonceaux, J.E.M. *et al.* (1992) Description and evaluation of the semiautomated 4-hour Rapid ID 32 Strep method for identification of streptococci and members of related genera. *J. Clin. Microbiol.*, **30**, 2657–2661.

Frost, W.D. (1921) Improved technique for the micro- or little plate method of counting bacteria in milk. *J. Infect. Diseases*, **28**, 176–184.

Fung, D.Y.C. (1991) Rapid methods in the food industry. In *Rapid Methods and Automation in Microbiology and Immunity* (eds Vaheri, A., Tilton, R.C. and Balows, A.). Springer, Berlin, pp. 503–511.

Fung, D.Y.C. (1992) New developments in rapid methods for food microbiology. *Trends Food Sci. Technol.*, **3**, 142–144.

Gannon, V.P.J., King, R.K., Kim, J.Y. and Thomas, E.J.G. (1992) Rapid and sensitive method for detection of Shiga-like toxin-producing *Escherichia coli* in ground beef using the polymerase chain reaction. *Appl. Environ. Microbiol.*, **58**, 3809–3815.

Gatti, S., Cevini, C., Bruno, A. *et al.* (1993) Cryptosporidiosis in tourists returning from Egypt and the island of Mauritius. *Clin. Infect. Diseases*, **16**, 344–345.

Gellhorn, A. (1991) Periodic physician recredentialing. *J. Am. Med. Assoc.*, **265**, 752–755.

Giesendorf, B.A.J., Quint, W.G.V., Henkens, M.H.C. *et al.* (1992) Rapid and sensitive detection of *Campylobacter* spp. in chicken products by using the polymerase chain reaction. *Appl. Environ. Microbiol.*, **58**, 3804–3808.

Gilchrist, J.E., Campbell, J.E., Donnelly, C.B. *et al.* (1973) Spiral plate method for bacterial determination. *Appl. Microbiol.*, **25**, 244–252.

Goldschmidt, M.C. (1991) Microbiological instrumentation for the food industry: a review. In *Rapid Methods and Automation in Microbiology and Immunology* (eds Vaheri, A., Tilton, R.C. and Balows, A.). Springer, Berlin, pp. 512–519.

Goulet, V., Rocourt, J., Courtieu, A.L. *et al.* (1993) Epidémie de listériose en France. *Bull. Épidémiol. Hebd.*, **4**, 13–14.

Goverde, R.L.J., Jansen, W.H., Brunnings, H.A. *et al.* (1993) Digoxigenin-labelled *inv*– and *ail*– probes for the detection and identification of pathogenic *Yersinia enterocolitica* in clinical specimens and naturally contaminated pig samples. *J. Appl. Bacteriol.*, **74**, 301–313.

Grant, K.A., Dickinson, J.H., Payne, M.J. *et al.* (1993) Use of the polymerase chain reaction and 16S rRNA sequences for the rapid detection of *Brochothrix* spp. in foods. *J. Appl. Bacteriol.*, **74**, 260–267.

Griffiths, M.W. (1993) Application of bioluminescence in the dairy industry. *J. Dairy Sci.*, **76**, 3118–3125.

Gruner, E., Von Graevenitz, A. and Altwegg, M. (1992) The API-ZYM system: a tabulated review from 1977 to date. *J. Microbiol. Meth.*, **16**, 101–118.

Habraken, C.J.M., Mossel, D.A.A. and Van den Reek, S. (1986) Management of *Salmonella* risks in the production of powdered milk products. *Neth. Milk Dairy J.*, **40**, 99–116.

Hansen, W. and Freney, J. (1993) Comparative evaluation of a latex aggutination test for the detection and presumptive serogroup identification of *Salmonella* spp. *J. Microbiol. Meth.*, **17**, 227–232.

Hawa, S.G., Morrison, G.J. and Fleet, G.H (1984) Method to rapidly enumerate *Salmonella* on chicken carcasses. *J. Food Protect.*, **47**, 932–936.

Hazeleger, W.C., Beumer, R.R. and Rombouts, F.M. (1992) The use of latex agglutination tests for determining *Campylobacter* species. *Lett. Appl. Microbiol.*, **14**, 181–184.

Heard, G.M. and Fleet, G.H. (1990) A convenient microtitre tray procedure for yeast identification. *J. Appl. Bacteriol.*, **68**, 447–451.

Hedberg, C.W., White, K.E., Johnson, J.A. *et al.* (1991) An outbreak of *Salmonella enteritidis* infection at a fast-food restaurant: implications for food handler-associated transmission. *J. Infect. Diseases*, **164**, 1135–1140.

Hedberg, C.W., Korlath, J.A., D'Aoust, J.Y. *et al.* (1992a) A multistate outbreak of *Salmonella javiana* and *Salmonella oranienburg* infections due to consumption of contaminated cheese. *J. Am. Med. Assoc.*, **268**, 3203–3207.

Hedberg, C.W., Levine, W.C., White, K.E. *et al.* (1992b) An international foodborne outbreak of shigellosis associated with a commercial airline. *J. Am. Med. Assoc.*, **268**, 3208–3212.

Huis in't Veld, J.H.J., Hartog, B.J. and Hofstra, H. (1988) Changing perspectives in food microbiology: implementation of rapid microbiological analyses in modern food processing. *Food Res. Int.*, **4**, 271–329.

Hutter, K.J. and Eipel, H.E. (1978) Flow cytometric determination of cellular substances in algae, bacteria, moulds and yeasts. *Antonie Van Leeuwenhoek*, **44**, 269–282.

Imperatrice, C.A. and Nachamkin, I. (1993) Evaluation of the Vitek EPS enteric pathogen screen card for detecting *Salmonella, Shigella* and *Yersinia* spp. *J. Clin. Microbiol.*, **31**, 433–435.

Ingram, M. (1963) Difficulties in counting viable clostridia in foods. *Rend. Cont. Ist. Sup. Sanità Roma*, **26**, 330–332.

In't Veld, P.H. and Notermans, S. (1992) Use of reference materials (spray–dried milk artificially contaminated with *Salmonella typhimurium*) to validate detection methods for *Salmonella*. *J. Food Protect.*, **55**, 855-858.

Islam, D. and Lindberg, A.A. (1992) Detection of *Shigella dysenteriae* type 1 and *Shigella flexneri* in feces by immunomagnetic isolation and polymerase chain reaction. *J. Clin. Microbiol.*, **30**, 2801–2806.

Islam, M.S., Hassan, M.K., Miah, M.A. *et al.* (1993) Use of polymerase chain reaction and fluorescent-antibody methods for detecting viable nonculturable *Shigella dysenteriae* type 1 in laboratory microcosms. *Appl. Environ. Microbiol.*, **59**, 536–540.

Jakobsen, M. and Lillie, A. (1992) Quality systems for the fish industry. In *Quality assurance in the fish industry* (eds Huss, H.H., Jakobsen, M. and Liston, J.). Elsevier, Amsterdam, pp. 515–520.

Jarecki-Khan, K., Tzipori, S.R. and Unicomb, L.E. (1993) Enteric adenovirus infection among infants with diarrhoea in rural Bangladesh. *J. Clin. Microbiol.*, **31**, 484–489.

Jarvis, B. (1985) A philosophical approach to rapid methods for industrial food control. In *Rapid Methods and Automation in Microbiology and Immunology* (ed. Habermehl, K.O.). Springer, Berlin, pp. 593–602.

Jensen, A. (1993) *Listeria monocytogenes* isolation from human faecal specimens: experiments with the selective media PALCAM and L-PALCAMY. *Lett. Appl. Microbiol.*, **16**, 32–35.

Jones, D.M., Sutcliffe, E.M. and Curry. A. (1991) Recovery of viable but non-culturable *Campylobacter jejuni*. *J. gen. Microbiol.*, **137**, 2477–2482.

Kampelmacher, E.H. (1983) Irradiation for control of *Salmonella* and other pathogens in poultry and fresh meats. *Food Technol.*, **37**, (nr. 6) 117–119, 169.

Kamphuis, H.J., Van der Horst, M.I., Samson, R.A. *et al.* (1992) Mycological conditions of maize products. *Int. J. Food Microbiol.*, **16**, 237–245.

Kapikian A.Z. (1993) Viral gastroenteritis. *J. Am. Med. Assoc.*, **269**, 627–630.

Kelly, M.T. and Latimer, J.M. (1980) Comparison of the AutoMicrobic system with API, Enterotube, Micro-ID, Micromedia systems and conventional methods for identification of Enterobacteriaceae. *J. Clin. Microbiol.*, **12**, 659–662.

Kennedy, J.E. and Oblinger, J.L. (1985) Application of bioluminescence to rapid determination of microbial levels in ground beef. *J. Food Protect.*, **48**, 334–340, 345.

Kerr, K.G., Hawkey, P.M. and Lacey, R.W. (1993) Evaluation of the API Coryne system for identification of *Listeria* species. *J. Clin. Microbiol.*, **31**, 749–750.

Kopecka, H., Dubrou, S., Prevot, J. *et al.* (1993) Detection of naturally occurring enteroviruses in waters by reverse transcription, polymerase chain reaction and hybridization. *Applied Environm. Microbiol.*, **59**, 1213–1219.

Kroll, R.G., Frears, E.R. and Bayliss, A. (1989) An oxygen electrode-based assay of catalase activity as a rapid method for estimating the bacterial content of foods. *J. Appl. Bacteriol.*, **66**, 209–217.

Kwaga, J., Iversen, J.O. and Saunders, J.R. (1990) Comparison of two enrichment protocols for the detection of *Yersinia* in slaughtered pigs and pork products. *J. Food Protect.*, **53**, 1047–1049, 1061.

Kwaga, J., Iversen, J.O. and Misra, V. (1992) Detection of pathogenic *Yersinia enterocolitica* by polymerase chain reaction and digoxigenin-labeled polynucleotide probes. *J. Clin. Microbiol.*, **30**, 2668–2673.

Lampel, K.A., Jagow, J.A., Trucksess, M.B. and Hill, W.E. (1990) Polymerase chain reaction for detection of invasive *Shigella flexneri* in food. *Appl. Environ. Microbiol.*, **56**, 1536–1540.

LeSaux, N., Spika, J.S., Friesen, B. *et al.* (1993) Ground beef consumption in noncommercial settings is a risk factor for sporadic *Escherichia coli* O157:H7 infection in Canada. *J. Infect. Diseases*, **167**, 500–502.

Lew, J.F., Moe, C.L., Monroe, S.S. *et al.* (1991) Astrovirus and adenovirus associated with diarrhoea in children in day care settings. *J. Infect. Diseases*, **164**, 673–678.

Long, E.G., White, E.H., Carmichael, W.W. *et al.* (1991) Morphologic and staining characteristics of a *Cyanobacterium*-like organism associated with diarrhoea. *J. Infect. Diseases*, **164**, 199–202.

Luby, S.P., Jones, J.L. and Horan, J.M. (1993) A large salmonellosis outbreak associated with a frequently penalized restaurant. *Epidemiol. Infect.*, **110**, 23–30.

Lund, A., Hellemann, A.L. and Vartdal, F. (1988) Rapid isolation of K88+ *Escherichia coli* by using immunomagnetic particles. *J. Clin. Microbiol.*, **26**, 2572–2575.

Lund, A.M., Zottola, E.A. and Push, D.J. (1991) Comparison of methods for isolation of *Listeria* from raw milk. *J. Food Protect.*, **54**, 602–606.

Macler, B.A. and Regli, S. (1993) Use of microbial risk assessment in setting US drinking water standards. *Int. J. Food Microbiol.*, **18**, 245–256.

Mafart, P., Cleret, J.J. and Bourgois, C. (1981) Optimization of 14C-lysine concentration and specific activity for the radiometric detection of micro-organisms. *Eur. J. Appl. Microbiol.*, **11**, 189–192.

Manninen, M.T., Fung, D.Y.C. and Hart, R.A. (1991) Spiral system and laser colony scanner for enumeration of microorganisms. *J. Food Safety*, **11**, 177–187.

Manning, P.R. and DeBakey, L. (1992) Lifelong learning tailored to individual clinical practice. *J. Am. Med. Assoc.*, **268**, 1135–1136.

Manofield, L.P. and Forsythe, S.J. (1993) Immunomagnetic separation as an alternative to enrichment broths for *Salmonella* detection. *Lett. Appl. Microbiol.*, **16**, 122–125.

Martins, S.B. and Selby, M.J. (1980) Evaluation of a rapid method for the quantitative estimation of coliforms in meat by impedimetric procedures. *Appl. Environ. Microbiol.*, **39**, 518–524.

Mattila, T. (1987) Automated turbidimetry – a method for enumeration of bacteria in food samples. *J. Food Protect.*, **50**, 640–642.

Mattila, T. and Alinehmas, T. (1987) Automated turbidimetry for predicting colony forming units in raw milk. *Int. J. Food Microbiol.*, **4**, 157–160.

Meyer, K.F. (1931) The protective measures of the state of California against botulism. *J. Prevent. Med.*, **5**, 261–293.

Miller, J.M. and Rhoden, D.L. (1991) Preliminary evaluation of Biolog, a carbon source utilization method for bacterial identification. *J. Clin. Microbiol.*, **29**, 1143–1147.

Mirhabibollahi, B., Brooks, J.L. and Kroll, R.G. (1991) Electrical detection of foodborne micro-organisms. *Trends Food Sci. Technol.*, **2**, 62–66.

Morinigo, M.A., Munoz, M.A., Martinez-Manzanares, E. *et al.* (1993) Laboratory study of several enrichment broths for the detection of *Salmonella* spp. particularly in relation to water samples. *J. Appl. Bacteriol.*, **74**, 330–335.

Mossel, D.A.A. (1982) Marker (index and indicator) organisms in food and drinking water. Semantics, ecology, taxonomy and enumeration. *Antonie Van Leeuwenhoek*, **48**, 609–611.

Mossel, D.A.A. (1983) Seventy-five years of longitudinally integrated microbiological safety assurance in the dairy industry in The Netherlands. *Neth. Milk Dairy J.*, **37**, 240–245.

Mossel, D.A.A. (1987) The microbiological examination of food and drinking water in the framework of health protection. In *Application of Biotechnology to the Rapid Diagnosis of Infectious Diseases* (ed. Balows, A.J.). Royal Society of Medicine, London, pp. 45–73.

Mossel, D.A.A. (1989) Adequate protection of the public against food-transmitted diseases of microbial aetiology. Achievements and challenges, half a century after the introduction of the Prescott–Meyer–Wilson strategy of active intervention. *Int. J. Food Microbiol.*, **9**, 271–294.

Mossel, D.A.A. (1991a) Food microbiology: an authentic academic discipline with substantial potential benefits for science and society. *J. A. O. A. C.*, **74**, 1–13.

Mossel, D.A.A. (1991b) Management of microbiological health hazards associated with foods of animal origin – contribution of the Plumb Strategy. *Arch. Lebensm. Hyg.*, **42**, 27–32.

Mossel, D.A.A. and Drion, E.F. (1954) Bacteriological requirements for and testing of sterilized milk and sterilized milk products. *Neth. Milk Dairy J.*, **8**, 106–114.

Mossel, D.A.A. and Struijk, C.B. (1992) *Prevention of the Transmission of Infections and Intoxinations by Foods: The Responsibility of the Veterinary Public Health Profession* (eds Bartlett, P.C. and Hewins, S.O.). American College Veterinary Preventive Medicine, San Antonio.

Mossel, D.A.A. and Van Netten, P. (1984) Harmful effects of selective media on stressed micro-organisms: nature and remedies. In *The Revival of Injured Microbes* (eds Andrew, M.H.E. and Russell, A.D.). Academic Press, London, pp. 329–369.

Mossel, D.A.A. and Van Netten, P. (1991) Microbiological reference values for foods: a European perspective. *J. A. O. A. C.*, **74**, 420–432.

Mossel, D.A.A. and Visser, M. (1960) The estimation of small numbers of microorganisms in opalescent non-alcoholic drinks by applying centrifugation. *Annales Institut Pasteur Lille*, **11**, 193–202.

Mossel, D.A.A., Van Netten, P. and de Pijper, M. (1991) A centrifugation/quadrantplate technique for the simplified differential-bacteriological examination of adequately heat-processed foods. *Lett. Appl. Microbiol.*, **13**, 115–117.

Mossel, D.A.A., Struijk, C.B., Jaisli, F.K. *et al.* (1992) Use of 24 hours' centrifugation/plating technique in a survey on the medical-microbiological condition of raw ham and hard raw-milk cheese originating from authenticated GMDP manufacture. *Arch. Lebensm. Hyg.*, **43**, 51–54.

Mossel, D.A.A., Corry, J.E.L., Struijk, C.B. and Baird, R.M. (1995) *Essentials of the Microbiology of Foods*. Wiley, Chichester (in press).

Muir, P., Nickelson, F., Jhetam, M. *et al.* (1993) Rapid diagnosis of enterovirus infection by magnetic bead extraction and polymerase chain reaction detection of enterovirus RNA in clinical specimen. *J. Clin. Microbiol.*, **31**, 31–38.

Nachamkin, I., Blaser, M.J. and Tompkins, L.S. (1992) Campylobacter jejuni: *Current Status and Future Trends*. American Society of Microbiology, Washington.

Nemes, P. and Altwegg, M. (1992) Rapid identification and susceptibility testing of routine urine isolates with COBAS MICRO semiautomated system. *J. Microbiol. Meth.*, **16**, 181–194.

Nesbakken, T., Kapperud, G., Dommarsnes, K. *et al.* (1991) Comparative study of a DNA hybridization method and two isolation procedures for detection of *Yersinia enterocolitica* O:3 in naturally contaminated pork products. *Appl. Environ. Microbiol.*, **57**, 389–394.

Nickerson, J.T.R. (1943) A modified little plate method for bacterial counts in vegetable freezing plants. *Food Res.*, **8**, 163–168.

Nilsson, L., Oliver, J.D. and Kjelleberg, S. (1991) Resuscitation of *Vibrio vulnificus* from the viable but nonculturable state. *J. Bacteriol.*, **173**, 5054–5059.

Ninet, B., Bannerman, E. and Bille, J. (1992) Assessment of the accuprobe *Listeria monocytogenes* culture identification reagent kit for rapid colony confirmation and its application in various enrichment broths. *Appl. Environ. Microbiol.*, **58**, 4055–4059.

Nir, R., Lamed, R., Gueta, L. and Sahar, E. (1990a) Single-cell entrapment and microcolony development within uniform microspheres amenable to flow cytometry. *Appl. Environ. Microbiol.*, **56**, 2870–2875.

Nir, R., Yisraeli, Y., Lamed, R. and Sahar, E. (1990b) Flow cytometry sorting of viable bacteria and yeasts according to galactosidase activity. *Appl. Environ. Microbiol.*, **56**, 3861–3866.

Nørrung, B., Sølve, M., Ovesen, M. and Skovgaard, N. (1991) Evaluation of an ELISA test for detection of *Listeria* spp. *J. Food Protect.*, **54**, 752–755, 761.

Notermans, S. and Kamphuis, H. (1990) Detection of moulds in food by latex agglutination: a collaborative study. *Food Agric. Immunol.*, **2**, 37–46.

Ogden, I.D. and Strachan, N.J.C. (1993) Enumeration of *Escherichia coli* in cooked and raw meats by ion mobility spectrometry. *J. Appl. Bacteriol.*, **74**, 402–405.

O'Hara, C.M., Rhoden, D.L. and Miller, J.M. (1992) Reevaluation of the API 20E identification system versus conventional biochemicals for identification of members of the family Enterobacteriaceae: a new look at an old product. *J. Clin. Microbiol.*, **30**, 123–135.

Park, R.W.A. (1990) Joint IUMS/ICFMH and UNESCO consultation on postgraduate teaching in advanced food microbiology with recommendation of a core curriculum. *Int. J. Food Microbiol.*, **11**, 107–117.

Parmar, N., Easter, M.C. and Forsyth, S.J. (1992) The detection of *Salmonella enteritidis* and *S.*

typhimurium using immunomagnetic separation and conductance microbiology. *Lett. Appl. Microbiol.*, **15**, 175–178.

Patchett, R.A., Back, J.P., Pinder, A.C. and Kroll, R.G. (1991) Enumeration of bacteria in pure cultures and in foods using a commercial flow cytometer. *Food Microbiol.*, **8**, 119–125.

Payne, M.J., Campbell, S. and Kroll, R.G. (1993) Lectin-magnetic separation can enhance methods for the detection of *Staphylococcus aureus, Salmonella enteritidis* and *Listeria monocytogenes. Food Microbiol.*, **10**, 75–83.

Perales, I. and Audicana, A. (1989) Evaluation of semi-solid Rappaport medium for detection of salmonellae in meat products. *J. Food Protect.*, **52**, 316–319.

Perales, I. and Mossel, D.A.A. (1989) Situacion actual y futuro de nuevos metodos rapidos para la comprobacion de la salubridad y de la calidad de los alimentos. Presented at the Symposium on "Nuevos metodos rapidos para el analisis de los productos lacteos y otros alimentos", 13–15 December 1989, Santander, Spain.

Persing, D.H. (1991) Polymerase chain reaction: trenches to benches. *J. Clin. Microbiol.*, **29**, 1281–1285.

Peterkin, P.I., Idziak, E.S. and Sharpe, A.N. (1991) Detection of *Listeria monocytogenes* by direct colony hybridization on hydrophobic grid-membrane filters by using a chromogen-labeled DNA probe. *Appl. Environ. Microbiol.*, **57**, 586–591.

Peterkin, P.I., Idziak, E.S. and Sharpe, A.N. (1992) Use of hydrophobic grid-membrane filter DNA probe method to detect *Listeria monocytogenes* in artificially contaminated foods. *Food Microbiol.*, **9**, 155–160.

Peterz, M. (1991) Comparison of Preston agar and a blood-free selective medium for detection of *Campylobacter jejuni* in food. *J. A. O. A. C.*, **74**, 651–654.

Peterz, M. and Steneryd, A.C. (1993) Freeze-dried mixed cultures as reference samples in quantitative and qualitative microbiological examinations of food. *J. Appl. Bacteriol.*, **74**, 143–148.

Pettipher, G.L., Watts, Y.B., Langford, S.A. and Kroll, R.G. (1992) Preliminary evaluation of COBRA, an automated DEFT instrument, for the rapid enumeration of microorganisms in cultures, raw milk, meat and fish. *Lett. Appl. Microbiol.*, **14**, 206–209.

Pfaller, M.A., Sahm, D., O'Hara, C. *et al.* (1991) Comparison of the AutoSCAN-W/A rapid bacterial identification system and the Vitek AutoMicrobic system for identification of Gram-negative bacilli. *J. Clin. Microbiol.*, **29**, 1422–1428.

Pierson, M.D. and Corlett, D.A. (1992) *HACCP*. Van Nostrand, Florence, Kentucky.

Poulis, J.A., de Pijper, M., Mossel, D.A.A. and Dekkers, P.P.A. (1993) Assessment of cleaning and disinfection in the food industry with the rapid ATP-bioluminescence technique combined with the tissue fluid contamination test and a conventional microbiological method. *Intl. J. Food Microbiol.*, **20**, 109–116.

Prescott, S.C. (1920) What should be the basis of the control of dehydrated foods? *Am. J. Public Health*, **10**, 324–326.

Ray, B. (1989) *Injured Index and Pathogenic Bacteria: Occurrence and Detection in Foods, Water and Feeds*. CRC Press, Boca Raton, Florida.

Reina, J., Hervas, J. and Borrell, N. (1993) Acute gastroenteritis caused by *Hafnia alvei* in children. *Clin. Infect. Diseases*, **16**, 443.

Roberts, T. (1985) Microbial pathogens in raw pork, chicken and beef: benefit estimates for control using irradiation. *Am. J. Agric. Economics*, **67**, 957–965.

Rodrigues, U.M. and Kroll, R.G. (1988) Rapid selective enumeration of bacteria in foods using a microcolony epifluorescence microscopy technique. *J. Appl. Bacteriol.*, **64**, 65–78.

Rodrigues, U.M. and Kroll, R.G. (1989) Microcolony epifluorescence microscopy for selective enumeration of injured bacteria in frozen and heat treated food. *Appl. Environ. Microbiol.*, **55**, 778–787.

Rodrigues, U.M. and Kroll, R.G. (1990) Rapid detection of salmonellas in raw meats using fluorescent antibody-microcolony technique. *J. Appl. Bacteriol.*, **68**, 213–223.

Rohner, P., Dharan, S. and Auckenthaler, R. (1992) Evaluation of the Wellcolex colour *Salmonella* test for detection of *Salmonella* spp. in enrichment broths. *J. Clin. Microbiol.*, **30**, 3274–3276.

Roszak, D.B., Grimes, D.J. and Colwell, R.R. (1984) Viable but nonrecoverable stage of *Salmonella enteritidis* in aquatic systems. *Can. J. Microbiol.*, **30**, 334–338.

Saha, S.K., Saha, S. and Sanyal, S.C. (1991) Recovery of injured *Campylobacter jejuni* cells after animal passage. *Appl. Environ. Microbiol.*, **57**, 3388–3389.

Salem, G. and Schantz, P. (1992) Toxocaral visceral larva migrans after ingestion of raw lamb liver. *Clin. Infect. Diseases*, **15**, 743–744.

Sallam, S.S. and Donnelly, C.W. (1992) Destruction, injury and repair of *Listeria* species exposed to sanitizing compounds. *J. Food Protect.*, **55**, 771–776.

Schmidt, H., Montag, M., Bockenmuhl, J. *et al.* (1992) Shiga-like toxin II-related cytotoxins in *Citrobacter freundii* strains from humans and beef samples. *Epidemiol. Infect.*, **109**, 397–403.

Scotter, S.L., Humphrey, T.J. and Henley, A. (1993) Methods for the detection of thermotolerant campylobacters in foods: results of an interlaboratory study. *J. Appl. Bacteriol.*, **74**, 155–163.

Shakespeare, A.P., Megson, G.M., Jones, P.A. *et al.* (1992) Evaluation of Mast-ID 15 system for identification of fresh clinical isolates of Enterobacteriaceae and *Acinetobacter*. *J. Clin. Pathol.*, **45**, 168–170.

Sharpe, A.N., Diotte, M.P., Peterkin, P.I. and Dudas, I. (1986) Towards the truly automated colony counter. *Food Microbiol.*, **3**, 161–184.

Silley, P. (1991) Rapid automated bacterial impedance technique (RABIT). *Soc. Gen. Microbiol. Quart.*, **18**, 48–52.

Silliker, J.H., Baird-Parker, A.C., Bryan, F.L. *et al.* (1988) *Micro-organisms in Foods 4. Application of the Hazard Analysis Critical Control Point (HACCP) System to ensure Microbiological Safety and Quality*. Blackwell, Oxford.

Skjerve, E. and Johnson, J. (1991) Safe food – industrial leadership. *Food Microbiol.*, **8**, 79–80.

Skjerve, E., Rørvik, L.M. and Olsvik, O. (1990) Detection of *Listeria monocytogenes* in foods by immunomagnetic separation. *Appl. Environ. Microbiol.*, **56**, 3478-3481.

Skovgaard, N. (1990) The need for continuous training in food factories. *Int. J. Food Microbiol.*, **11**, 119–125.

Slade, P.J. (1992) Monitoring *Listeria* in the food production environment. *Food Res. Int.*, **25**, 45–56, 203–214, 215–225.

Smulders, F.J.M., Barendsen, P., van Logtestijn, J.G. *et al.* (1986) Lactic acid. Considerations in favour of its acceptance as meat decontaminant. *J. Food Technol.*, **21**, 419-436.

Sorrells, K.M. (1981) Rapid detection of bacterial content in cereal grain products by automated impedance measurements. *J. Food Protect.*, **44**, 832–834, 838.

Stager, C.E. and Davis, J.R. (1992) Automated systems for identification of microorganisms. *Clin. Microbiol. Rev.*, **5**, 302–327.

Stannard, C.J., Petit, S.B. and Skinner, F.A. (1990) *Rapid Microbiological Methods for Foods, Beverages and Pharmaceuticals*. Blackwell, Oxford.

Starbuch, M.A.B., Hill, P.J. and Steward, G.S.A.B. (1992) Ultra sensitive detection of *Listeria monocytogenes* in milk by the polymerase chain reaction (PCR). *Lett. Appl. Microbiol.*, **15**, 248–252.

Stewart, G.N. (1899) The changes produced by the growth of bacteria in the molecular concentration and electrical conductivity of culture media. *J. Exp. Med.*, **4**, 235–243.

Szabo, R., Todd, E., McKenzie, J. *et al.* (1990) Increased sensitivity of the rapid hydrophobic grid membrane filter enzyme-labeled antibody procedure for *Escherichia coli* O157 detection in foods and bovine feces. *Appl. Environ. Microbiol.*, **56**, 3546–3549.

Tate, C.R., Miller, R.G. and Mallinson, E.T. (1992) Evaluation of two isolation and two non isolation methods for detecting naturally occurring salmonellae from broiler flock environmental drag-swab samples. *J. Food Protect.*, **55**, 964–967.

Tauxe, R.V. (1991) *Salmonella*: a postmodern pathogen. *J. Food Protect.*, **54**, 563–568.

Taylor, J.L., Dwyer, D.M., Groves, C. *et al.* (1993) Simultaneous outbreak of *Salmonella enteritidis* and *Salmonella schwarzengrund* in a nursing home. Association of *S. enteritidis* with bacteremia and hospitalization. *J. Infect. Diseases*, **167**, 781–782.

Tebbutt, G.M. and Midwood, C.A. (1990) Rapid and traditional methods of assessing cleaning standards in food premises. *Environ. Health*, **98**, 235–237.

Thomas, L.V., Gross, R.J., Cheasty, T. and Rowe, B. (1990) Extended serogrouping scheme for motile, mesophilic *Aeromonas* species. *J. Clin. Microbiol.*, **28**, 980–984.

Turpin, P.E., Maycroft, K.A., Rowlands, C.L. and Wellington, E.M.H. (1993) Viable but non–culturable salmonellas in soil. *J. App. Bacteriol.*, **74**, 421–427.

Van Damme-Jongsten, M., Rodhouse, J., Gilbert, R.J. and Notermans, S. (1990) Synthetic DNA probes for detection of enterotoxigenic *Clostridium perfringens* strains isolated from outbreaks of food poisoning. *J. Clin. Microbiol.*, **28**, 131–133.

Van der Giessen, A.W., Dufrenne, J.B., Ritmeester, W.S. *et al.* (1992) The identification of *Salmonella enteritidis*-infected poultry flocks associated with an outbreak of human salmonellosis. *Epidemiol. Infect.*, **109**, 405–411.

Van Netten, P., Van der Zee, H. & Mossel, D.A.A. (1984) A note on catalase enhanced recovery of acid injured cells of Gram-negative bacteria and its consequence for the assessment of the lethality of L-lactic acid decontamination of raw meat surfaces. *J. Appl. Bacteriol.*, **57**, 169–173.

Van Netten, P., Mossel, D.A.A. and Van de Moosdijk, A. (1987) Rapid detection of excessive cfu counts of Enterobacteriaceae, *S. aureus* and yeasts by nitrate reduction/glucose dissimilation (NGD) monitoring in selective media. *J. Food Sci.*, **52**, 751–756.

Van Netten, P., Van Gaal, B. and Mossel, D.A.A. (1991) Selection, differentiation and counting of haemolytic *Listeria* spp. on PALCAM medium. *Lett. Appl. Microbiol.*, **12**, 20–22.

Van Poucke, L.S.G. (1990) *Salmonella*-TEK, a rapid screening method for *Salmonella* species in food. *Appl. Environ. Microbiol.*, **56**, 924–927.

Vermunt. A.E.M., Franken, A.A.J.M. and Beumer, R.R. (1992) Isolation of salmonellas by immuno-magnetic separation. *J. Appl. Bacteriol.*, **72**, 112–118.

Visser, M.R., Bogaards, L., Rozenberg-Arska, M. *et al.* (1992) Comparison of the autoSCAN-W/A and Vitek Automicrobic systems for identification and susceptibility testing of bacteria. *Eur. J. Clin. Microbiol. Infect. Diseases*, **11**, 979–984.

Vugia, D.J., Mishu, B., Smith, M. *et al.* (1993) *Salmonella enteritidis* outbreak in a restaurant chain: the continuing challenges of prevention. *Epidemiol. Infect.*, **110**, 49–61.

Wachsmuth, I.K., Evins, G.M., Fields, P.I. *et al.* (1993) The molecular epidemiology of cholera in Latin America. *J. Infect. Diseases*, **167**, 621–626.

Wang, G.I.J. and Fung, D.Y.C. (1986) Feasibility of using catalase activity as an index of microbial loads on chicken surfaces. *J. Food Sci.*, **51**, 1442–1444.

Wang, R.F., Slavik, M.F. and Cao, W.W. (1992) A rapid method for direct detection of low numbers of *Campylobacter jejuni*. *J. Rapid Methods Automat. Microbiol.*, **1**, 101–108.

Weenk, G., Fitzmaurice, E. and Mossel, D.A.A. (1991) Selective enumeration of spores of *Clostridium* species in dried foods. *J. Appl. Bacteriol.*, **70**, 135–143.

Weenk, G.H., Struijk, C.B. and Mossel, D.A.A. (1994) Reliable and convenient methods for the enumeration of clostridia in foods. *Proc. Int. Symp. Food Microbiology and Hygiene*, Bingen (in press).

Westblom, T.U. and Milligan, T.W. (1992) Acute bacterial gastroenteritis caused by *Hafnia alvei*. *Clin. Infect. Diseases*, **14**, 1271–1272.

Widjojoatmodjo, M.N., Fluit, A.C., Torensma, R. *et al.* (1992) The magnetic immuno polymerase chain reaction assay for direct detection of salmonellae in fecal samples. *J. Clin. Microbiol.*, **30**, 3195–3199.

Wieneke, A.A. (1991) Comparison of four kits for the detection of staphylococcal enterotoxin in foods from outbreaks of food poisoning. *Int. J. Food Microbiol.*, **14**, 305–312.

Wilson, G.S. (1955) Symposium on food microbiology and public health: general conclusion. *J. Appl. Bacteriol.*, **18**, 629–630.

Wilson, G. (1964) The public health aspect of food microbiology. *Chem. Ind.*, 1854–1859.

Winter, F.H., York, G.K. and El-Nakhal, H. (1971) Quick counting method for estimating the number of viable microbes on food and food processing equipment. *Appl. Microbiol.*, **22**, 89–92.

Wirtanen, G., Mattila-Sandholm, T., Manninen, M. *et al.* (1991) Application of rapid methods and ultrasound imaging in the assessment of the microbiological quality of aseptically packed starch soup. *Int. J. Food Sci. Technol.*, **26**, 313–324.

Wolcott, M.J. (1991) DNA-based rapid methods for the detection of foodborne pathogens. *J. Food Protect.*, **54**, 387–401.

Yolken, R.H. (1991) New methods for the diagnosis of enteric infections. *World J. Microbiol. Biotechnol.*, **7**, 150–156.

York, M.K., Brooks, G.F. and Fiss, E.H. (1992) Evaluation of the autoSCAN-W/A rapid system for identification and susceptibility testing of Gram-negative fermentative bacilli. *J. Clin. Microbiol.*, **30**, 2903–2910.

2 Development and evaluation of membrane filtration techniques in microbial analysis

A.N. SHARPE

2.1 Brief history of membrane filters

Membrane filters (MFs) were used before agar was adopted by microbiologists, when Fick (1855) made nitrocellulose membranes as substitutes for animal membranes in dialysis experiments (Presswood, 1981). However, it was some time before a useful preparation method was developed (Zsigmondy and Bachmann, 1918), leading to commercial production by Sartorius-Werke AG (Gottingen, Germany). Before World War II, MFs were used primarily for sterilization. However, devastation of laboratories and contamination of water supplies by bombing raids led to their application for determining water safety by German hygiene institutes. After the war, the first MF methods for culturing organisms were published (Muller, 1947a, b).

Until 1963, MFs were predominantly manufactured from nitrocellulose or cellulose acetate esters. Today they are available in nylon, polyvinyl chloride, polysulfone, polycarbonate, polyester, and even metals. Most MFs, being prepared by deposition from solvent, have tortuous structures, with pore length to width ratios of 750 or more (Presswood, 1981). They have been valuable in techniques based on the growth of organisms to form colonies; for this, they are usually placed on agar plates or soaked pads of nutrients. More recently, the Nuclepore polycarbonate and polyester MFs (in which straight cylindrical pores are produced by etching the tracks of nuclear particles) have been introduced. Because these are extremely thin (10 μm) and are rendered invisible by the proper mountant, they have permitted development of sensitive techniques based on direct microscopic observation of microorganisms. Even more recently, MFs (particularly nitrocellulose and nylon) have found application as a vehicle in a variety of immunological and DNA-based 'blotting' techniques; however, this chapter will not cover that aspect, and will be confined to microscopic and growth techniques.

Some of the potential advantages and disadvantages of MFs, when used in modifications of conventional procedures, are listed below.

1. Filtration of larger volumes of sample through MFs can improve limits of detection, a feature exploited dramatically in water microbiology, in the direct epifluorescent filter technique (DEFT), and to a lesser extent in hydrophobic grid membrane filter (HGMF) methods.
2. Inoculation by filtration eliminates the possibility of injuring organisms by exposing them to hot agar.
3. Filtration removes most of any growth inhibitors or interfering substances in samples.

4. Since microbial cells are confined to upper surfaces, and receive nutrients by diffusion from beneath, MFs may be transferred from one type of nutrient surface to another without disturbing the individuality of the developing growths. Thus, MF techniques can provide injured cells with a full recovery period on a benign medium before exposing them to selective conditions, without losing the quantitative nature of the analysis.
5. The same property allows MFs to be transferred across media so that biochemical characteristics of growths can be determined.
6. Although colonies tend to be slightly smaller, the small growth area of conventional MFs (typically 35 mm for 47 mm circles) leads to relative colony crowding compared with plates. Thus, the overlap error introduced when one attempts to count crowded colonies may be worse on normal MFs. However, this does not apply to the HGMF, on which dense growth is encouraged, and which yields linear counts over four orders of magnitude.
7. One type of MF, the HGMF, permits the use of electronic counters which combine a satisfactory degree of reliability and versatility in the automated count.
8. Membrane filters, through their large surface area, can affect microbial systems by adsorbing selective agents, proteins, and metabolic products. The range of materials in which MFs are available allows surface properties to be exploited to advantage.
9. Analyses carried out using MFs (but particularly HGMFs) may be less expensive than conventional methods.

These attractions should be considered together with potential disadvantages:

1. MFs and HGMFs increase the direct cost of single agar plates. Their attraction thus depends on, for example, improved sensitivity, reduced analysis time, or on the potential for automation or their ability to reduce labor or the number of dilution steps or platings carried out.
2. Inoculation of MFs, if by filtration, increases the amount of labor required at the start of an analysis.
3. Some sample suspensions do not naturally filter easily through MFs, and preparing them in a suitable form may introduce additional steps to the analysis.

2.2 Early uses of membrane filters

Leaving aside the enormous consumption of MFs for clarification and sterilization in the electronics, pharmaceutics, and other industries, the major application of MFs is in the field of water microbiology. It is hard to imagine how public health authorities could accommodate the numbers of samples and statutory limits of detection necessary to maintain current levels of control over drinking-water supplies, sanitary operations, recreational waters, and so on, without the MF. There is an enormous body of literature devoted to MFs and water quality, which is mentioned only where it bears relevance to membrane filter food microbiology.

Up to 1980, although a few food microbiologists explored the potential advantages, MF methods in general excited little interest amongst regulators, authors of standard texts, and others in a position to encourage their development. The situation is changing, with increasing popularity of the direct plate, direct epifluorescent filter, and hydrophobic grid-membrane filter techniques, and particularly as molecular methods using antibodies and nucleic acid probe procedures become more practical and popular.

There are numerous descriptions in the literature of the use of MFs for rapid analyses of beers or wines, either by direct microscopy of the filtered sample, or after incubation to develop microcolonies. A staining procedure for direct microscopic counts on MFs was described by Ehrlich (1955) as applicable to 'liquids in which the concentration of microorganisms is relatively low', with discussion of the obscuration of bacteria by debris, though it is not clear just what liquids the technique was intended to be applied to. Schiemann (1955) used MFs in the control of brewery operations, both by a culture method and as a direct microscopic method. Haas (1956) filtered an unspecified volume of beer through a variety of MFs, examining the filter after 36–48 h incubation. Middlekauff *et al.* (1967) described a 2–3 h technique for detecting yeast colonies in beer; 10 l of beer were filtered through 0.45 µm MFs, incubated, and examined for microcolonies under a 400× power microscope. Nobile (1967) trapped yeast cells from beer and wines on 1.2 µm MFs but noted that bacteria passed through. Richards (1970) described filtration of beer samples (volume unspecified) through Sartorius MFs, incubation for 10–20 h, and staining with safranin followed by counting of the microcolonies. Bourgeois *et al.* (1973) concentrated *Saccharomyces carlsbergensis* cells from beer and detected them by uptake of [^{14}C]lysine. Later, the work was extended to other *Saccharomyces* spp., *Candida* spp., and to *Escherichia coli* and *Lactobacillus casei* (Mafart *et al.*, 1978, 1981). Combined filtration and epifluorescence microscopy was used by Portno and Molzahn (1977) and Paton and Jones (1975) for enumeration of yeasts in beer. Ingledew and Burton (1979) described filtration of yeast suspensions onto MFs during investigation of the chemical preservation of spent brewers' yeast. Ingledew and Burton (1980) also described a practice of shipping inoculated MFs in multiplant brewery control, and use of skim-milk powder in the filtrant (or an after-rinse) to deposit organic material with the cells to protect them from desiccation while in the mail. Apart from commenting that variations in flow rate existed between lot numbers of filters, these authors noted no difficulty in filtering 5–7 ml of 4% skim-milk (0.2 g of dry material) through 0.45 µm Millipore MFs.

Use of MFs for analysis of sugar liquids began almost at the same time as for beers and wines. Moroz (1957) filtered up to 10 g of cane syrup (after dilution), for determination of yeasts, but commented that blackstrap molasses was more difficult to filter. Coleman and Bender (1957) filtered 20 ml samples of liquid sugars used in soft drink manufacture. Halden (1957) and Halden *et al.* (1958, 1960) filtered solutions containing the equivalent of as much as 100 g of dry sugar (or 150 g of liquid sugar) through 0.45 µm MFs to detect low levels of bacteria, yeasts and molds. Halden *et al.* also noted (1960) that yeast counts on 0.45 µm MFs and 0.80 µm MFs were not significantly different, but the latter permitted use of larger sample sizes. Nakhodkina and Dvornikova (1971) commented that MF methods for sugar had shorter analysis times, reduced requirements for sterile media and dishes, and the MFs could be stained, fixed and stored for long periods.

One can understand early interest in the application of MFs in the microbio- logical analysis of beers, wines, and sugar liquids, since, notwithstanding the heavy loads of particulate unfermented starches, pectins and other materials beer worts, wine musts and syrups carry before clarification, they are easily perceived as inherently water-like and filterable. Indeed, filtration through depth or even membrane filters is, nowadays, a frequent precedent to bottling. However, considering the many potential advantages of MF techniques, MFs seem, until very recently, to have been largely ignored as vehicles for analytical microbiology of foods. It seems that food microbiologists tended to overlook membrane filters and any advantages they might carry, assuming that food suspensions would not filter adequately through them. The assumption is not unfounded, of course. An indication of filterability difficulties is seen, for example, in the use of combined protease and surtactant to assist the filtration of milk (Godberson, 1974). But, that it is by no means a real impediment is evidenced by the work of Bourgeois *et al.* (1984), in which 300 ml volumes of milk were filtered through 0.8 µm MFs to detect spores of *Clostridium tyro- butyricum.*

Much of the early literature concerns milk which, because of the number of tests carried out each year, accounts for a major portion of analytical food micro- biological effort. Barber *et al.* (1954) discussed uses of MFs for controlling sanitary operations in dairies, but felt they would be too easily clogged by other materials, such as milk. Fifield *et al.* (1957) used a filtration apparatus prewarmed with 0.1% Triton X-100 solution at 45°C to filter up to 10 ml of milk or 1 ml of cream. In 1958 a Reference Committee reported to the Subcommittee on Standard Methods for the Examination of Dairy Products of the American Public Health Association, on an interlaboratory study of 'Membrane Filters for Coliform and Total Counts in Milk'. Remarkably, although several brands of MF were investigated, no attempt to solubilize milk samples appears to have been made. The Committee concluded that, '. . . while it is true the MF technic does offer certain advantages, including transportation of samples, use of filters in the field, shorter incubation period . . . (etc.) . . . these advantages are partly, if not completely, nullified unless reasonably comparable determinations can be reported on milk by the MF method. The colloidal state, even in the diluted milk, is the chief obstacle to be overcome . . .' and that uniform agreement between laboratories should be a goal. For certain, the use of a surfactant in the protocol would have led to a very different conclusion.

Nutting *et al.* (1959) mixed 0.5 g of ice cream with 20 ml of warm Triton X-100, after the method of Fifield *et al.* (1957). The surfactant increased filterability by 5–20 times, over that of untreated aliquots, but only 0.1 ml of chocolate ice cream could be filtered. Reusse (1959) filtered 5 ml of milk, or 1 ml of cream (mixed in 50 ml of Triton X-100 solution) through 0.4 µm MFs, followed by incubation, in order to detect *Brucella*. Busta and Speck (1965) used membrane filtration during the enumeration of *Bacillus stearothermophilus* to remove inhibitors present in milk; up to 1.0 ml of milk was mixed with warmed Triton X-100 solution in the filter funnel and drawn through the MF. Graves and Schipper (1966) filtered very dilute milk suspensions through Millipore MFs, followed by 48 h incubation, and indicated that the MF method gave a more complete picture of the normal udder flora than a standard 18 h incubation technique. Duitschaever and Ashton (1968) filtered unspecified volumes of milks through (unspecified) Millipore MFs, comparing MF microscopic counts with direct microscopic counts. The MF method gave higher counts for somatic cells, but lower counts for bacteria. Revival of injured bacteria in milk by a preliminary incubation on trypticase soy agar, and a procedure for differentiating between injured and uninjured cells by use of MFs, was described by Goff *et al.* (1972). Claydon (1975) described resuscitation as an advantage of an MF technique for milk analysis.

Most other applications of MFs in milk analysis exploited their ability to concentrate organisms, to permit development of very rapid analyses based on direct microscopic counts. Thus, Merrill (1963) filtered 0.2 ml aliquots of milks (10 ml of 1 ml in 50 ml of Triton X-100 surfactant) through (brand unspecified, presumably 25 mm diam.) MFs. The preparations were wet-stained for direct microscopy to give an approximately 200-fold increase in sensitivity over the standard direct count. From 1977 on, the milk testing area has seen a rapidly developing interest in the DEFT. In this technique, milk samples are treated with trypsin and Triton X-100, concentrated on a polycarbonate membrane, stained with acridine orange, and examined under the epifluorescent microscope. The technique is ideally suited to rapid (25 min) analysis of the bacteriological quality of milk and has been adopted by several European Milk Marketing Boards.

A few other studies of foods are to be found outside the dairy industry. Murphy and Silliker (1955) and Silliker *et al.* (1957) exploited the retentive properties of MFs in a non-destructive sampling method for meats; MFs were pressed against the meat and transferred to agar plates for incubation. Kirkham and Hartman (1962) used MFs for resuscitation in the analysis of *Salmonella* in egg albumen; samples were hydrolyzed with pepsin, ficin, and protease, clarified with Celite, after which aliquots equivalent to 10 g of albumen were filtered through 0.45 µm MFs, incubated and examined for suspect colonies. The MF method gave a better recovery than the most probable number (MPN) method.

Frazier and Gneiser (1968) described a short-time MF method for estimating rinsings or swabbings from the surfaces of food or food-processing equipment; at least 250 ml of suspension were filtered through 0.45 μm Millipore HA-type MFs, rinsed, and incubated for 8 h or more before being stained and examined at 315× power under a microscope. Meynell and Meynell (1970) commented on better recovery of bacteria by MFs compared with pour plates because of the avoidance of thermal death. Winter *et al.* (1971) filtered 10 ml volumes of wash water from vegetables, or swab samples of food processing surfaces, first through 8.0 μm MFs to remove large particles, then through 0.45 μm filters. The MFs were incubated for 4–6 h before being stained and examined for micro-colonies.

Dempster *et al.* (1973) filtered 5 ml of swab-rinse suspensions of cooked meats through 0.45 μm MFs, followed by a 2 h resuscitation period and incubation for 18 h. The direct plating procedure of Delaney *et al.* (1962) which exploited the stainability of an MF for indole formed during colony growth (used initially for enumeration of *Escherichia coli* biotype 1 in water) was adapted for use with foods with great success by Anderson and Baird-Parker (1975). The food suspension is not filtered but simply spread over a large diameter MF laid on the surface of a predried (i.e. absorbent) agar plate, into which it seeps over the course of minutes or hours. Ercolani (1976) filtered up to 3 l of wash water from vegetable suspensions, first through coarse cheesecloth to remove large particles, then through 47 mm 0.45 μm Millipore MFs. More recently, Richards (1979) found the Millipore Coli-Count Samplers as good as the MPN method for detecting fecal coliforms in crabs. The 1976 *Compendium of Methods for the Microbiological Examination of Foods* (APHA, 1976) mentioned MFs as being useful for 'certain foods' (without specifying which these might be), to improve the accuracy of quantitative microbiological analyses by allowing the testing of relatively large samples. However, the latest compendium (APHA, 1992) has large sections devoted to MF and HGMF methods for total count, coliforms, *E.coli* and *Salmonella*.

A few other areas might be considered relevant to food microbiology, from the viewpoint of sample preparation, filtration and other factors. Kenner *et al.* (1957) isolated *Salmonella* after membrane filtration of selective enrichments of feces, followed by serological identification of colonies. Jannasch (1958) commented on the effect the amount of organic matter may have on the deviation between MF direct microscopic counts and the Standard Plate Count. MacLeod *et al.* (1966) described a rapid radiometric method for enumerating bacterial cells involving collection on a Millipore HA filter, allowing them to take up ^{32}P labeled phosphate, followed by radiometric detection. A claimed detection limit of 23 cells ml[-1] was later (1970) revised to 5×10^4 cells ml[-1]. Crum (1971) described the collection of *Pseudomonas aeruginosa* from jet fuels, cutting liquids and lubricants on 0.45 μm Millipore HA filters. Organisms were enumerated radiometrically after allowing them to take up ^{32}P from labeled phosphate buffer. Correlations with growth data were adequate. This work was undertaken with a view to marketing test kits, but it appears little came of it. Schrot *et al.* (1973) described the detection of as few as 100 cfu of *Pseudomonas*, *Klebsiella*, *Shigella*, and other species in blood, by filtering lysed blood through Millipore MFs, followed by radiometric determination of $^{14}CO_2$ release during incubation on a labeled substrate. Green *et al.* (1974) quantified antibody for radio-immunoassay by collecting gonococcal cells on cellulose acetate or polycarbonate MFs and incubating filters with ^{121}I-labeled antisera. Petersen *et al.* (1973) filtered 50 ml of handwashings through 0.45 µm Millipore MFs in order to collect and count organisms released by handwashing. Hobbie *et al.* (1977) discussed the value of Nuclepore polycarbonate MFs for microscopic examinations. Oxborrow *et al.* (1976) used MFs to purify microbial cells for identification by pyrolysis gas chromatography.

Two rather interesting instruments utilized filtration through an MF. An instrument referred to as Vitatect II (Vitatect Corp., Alexandria, VA), in which samples from water monitoring or clinical infection studies were filtered through a moving MF tape prior to the detection/estimation of microorganisms by measuring adenosine triphosphate, was described by Picciolo *et al.* (1977), with indications that food applications would be investigated. However, it is not known whether this came about. A semi-automated fluorescent antibody detection system for *Salmonella*, incorporating a membrane filtration step (Automated Bioassay System or ABS, Organon Inc.) was evaluated by Thomason *et al.* (1975). The instrument was tested with creek water, powdered egg, meatmeal, poultry, sausage and candy samples; it was also evaluated by Munson *et al.* (1976), for milk powder, dried yeast and frogs legs. In this instrument the filtrant was normally taken from the enrichment culture. Thus, the food would have been diluted considerably, which may have aided filtration. The instrument was developed in anticipation that the US Food and Drug Administration would mandate heavily on *Salmonella* in foods, and was tested extensively by that agency. In the event, the anticipated ruling did not happen and only two instruments were made. A manual technique, also based on fluorescent antibody staining of enrichment cultures which had been filtered through 0.4 µm polycarbonate MFs, was described by Early and Patterson (1982). The MFs were then stained and examined under the fluorescent microscope. Membrane filtration improved the correlation between the fluorescent antibody method and the conventional cultural procedure.

Finally, it may be of interest that MFs have not shown much promise for detecting or enumerating microorganisms in air, despite its filterability. Bacteria trapped on dry conventional MFs die quickly, probably through desiccation. Their survival is better on gelatin filters; however, MFs have generally yielded recoveries only 66–79% of the (standard) Anderson Sampler (Fields *et al.*, 1974; Hambraeus and Benediktsdottir, 1980; Lundholm 1982). For controlling air quality around food processing operations, the various commercial slit-samplers, Anderson Sampler, or simple settling plates, appear to be preferable.

Thus, the literature has not bulged with food-related applications of MFs, even though indications from uses outside of water, fermented beverage and milk analysis, suggest that problems of filterability of samples have rarely been severe. The real extent of filterability problems, and the solutions where problems exist, were extensively studied during the development of HGMF methods. Food microbiology might, however, have benefited considerably from a serious investigation of the attractions of MF methods at a much earlier stage.

2.3 Inter-brand variability of membrane filters

Although there exists a large body of literature concerning differences between MFs in their ability to recover microorganisms, there is no need for concern that this affects their performance in food microbiology. The literature is confined to detection of (principally fecal) coliforms in water by early types of MFs. The loss of count is due to dehydration when sensitive organisms fail to penetrate significantly into the structure of the MF (Sladek *et al.*, 1975). Some MFs naturally have wider surface pores, into which organisms are drawn, and are protected from excessive drying. Most manufacturers of low-recovery MFs took steps to correct the problem; the Millipore Corporation, for example, adopted a different method of manufacture for its HC-type filters which are used for fecal coliform analyses on water. In any case, it should be noted that suspensions prepared for microbiological analysis of foods contain debris, which is deposited on the MF with the microorganisms. Food debris forms a mat which protects the organisms at least as well as they would be on a 'good' MF; in fact, it eliminates differences in performance between MF brands (Farber and Sharpe, 1984; Farber *et al.*, 1985).

2.4 Current status of membrane filter methods

The outlook for MF methods is improving. Even by 1983, the DEFT was being used in over 20 laboratories in Europe, and by the Jersey Milk Marketing Board for the examination of farm supplies on a payment-by-quality arrangement. It also appears to be gaining popularity as a rapid screening method for tanker milk at dairy factories, and there are instruments available to automate the operations. Manually, however, it requires a significant amount of sample preparation, and in some countries seems to be regarded as tedious compared with, say, analysis of milk for pyruvate by Autoanalyzer (Sharpe, 1987).

In a comparative study the International Commission on Microbiological Specifications for Foods (ICMSF) compared the Anderson–Baird-Parker direct plate technique (DP) with the MPN technique for determination of *E.coli* in meat products (Rayman *et al.*, 1979), and recommended the technique for use with meats. The HGMF has received First or Final Action approvals by the

Association of Official Analytical Chemists (AOAC), for aerobic colony count, total coliform, fecal coliform, *E.coli* and *Salmonella* determination in foods. The Millipore Corporation markets a range of dip-type samplers for monitoring microorganisms in liquids, and various companies market kits of nitrocellulose membrane microfiltration apparatus and reagents for RNA/DNA hybridization tests, which have potential use in food microbiology. The future for MF methods in food microbiology appears bright, even if it must be said of today that the majority of food microbiologists probably do not even possess an apparatus for membrane filtration.

2.5 Analyses based on membrane filters

2.5.1 Campylobacter *species*

Being small (0.2–0.8 µm wide) bacteria, the heat-tolerant campylobacters are able to pass through conventional 0.45 or 0.65 µm MFs. Thus, filtration through 0.65 µm MFs can greatly increase the selectivity of campylobacter detections, although 80% of organisms may be retained on the filter. Park *et al.* (1981) filtered 5 ml of enrichment culture through 0.65 µm MFs to isolate *C.jejuni* from fresh chicken. However, some strains of *Pseudomonas* and *Proteus* spp. are also filterable (APHA, 1984). In one method of detection, membrane filtration through a 0.65 µm pore size MF is recommended if non-campylobacters are predominant after examination of the enrichment culture by phase-contrast microscopy (APHA, 1984).

   In the method of Steele and McDermott (1984), a 0.45 or 0.65 µm MF (47 mm diam.) is laid on the surface of a Columbia agar plate (plus cycloheximide and defibrinated horse blood) and the sample suspension (approximately 500 µl) placed on top and allowed to seep through for 30–45 min. The MF is then removed and the plate incubated.

   Bolton *et al.* (1988) compared membrane filtration using the method of Steele and McDermott (1984) against Skirrow's agar, Preston agar, modified CCD agar and Fennell's medium, for isolation of *Campylobacter* spp. from feces. A 0.45 µm MF technique was the only one to isolate all catalase-negative campylobacter strains, though it was less effective overall. Further evaluation showed that more strains of *C.jejuni* and *C.coli* were isolated by using 0.65 µm MFs.

   Albert *et al.* (1992) compared the MF technique against a blood-free medium technique in which blood is replaced by charcoal (Bolton *et al.*, 1984). While *Campylobacter* spp. were isolated 213 times out of 225 positive patients by using the MF technique, and only 75 times by using the blood-free medium, the authors concluded that the combined use of a selective medium and a non-selective medium with the filter technique was better than either medium alone for isolation of *Campylobacter* spp. from feces.

Baggerman and Koster (1992) concluded that the time for isolation of *C.jejuni* and *C.coli* from fresh and frozen foods can be reduced from the traditional 4 to 3 days by using the Steele and McDermott MF technique. A further reduction to 2 days can be achieved by using the Falcon MF unit (Falcon 7102 membrane filter unit, 0.45 μm pores, with cellulose nitrate filter, Becton and Dickinson Labware, Oxland, CA) for combined enrichment and isolation. The 2-day period required for confirmation could be reduced to 2 minutes by using the Microscreen latex agglutination test (Mercia Diagnostics Ltd, Shalford, Surrey).

In an interlaboratory study (Scotter *et al.*, 1993) of detection of *Campylobacter* on minced chicken skin, the (ISO 1989) technique was significantly superior at levels of 2 or 20 cells g^{-1}, but the extensive manipulations involved were deemed not likely to be well received in a busy laboratory. A non-selective blood agar medium in combination with the MF technique or pre-enrichment of samples with gradual addition of antibiotics to suppress competing organisms, were preferred.

2.5.2 Escherichia coli *biotype 1 count*

Determination of *E.coli* has been one of the cornerstones of MF microbiology, and for this reason it is covered in some detail, although the rivalry has probably become irrelevant following the introduction of glucuronidase-based tests (e.g. see Opti-MUG, below). However, the methods still have much to recommend them.

The MF procedure first described for enumeration of *E.coli* in foods is known as the DP method (Anderson and Baird-Parker, 1975). The DP procedure is derived from a method described by Delaney *et al.* (1962) for detection of *E.coli* in rivers and sewage. Closely related to it are HGMF procedures described by Sharpe *et al.* (1983b), and by Entis (1983, 1984a,b) and AOAC (1985), differing from each other mainly in the procedure used to stain for indole. In all these methods, *E.coli.* colonies are demonstrated by means of the acid-catalyzed condensation reaction of 4-dimethylaminobenzaldehyde (DAB) with indole (Ehrlich reaction) to produce a red stain in or around the colony.

Correct matching of staining procedure to MF material is essential in these methods. Retention (in the colony or in the membrane) of indole and the colored products of the Ehrlich reaction depends on the filter material. Indole is strongly adsorbed by cellulosic MFs (Millipore HAWP, Oxoid Nuflow, etc.) used in the DP procedure, and yields a very intense stain with the procedure described. Indole is not adsorbed by the polysulfone material of the ISO-GRID HGMF and a more sensitive staining procedure is used which detects it in the colonies rather than the surrounding membrane (Sharpe *et al.*, 1981). HGMF staining procedures use the indole reagent (DAB) at one-tenth of the concentration called for by the DP procedure. In the DP method, indole-negative colonies stain yellow–brown, with an intensity which sometimes makes classification difficult. With a lower concentration of the reagent, this yellow–brown staining is eliminated. Colonies show pink–red against a white background, with less risk of incorrectly classifying colonies as indole-positive. In all techniques precise incubation control is essential, since a slightly low temperature reduces specificity, while *E.coli* may not grow at a slightly higher temperature. A water-jacketed incubator is preferable; if an air incubator is used, it should be checked for its ability to maintain $44.5 \pm 0.5°C$.

Both the DP and HGMF methods require only one day to execute, if one accepts the assumption that growth at $44.5°C$ on tryptone bile agar (TBA), with production of indole, is sufficient to identify *E.coli*. If the assumption is not acceptable, at least one extra day and extra labor will be required. Since drying during the staining procedure increases the concentration of hydrochloric acid in the MFs to approximately 7M (Sharpe *et al.*, 1981) and effectively sterilizes them, any sub-culturing needed for confirmation must be done prior to staining, either by picking out colonies or by making a replicate or blot-transfer of the MF.

The assumption that indole production is sufficient to identify colonies of *E.coli* is not entirely correct, since about 3–5% of indole-positive isolates do not confirm as *E.coli* (Anderson and Baird-Parker, 1975; Rayman *et al.*, 1979; Yoovidhya and Fleet, 1981). The time-saving usually outweighs this disadvantage. The techniques will not enumerate those *E.coli* incapable of producing indole; however, only about 1% of *E.coli* strains fall in this category (Ewing, 1972). At the same time, the MF methods enumerate anaerogenic and late lactose-fermenting strains of *E.coli* biotype 1 which are missed by the MPN method, and many of which are enteropathogenic (Arbuzova, 1970; Molchenov, 1970; Mehlman *et al.*, 1974). According to Ewing (1972), these comprise as much as 10% of *E.coli* strains. Anderson and Baird-Parker considered indole formation to be more characteristic of the organism than lactose fermentation; during testing on 843 samples of food, 95% of presumptive colonies were identified as *E.coli* and a further 3.4% as fecal coliforms.

Performance of the DP and the HGMF methods has been examined in several studies. Both techniques performed excellently in the majority of their evaluations, and disappointingly in one. Since MF methods can incorporate a resuscitation step, during which stressed organisms recover, counts of *E.coli* which have been frozen or otherwise stressed are at least equivalent to more traditional procedures incorporating cumbersome liquid or agar overlay resuscitation steps. Holbrook *et al.* (1980) found that 90% of freeze-stressed cells were recovered by DP after a 4 h resuscitation period, compared with about 20% without. While the temperature for resuscitation is important, it is not critical; whereas 5 h was needed to get maximum recovery at $20–30°C$ with freeze-stressed cells, 2 h was sufficient at $37°C$.

The HGMF technique was evaluated in an AOAC collaborative study against the AOAC (AOAC, 1984) or American Public Health Association (APHA, 1976) pour plate count methods for raw milk, raw poultry meat, whole-egg powder, spices and flours. Counts from the methods were statistically equivalent on milk, poultry, flours and spices, but were significantly higher by the HGMF method for whole-egg powder (Entis, 1984a,b). The HGMF method was accorded Official First Action status (AOAC, 1985). Subsequently, however, the method was superceded by the more efficient Opti-MUG procedure which also gained AOAC official action status.

The DP method was evaluated by Yoovidhya and Fleet (1981) against the A-1 MPN procedure (Andrews and Presnell, 1972) and the Standards Association of Australia (SAA) MPN procedure for enumerating *E.coli* in Sydney rock oysters. The DP procedure gave counts which were not significantly different from the SAA method, although mean counts were slightly lower. At high count levels, its reproducibility was comparable to that of the SAA method; however, at levels lower than 1 cell g^{-1} it was less reproducible, and it was necessary to improve the sensitivity of the DP method by the use of 110 mm circles of MF, laid in large diam. plates, to match the detection level of 2–5 cells g^{-1}. The method was accepted by the SAA for the enumeration of *E.coli* in foods (AS1766.2.12), including oysters, and has been used since 1981 by the New South Wales Department of Health for oyster analyses. The 'Australian' DP method was given a less enthusiastic report by Motes *et al.* (1984) after comparison with a 'French' MPN method (Mackenzie *et al.*, 1948), a 'British' roll-tube method (Reynolds and Wood, 1956) and the 'American' APHA (1970) MPN method for the enumeration of *E.coli* in estuarine waters, oysters, mussels and clams. The DP counts were significantly lower than those obtained with the APHA procedure for all sample types; only the roll-tube method compared favorably with the APHA MPN method for mussels and waters.

The DP method (with resuscitation step) was evaluated for use with raw meats in an international comparative study for the ICMSF (Rayman *et al.*, 1979). It yielded higher counts than the MPN procedure of the APHA (APHA, 1976) for frozen samples, though counts for samples which were not frozen did not differ significantly, and it detected 823 out of the 868 positive samples, whereas the MPN method detected 797. It was concluded that the DP procedure was preferable to the MPN method on account of lower variability, better recovery from frozen samples, decreased media requirement, and decreased analyst time. The DP and an HGMF method were compared against the MPN procedure MFA-19 of the Health Protection Branch (Sharpe *et al.*, 1983a) for a variety of foods inoculated with a range of *E.coli* strains. Both MF methods gave higher recoveries than the MPN procedure and both were concluded to be preferable to the MPN procedure. None of the methods performed satisfactorily with bean and alfalfa sprouts containing high levels of *Klebsiella* spp. which also produced indole. In an ICMSF study in which an HGMF method was compared against the DP procedure for a range of naturally contaminated foods (Sharpe *et al.*, 1987), five of the eight participating laboratories obtained higher counts by DP procedure. As a result of this the HPB indole staining procedure was later changed to that of the AOAC method (i.e. 15–20 min on the bench, rather than 5 min in a 35°C incubator).

Although the DP and HGMF procedures are similar in principle, differences exist which may make one or the other more attractive in individual situations. Limits of detection (LOD) of the techniques may often settle the choice. Microbial standards for foods have often been set on the basis of the LOD of the MPN method. Since each of the first dilution MPN tubes can receive 1.0 ml of inoculum, the LOD of the DP method is inferior to that of the MPN method unless at least two 1.0 ml aliquots of suspension are used. In the HGMF methods, inoculation is usually by vacuum filtration; for suitable foods this allows analysis of a larger aliquot, and the improved LOD may be attractive when the acceptance level for *E.coli* contamination is low. At the same time, the HGMF's larger numerical range usually eliminates the need to make serial dilutions in order to cover an adequate counting range.

2.5.3. Pediococcus *spp.*

A 4 h microscopic method for detecting these spoilage organisms in brewers pitching yeast was described by Whiting *et al.* (1992). Usually, they are present only in very low concentrations compared with the yeast cells. The method uses preliminary centrifugation of suspensions at $5100 \times g$, followed by filtration through 0.45 µm filters, staining by means of a fluorescein-conjugated monoclonal antibody to the target organism, and microscopy at $1000\times$ magnification. The limit of detection of the method is two orders of magnitude lower than previous microscopy methods, and at least one order of magnitude lower than the lowest pediococcal level known to cause an off-flavor in beer.

2.5.4 Epifluorescence microscopy

The DEFT count described by Cousins *et al.* (1979) and Pettipher *et al.* (1980) rapidly gained acceptance as a rapid, sensitive method for enumerating viable bacteria in milk and milk products. The count is completed in 25–30 min and detects as few as 6×10^3 bacteria ml^{-1} in raw milk and other dairy products, three to four orders of magnitude better than direct microscopy (Pettipher and Rodrigues, 1981). Because it is a microscopic technique, the DEFT gives some evidence of the presence of yeasts, molds, bacteria or other contaminants. Stained preparations are stable for up to one month, offering the possibility of examination at a central facility.

The sensitivity of the DEFT results from the concentration of cells by membrane filtration before staining. Its ability to distinguish live and dead bacteria comes from use of the nucleophilic fluorochrome *acridine orange*, which fluoresces at different colors in cells during different phases of growth (Hobbie *et al.*, 1977). However, the actual mechanism is complex and the degree and color of fluorescence achieved depend on factors such as the amount of RNA present in cells (RNA fluoresces red, DNA green) and the permeability of the cell wall to acridine orange (Pettipher, 1989). Generally, viable cells fluoresce orange–red, while dead ones fluoresce green. This rule applies only to bacteria, and there are exceptions; for example, heat-killed cells may fluoresce orange. Streptococci (e.g. in starter cultures) may be problematic in keeping their red fluorescence even when dead, but treatment with lysozyme/EDTA to damage cell walls may prevent this.

Performance of the DEFT and its agreement with the standard plate count were verified by the National Institute for Research in Dairying (UK), and the count is used by the several Milk Marketing Boards to assess farm milks for payment purposes. Kits of reagents and other disposables for the procedure are available (Micro-Measurements Ltd, Saffron Walden, Essex). It appears that the technique can be used successfully on a variety of other foods; for frozen meat and fish, for example, DEFT and plate count data fitted the regression line $y = 0.47 + 0.88x$ with a correlation coefficient of 0.94 (Pettipher and Rodrigues, 1982). However, for some dried products, the DEFT count exceeded the plate count by factors of three to twenty.

For raw milk, the MCC or membrane clump count (i.e. the count of *clumps* of bacteria on the membrane without reference to the number of organisms each clump contains) agrees most closely with the standard plate count. The hygienic quality of milk is, however, most closely related to the total number of viable bacteria. Thus, the membrane single cell count (the total count of orange-fluorescing bacterial cells) may be a more realistic measure of its bacteriological activity (Pettipher *et al.*, 1980), and this may also apply to foods.

The initial investment for fluorescence microscopy equipment is somewhat high; thereafter the count is inexpensive to perform. An automated counting system is available (Micro-Measurements Ltd). It removes the problem of eye fatigue and reportedly gives good agreement with visual counts (Pettipher and Rodrigues, 1982). Very recently, a completely automated DEFT system was announced (Cobra 2024, Biocom, France). It features three computers, attending to the sample preparation, staining, filtration, drying and image analysis stages. The manufacturers claim a sample throughput rate of 150 samples h^{-1}, repeatability better than the (manual) DEFT between 20000 and 1 million cfu ml^{-1} with a lower counting limit better than 20000 cfu ml^{-1}. Cost is approximately US$120000.

For the milk analysis, 2.0 ml of milk (or a dilution in one-quarter strength Ringer's solution, if necessary) is incubated with 0.5 ml trypsin solution and 2.0 ml Triton X-100 solution at $50 \pm 1°C$ for 10 min. The treated milk is filtered through a prewarmed Nuclepore MF. The membrane is overlaid with 2 ml acridine orange solution for 2 min, sucked clear and rinsed with citrate–NaOH buffer and then isopropanol. After air-drying, the membrane is mounted in non-fluorescing immersion oil beneath a cover slip and examined under the epifluorescence microscope. At least 15 fields may be counted.

For foods, the method is similar, but sample suspensions are preferably prepared by Colworth Stomacher to minimize the debris content, then prefiltered through 5 μm pore size mesh prefilters, before receiving the surfactant-enzyme treatment.

2.6 HGMF

The HGMF is the basis of the most highly developed of the membrane filter-based analytical systems. The HGMF was developed by the Canadian Health Protection Branch as a vehicle for automated counting (Sharpe and Michaud, 1974; Sharpe, 1982; Sharpe and Peterkin, 1988). The ISO-GRID HGMF™, the automated counter (HGMF Interpreter) and related system items are sold worldwide by QA LifeSciences Inc (6645 Nancy Ridge Dr., San Diego, CA, USA), and Richard Brancker Research Ltd (27 Monk St., Ottawa, Ont, Canada). In addition to techniques for total aerobes, several techniques for detection of pathogens have been developed.

Figures 2.1 and 2.2 show the 1600 grid-cell ISO-GRID HGMF (QA LifeSciences Ltd) and the typical appearance of incubated filters. The black hydrophobic print divides the MF into discrete growth compartments (the grid-cells) which are usually (though not always) inoculated by filtration. During incubation, microbial growth within grid-cells results in the development of square 'colonies' over the HGMF surface. The lateral restraint isolating growths to particular grid-cells results entirely from the hydrophobicity of the grid-lines; there is no other mechanical barrier. On first acquaintance, the HGMF's most striking feature is its enormous numerical range compared with familiar techniques, and the way counts can reliably be made on HGMFs which experience with agar plates would suggest are too crowded for use. However,

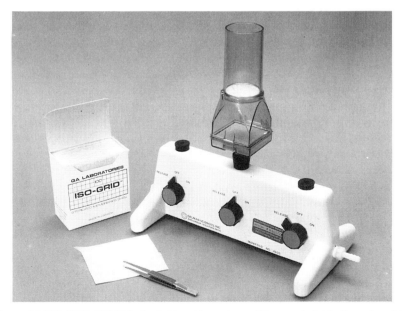

Figure 2.1 ISO-GRID HGMF hydrophobic grid membrane filters are individually packaged in boxes of 100. Border marks divide HGMFs into 2, 5 or 10 portions to assist manual counts and alignment under a replicator.

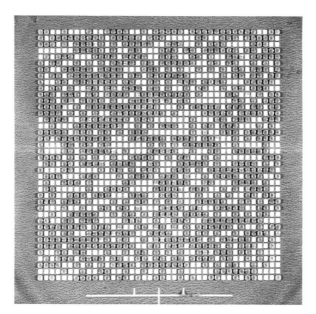

Figure 2.2 Coliforms on an ISO-GRID-HGMF, incubated on mFC agar.

for a laboratory making many bacterial counts each day, its advantages are likely to be realized in reduced technical time for making and plating dilutions, in counting plates and recording data, and in considerably lower costs per analysis.

There are HGMF-based analyses for many common foodborne microorganisms; this permits a unified approach to laboratory procedures. Many HGMF techniques (Aerobic Plate Count, coliform, fecal coliform, *E.coli* and *Salmonella*) enjoy AOAC Official Action and HPB Method or Laboratory Procedure status, and HGMF procedures are described in the *Compendium of Methods for the Microbiological Examination of Foods* (APHA, 1992). The book by Sharpe and Peterkin (1988) reviews most HGMF and MF procedures, including fecal streptococci, *Staphylococcus aureus*, *Vibrio parahaemolyticus*, yeasts and molds, *Clostridium perfringens*, *Pseudomonas aeruginosa* and *Yersinia enterocolitica*.

Sample preparation consists of dispersing by Stomacher, digesting with an enzyme if necessary, vacuum filtering, and laying the HGMF on growth medium. Serial dilutions are usually unnecessary. A unique apparatus called Spreadfilter relies on the hydrophobic border of the HGMF to retain sample volumes up to 10 ml. One uses the pipette to dispense the sample onto the HGMF and spread over the HGMF. The rotatable filter head makes this easy. The Spreadfilter can be immediately reused without sterilization; it is ideal for plating enrichment cultures on HGMFs.

Cost per analysis by HGMF compares very favorably with that of other techniques. For example, Chain and Fung (1991) estimated the cost of doing aerobic plate count by ISO-GRID HGMF was US\$3.33, compared with US\$13.62 by standard APC, and US\$8.22 by both Petrifilm and Redigel.

2.6.1 Numerical range and accuracy of the HGMF

The ISO-GRID HGMF possesses a large numerical range, and provides accurate counts at high growth densities when, within grid-cells, growths manifestly overlap. The explanation lies in the way it exploits the MPN mathematic, rather like a 1600-tube, single-dilution MPN test. The HGMF grew from a theoretical re-examination of food microbiological analysis, which showed that MPN systems are better suited to electronic counting than are random growths on Petri dishes (Sharpe, 1978) and that, given sufficient growth compartments, these methods can be more precise than plate counts. By arranging that counters (human or electronic) count only those HGMF grid-cells containing the growth of interest, the colony overlapping which degrades conventional plate or MF counts is actually exploited. Under normal circumstances an HGMF comprising 1600 grid-cells accommodates inocula having a concentration range of 4-log cycles. Accepting this calls for an act of faith in mathematics, but the theory is amply upheld by experiment (Sharpe and Michaud, 1974, 1975, 1978; Sharpe *et al.*, 1979, 1981, 1983a,b, 1986; Sharpe, 1981; Brodsky, 1982; Brodsky *et al.*, 1982a; Entis 1983, 1984a,b, 1986; Entis and Boleszczuk, 1983, 1986; AOAC, 1985, 1986; Lin *et al.*, 1984)

The ability of HGMFs to yield 'linear counts' at levels greater than suggested by the number of grid-cells poses no difficulty once the principle is understood. However, the conceptual difference between agar plates (or ordinary MFs) and HGMFs must be emphasized. The user sees microbial growths on an HGMF, just as on a plate or MF, but must avoid all tendency to assume that they should be counted individually. The user simply scores every grid-cell which contains growth of the target organism as *one*, no matter how many individual colonies may be visible within it. The wide numerical range is only manifested after inoculation by filtration, which ensures that each HGMF grid-cell receives the same inoculum volume as its neighbors. With this condition fulfilled, the equation relating an HGMF score to the *most probable number of growth forming units* (MPNGU) inoculated onto it, is

$$MPNGU = -N \log_e \frac{(N-X)}{N}$$

where N is the number of grid-cells on the HGMF (1600 for the ISO-GRID HGMF), and X is the number of positive grid-cells. This equation holds throughout the range of positive grid-cells until the HGMF becomes saturated. The absolute upper 'counting' limit of the HGMF occurs when all but one of its grid-cells (1599 for the ISO-GRID HGMF) are positive for the organism of interest. Thus the maximum MPNGU value for an ISO-GRID HGMF is 1.2×10^4. In practice, the precision of counts taken at this level will be unacceptably poor because the wide spread between upper and lower 95% confidence limits when there are very few negative (empty) grid-cells mirrors the spread when there are very few positive ones. The maximum recommended counting level is currently 95% of saturation (1520 positive grid-cells on the ISO-GRID HGMF) for which the MPNGU is about 4800 (Sharpe et al., 1983b). As with MPN, plate, or other MF counts, very high levels of competing flora may make enumeration of the target organism difficult by reducing the color or stain intensity in positive HGMF grid-cells. The effect depends on local conditions and is occasionally a problem.

2.6.2 Counting and scoring HGMFs

While electronic counters are now available, early users of HGMFs were forced to count them manually. HGMFs bearing moderate numbers of positive grid-cells are no more difficult or tedius to count than conventional plates. However, they can be forbidding when full of growth. The problem is in the confusion the human eye experiences when attempting to focus on individual grid-cells against the background of the extended grid pattern. The manual 'Linecounter' a viewer equipped with a horizon bar and electronic tally, greatly assists the manual count (Sharpe et al., 1983b). Linecounters (with microprocessor-based touchpads for automatic MPNGU calculation) are available from QA LifeSciences. Using this, analyst variability was generally less than 2.5% between 41 analysts from six Health Protection Branch laboratories when counting HGMFs containing red (indole-stained) colonies of E.coli (Sharpe et al., 1983b). Similar variability was obtained during a comparative study of E.coli counts from foods carried out for the ICMSF (Sharpe et al., 1987). These figures compare excellently with those obtained by analysts counting conventional plates (Fruin and Clark, 1977).

Today, computerized HGMF counters are available, which count an HGMF in approximately 2 s, and record the results in a file suitable for import to a database program. Using a histogram of reflected intensities from the grid-cells, the instrument normally identifies and counts positive grid-cells without assistance from the operator. The earlier (MI-100) Interpreter has recently been superceded by the MI-200, which uses a color TV system (Figure 2.3).

2.6.3 Some well-established HGMF-based techniques

2.6.3.1 Aerobic colony count. A method based on incubation on agar containing Fast Green FCF dye received official action status of the AOAC several years ago (Entis, 1986; Entis and Boleszczuk, 1986). The method involves incubation for 1–2 days. Colonies are green to blue. Because the range of hues produced, and a tendency for food debris to take up the dye, do not provide the best conditions for automated counting using the HGMF interpreter, an improved technique, based on triphenyl tetrazolium chloride (TTC) was developed

Figure 2.3 The newly introduced MI-200 HGMF Interpreter uses a colour TV camera.

(Parrington *et al.*, 1993). Because TTC is colorless, it cannot produce a background color to interfere with recognition; the quality of the optical signal is greatly improved for automated counting. To avoid potential problems of growth inhibition by TTC, a standard medium such as trypticase soy agar is used for incubation; after 24 or 48 h (depending on the type of sample), the HGMF is simply raised momentarily with forceps while 2 ml of 0.1% TTC solution is pipetted on the agar surface. After 15 min on the bench, the HGMF is ready for counting. In some situations, agars containing TTC may be satisfactory (McNab *et al.*, 1991).

2.6.3.2 Total coliform count.   The method uses overnight incubation at 35°C on mFC agar (Difco). Coliforms are dark blue. Collaborative studies of this method showed a better recovery, repeatability and reproducibility than the FDA BAM method, and it now has AOAC official final action status for all foods. There are significant savings in cost, analysis time and person hours over the MPN method. An HGMF method has also been proposed as an alternative to the most probable number method for determination of coliforms in surface waters (McDaniels *et al.*, 1987).

2.6.3.3 Fecal coliform count.   Incubation is overnight at 44.5°C on mFC agar. A resuscitation period is included for frozen, desiccated or otherwise stressed organisms. It has AOAC official final action status.

2.6.3.4 E.coli biotype 1 count.   One demonstrates indole production at 44.5°C. A resuscitation period is included if organisms have been stressed. The method has AOAC official final action status for all foods. Very significant savings in cost, analysis time and person hours compared with the MPN method have been detailed (Sharpe and Peterkin, 1988).

2.6.3.5 E.coli O157:H7.   This very selective method is based on incubation of the HGMF at 43°C on HC agar (Oxoid), followed by rinsing away most of the colony material and staining by means of horse radish peroxidase-labeled anti-*E.coli* O157 monoclonal antibody (Todd *et al.*, 1988). Staining is simple and takes less than 2 h; grid-cells containing any *E.coli* O157 turn purple even with a massive background flora. Without enrichment the technique detects and counts 10 cells g^{-1} food; with enrichment the sensitivity can be higher. The HGMF replicator may be used before the staining step, to copy the HGMF in case further confirmation is needed – for example, a second immunological step using H7 antibody to confirm O157:H7. The method is much less cumbersome that an earlier HGMF method described by Doyle and Schoeni (1987).

2.6.3.6 Opti-MUG coliform/E.coli *count.* A clever and efficient double procedure (Entis, 1989; Entis and Boleszczuk, 1990) uses preliminary incubation on lactose–monensin–glucuronate (LMG) agar for 24 h at 35°C. Coliform colonies are blue and are counted. Any HGMF positive for coliforms is transferred to buffered MUG agar (BMA) for 2 h at 35°C and examined under longwave UV light. *Escherichia coli* colonies fluoresce blue–white. The procedure has AOAC official action status.

2.6.3.7 Salmonella *spp.* The updated 42 h AOAC-HGMF method uses overnight non-selective enrichment, and 6 h selective enrichment. A portion of the selective enrichment on an HGMF is incubated overnight at 42°C on EF-18 agar (QA LifeSciences). Green (lactose negative) *Salmonella* colonies are sub-cultured for confirmation. The method detects 1 cell per 25 g with a claimed false-positive rate of less than 0.5%. It has AOAC official action status for foods (Entis, 1990). Peterkin *et al.* (1992a) used both EF-18 and Hektoen agars to *enumerate* nalidixic acid-resistant salmonellae in chicken ceca, and follow their decline, during studies of the elimination of the organism from flocks by feed modification.

Techniques have been developed, based on replicating libraries of up to 400 cultures per HGMF, exposing them to permutations of diagnostic reaction conditions such as enzyme-labeled antibodies or DNA probes, and evaluating the growths or stains by means of the HGMF Interpreter (Peterkin *et al.*, 1989; Sharpe *et al.*, 1989a,b). The techniques permitted evaluation of the specificity of a large number of anti-*Salmonella* antibodies, in far more depth than would normally be practicable, and identified two which appeared extremely promising for *Salmonella* detection (Todd *et al.*, 1991). Preliminary experiments (unpublished results) suggest that these monoclonal antibodies are sufficiently sensitive and specific to permit enumeration of salmonellae in foods by direct plating of the sample suspension, followed by immunological staining.

2.6.3.8 Vibrio parahaemolyticus *count.* This overnight count is based on non-fermentation of sucrose at 42°C and supercedes an earlier method described by Entis and Boleszczuk (1983). Green growths are *V. parahaemolyticus*. The method gave significantly better recovery of *V. parahaemolyticus* from seafoods than the FDA BAM procedure (Entis and Bolezczuk, 1983). A recent study (DePaola *et al.*, 1988) showed the HGMF method to be superior, in all aspects, to the BAM and two other methods. An interesting, alternative method based on the use of radioactive DNA probes on the HGMF was described by Kaysner *et al.* (1990). The method also was able to differentiate *V. parahaemolyticus* from *V. vulnificus*, which are difficult to differentiate biochemically. The complete procedure requires 4 days.

2.6.3.9 Fluorescent pseudomonads.   These organisms are in large part responsible for spoilage of poultry. They are difficult to enumerate on ordinary agar plates because their initial concentration is too low. By filtering larger volumes (e.g. 10 ml) of suspensions through HGMFs, they can be enumerated in 48 h on either Flo Agar (BBL) or the S1-medium of Gould *et al.* (1985), followed by examination under long-wavelength UV light (Knabel *et al.*, 1987).

2.6.3.10 Lactic bacteria.   HGMFs have been found useful in a number of analyses related to the quality of yoghurt and fermented meat products. The first counting of lactic populations (*Streptococcus*, *Lactobacillus* and *Leuconostoc* spp.) on MRS agar was described by Chu *et al.* (1987). Later, Holley and Millard (1988) used MRSD medium with additions of L-arginine, phenol red and polymyxin B, followed by anaerobic incubation (32°C, 48 h), to differentiate between pediococci and lactobacilli in fermented meat and starter cultures. Millard *et al.* (1989, 1990) then described the use of tryptone–phytone–yeast extract containing erioglaucine pH indicator, and incubation at 37°C for 72 h, to permit counting of *Streptococcus salivarius* and *Lactobacillus delbruekii* on the same HGMF. Ryser and Richard (1992) described a method using colonies laid out in a regular grid pattern, to detect bacteriocin-producing cultures; the method was reported to be similar to or better than cross-streak or agar disk methods.

2.6.3.11 Listeria monocytogenes.   *L. monocytogenes* was enumerated in pasteurized eggs and mayonnaise by plating on selective veal agar with addition of Fast Green, yeast extract, aesculin and lithium chloride and incubating at 35°C for 72 h (Erickson and Jenkins, 1991, 1992). Suspect colonies were confirmed biochemically. The authors noted that the method was superior in sensitivity and precision to the USDA method, especially for recovering acid-stressed *L. monocytogenes* cells. An alternative procedure using a chromogenically-labeled DNA probe to detect *L. monocytogenes* in foods was described by Peterkin *et al.* (1992b). Earlier (1989), these authors made use of the HGMF in screening libraries of cultures to obtain the specific DNA probe.

2.6.3.12 Aeromonas hydrophila.   Erickson and Jenkins (1992) enumerated this organism by plating on trypticase soy agar (TSA-AH) with bile salts and glucose, and incubating at 35°C for 24–28 h. Only aeromonads, coliforms and fecal streptococci grew on the anaerobic medium. After incubation, cytochrome oxidase-positive colonies were detected by Kovac's reagent.

2.6.3.13 Yersinia enterocolitica.   Erickson and Jenkins (1992) enumerated this organism in pasteurized egg by plating on *Yersinia* selective agar (YSA-Difco) with antimicrobial supplement CN (Difco) and incubating at 30°C for 72 h. Typical *Y. enterocolitica* colonies were picked for biochemical confirmation.

2.6.3.14 Yeasts and molds.   A preliminary publication (Brodsky *et al.*, 1982b) described the use of potato dextrose agar containing safranin as useful for enumerating yeasts and molds by HGMF. Later, Lin *et al.* (1984) indicated that trypan blue agar was as reliable, and more convenient for enumerating these organisms in cheeses, spices and dry products, or salad samples.

2.6.4 HGMFs in very practical HACCP situations

The HGMF system has been used extensively by Agriculture Canada (Dr R.E. Clarke's group in Guelph) to investigate microbial loads in abattoirs, as part of the Canadian Food Safety Enhancement Program. Laboratories based on HGMFs were set up directly on abattoir floors to study HACCP in poultry plants and were supplied with prepoured TTC agar plates and the necessary diluents (Figures 2.4 and 2.5). Chickens were rinsed in plastic bags in a

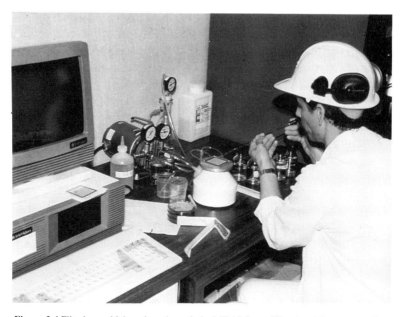

Figure 2.4 Filtering a chicken rinse through the MF-10 Spreadfilter in a laboratory on the abattoir floor.

Figure 2.5 Abattoir staff counting HGMFs by means of the MI-100 Interpreter.

commercial paint shaker which handled six samples at a time, returned to the line, the rinses were filtered onto HGMFs by using a Spreadfilter, and the HGMFs were plated on plates containing TTC. After incubation, the HGMFs were read in the computerized MI-100 HGMF Interpreter by the inspector or factory staff, who were instructed simply to put the Petri dish in the instrument and record the result. Repeatability of the HGMF analyses was excellent, and because they were done on the spot, the results were highly pertinent. The cost of counting grids this way was only one-tenth that of sending them to the laboratory for aerobic plate count. In one study alone, over 6000 birds were tested, with dramatic financial savings (McNab *et al.*, 1991). Similar studies have been carried out in beef abattoirs under the direction of Dr K. Jericho (Agriculture Canada, Calgary).

Other workers have also used HGMFs to determine contamination levels relevant to abattoir practices. For example, Charlebois *et al.* (1991) examined fecal coliform levels on beef carcasses, by incubating for 24 h at 44.5°C on mFC agar. Fliss *et al.* (1991) examined levels of total aerobic counts and coliforms, for various meat species in Tunisian abattoirs and markets.

2.7 Millipore samplers

The interesting range of dip-type Samplers marketed by the Millipore Corporation uses membrane filters sealed over an absorbent, nutrient-releasing paddle. Four types – Total Plate Count, SPC (standard plate count) Total Count, Coli-Count, and Yeast and Mold Samplers – are available. These may be used alone, or in conjunction with a Swab Test Kit. Leaflets supplied for the system contain all necessary details of sample preparation, immersion, incubation and reading the devices. A particularly useful feature of the Application leaflet is a page containing reproductions of incubated Samplers at different levels of saturation. Estimates of colony numbers can be made simply by comparing a test Sampler with the photographs.

The Sampler was conceived as a means of monitoring supermarkets, nursing homes and food processing plants, or evaluating the effectiveness of cleaning and disinfecting procedures. While studying the conditions under which testing should be carried out, the company noted that the results of monitoring were favorably received when sampling was done in the presence of, and particularly by, the person involved in cleaning, indicating the desirability of an extremely simple test which can be carried out by unskilled personnel. The goal was a two-piece test which would fit into a shirt pocket; the Swab Test Kit and Samplers evolved more or less together (personal communication, A. Haffenraffer, Millipore Corp.). The swab and dip tester were packaged in similar vials (of shirt-pocket size). The swab has a 5 cm long stem for easy handling, and its vial contains sterile buffer solution. A simple '40-linear inch' swabbing procedure, consisting of the drawing of five letter 'Ms' (all legs being 2 in or 5 cm) at random points on equipment surfaces, was developed. In normal use the swab is returned to its vial, microorganisms on it are dispersed by vigorous shaking, and the swab is discarded.

The Sampler outer surface is a standard pore size MF, heat sealed to the outer edges of the paddle. It carries a black grid to facilitate counting. Beneath the MF, an absorbent pad containing nutrient medium is formulated to absorb 1 ml of sample. The Samplers are color coded: red for water or other bacteria which may be stressed (Total Count), white for unstressed organisms (SPC Total Count), yellow for yeasts and molds, and blue for coliforms. In the complete test kit the Sampler paddle is introduced into the buffer solution of the swab unit for 30 s, then returned to its vial for incubation. Alternatively, it may be used alone on other solutions or homogenates. A small amount of care is required to immerse the Sampler without undue agitation, which might prevent organisms attaching properly to it. Reproducibility of fluid absorption is relatively good. For example, mean uptake of water was 1.04 g, varying between 0.98 and 1.12 g (Hedberg and Connor, 1975). For fresh and frozen crab homogenates, mean uptakes of 0.991 and 0.986 g with standard deviations of 0.032 and 0.034, respectively, were reported (Richards, 1979). While this method of inoculation is not as precise as that obtainable by careful pipetting, it is quite adequate for the purpose.

When interpreting the results of evaluations of Samplers, one should keep in mind the intent of the devices. They are not designed as absolute tests of micro-biological quality, but merely for monitoring areas and critical control points, quickly, conveniently, with minimal requirements of space and skill. It is a bane of clever devices that more becomes expected of them than is reasonably possible. For example, when an evaluation of Coli-Count Samplers for counting total and fecal coliform organisms in recreational waters gave ratios of counts ranging from 0.57 to 1.10 of a MF control count, the authors (Hedberg and Connor, 1975) concluded that, though useful for semi-quantitative analyses, they were not suitable for standard examinations of waters.

Coli-Count Samplers were evaluated against an MPN method and a violet–red bile agar pour plate method for the monitoring of fecal coliforms, in 25 fresh samples of the Eastern Oyster (Richards, 1978). Relatively low counts by Samplers at low levels of presumptive coliform counts in 10^{-1} dilutions were attributed to blockage of the MF membrane by the oyster homogenate. These suspensions had been prepared by bladed blender; stomachered suspensions would probably have behaved better. At higher levels the Sampler counts compared well with the other two methods. The Sampler method was deemed an acceptable alternative to the standard MPN method for monitoring coliform levels in oysters. Subsequently, the Samplers were examined for monitoring fecal coliforms in the Blue Crab; although the counts were generally somewhat lower, they did not differ significantly from the MPN counts, and it was concluded that the Coli-Count Samplers might be acceptable for determining fecal coliform counts at levels greater than 10 g^{-1} in crab meat, and should be useful for enumerating the organisms at critical control points throughout crab processing operations (Richards, 1979).

The Yeast and Mold and Total Count SPC Samplers were evaluated against standard (47 mm) MF methods, for potential use in brewery operations (Ingledew and Burton, 1981). Six replicate analyses were made by each method. For eight species of yeasts and molds, counts by Sampler ranged from 10.5 to 52.5% of those of the standard method. For seven species of bacteria the recoveries ranged from 15.3 to 94.9% of the standard. In addition, standard deviations among some of the replicate Sampler counts were relatively high. The authors concluded that, in their present form, the Samplers could not be recommended for use in the brewery.

The SPC Samplers were compared against standard plate counts (APHA, 1980) for monitoring bacterial counts in cannery cooling waters (Rey *et al.*, 1982). Samplers were counted using a stereomicroscope, standard plates using a Quebec-type colony counter. With 24 h of incubation, counts from the Samplers were very low. However, after 48 h of incubation, the counts were consistently higher than those from the standard method. The authors noted that colonies from chlorinated cooling waters were generally minute on both standard plates and Samplers, and that use of a stereomicroscope for examining Samplers may have facilitated their detection. Other than making the caveat that counts should be made in duplicate to counteract a lower reproducibility, these authors concluded that, as a result of favorable recovery rates, the compactness of the system, and its general simplicity and low requirement for operator skill, the sampler method is a convenient alternative to the standard plate count for routine audit of cannery cooling waters. While on the subject of water it may be of interest that the American Society for Testing and Materials proposed the use of the Samplers as a field technique for achieving maximum counts of bacteria in water (ASTM, 1983).

Devenish *et al*. (1985) evaluated the Total Count and SPC Swab Test Kits against a pour plate technique for the recovery of heat and chlorine-stressed cells. Hand shaking was as effective as vortex stirring at dislodging bacteria from the swabs, but bacteria were unstable in the buffer, even at 4°C, indicating the desirability of completing both swabbing and inoculation in the field, before returning Samplers to the laboratory for incubation and enumeration. Following incubation at 35°C for 24 or 48 h, colonies on the Sampler paddles were enumerated using a stereomicroscope at low magnification. Over the range 20–300 *E.coli* cells ml^{-1}, data fitted the line: Total Count Sampler = –0.255 + (0.898 × Pour Plate), with a correlation coefficient of 0.997. At inoculation levels of more than 150 cfu, counts from Total Count Samplers were increasingly lower than expected, due to overcrowding of colonies on the small area. The authors noted that, due to a certain amount of drying out of the Sampler paddle during incubation, colony sizes were limited and confluent growth was never a problem. However, the stereomicroscope was necessary for accurate enumeration.

These workers also determined the recovery of stressed cells using Standard Methods and violet–red bile agars. In two separate trials, SPC Samplers recovered 126–724% of heat-stressed and 37–117% of chlorine-stressed *E.coli* cells, respectively, corresponding figures for Total Count Samplers being 0.2–100 and 279–383%. This suggests that Total Count Samplers were able to recover chlorine-stressed, but not heat-stressed cells, whereas the opposite was true for SPC Samplers. MF techniques should generally provide better recovery of stressed organisms than do pour plates, since additional heat stress is avoided (Stapert *et al*., 1962). The authors concluded that with a new formulation to achieve better recovery of cells stressed by either heat or chlorine, the kits could replace the more tedious and time-consuming standard method currently used for monitoring sanitary conditions related to public health.

References

Albert, M.J., Tee, W., Leach, A., Asche V. and Penner, J.L. (1992) Comparison of a blood-free medium and a filtration technique for the isolation of *Campylobacter* spp. from diarrhoeal stools of hospitalized patients in central Australia. *J. Med. Microbiol.*, **37**, 176–179.

Anderson, J.M. and Baird-Parker, A.C. (1975) A rapid and direct plate method for enumerating *Escherichia coli* biotype 1 in food. *J. Appl. Bacteriol.*, **39**, 111–117.

Andrews, W.H. and Presnell, M.W. (1972) Rapid recovery of *Escherichia coli* from estuarine water. *Appl. Microbiol.*, **23**, 521–523.

AOAC (1984) *Official Methods of Analysis*, 14th edn (ed. Williams, S.) Association of Official Analytical Chemists, Arlington, Virginia.

AOAC (1985) Total coliform, fecal coliform and *Escherichia coli* in foods, hydrophobic grid membrane filter method. *J. A. O. A. C.*, **68**, 404.

AOAC (1986) Aerobic plate count in foods, hydrophobic grid membrane filter method. First action.

J. A. O. A. C., **69**, 376–378.

APHA (1970) *Recommended Procedures for the Examination of Seawater and Shellfish*, 4th edn, American Public Health Association, Washington, DC.

APHA (1976) *Compendium of Methods for the Microbiological Examination of Foods* (ed. Speck, M.L.). American Public Health Association, Washington, DC.

APHA (1980) *Standard Methods for the Examination of Water and Waste Water*. 15th edn, American Public Health Association, Washington, DC.

APHA (1984) *Compendium of Methods for the Microbiological Examination of Foods* (ed. Speck, M.L.). 2nd edn, American Public Health Association, Washington, DC.

APHA (1992) *Compendium for the Microbiological Examination of Foods* (eds Vanderzant, C. and Splittstoesser, D.F.). American Public Health Association, Washington, DC.

Arbuzova, V.A. (1970) Biological characteristics of enteropathogenic *Bacillus coli* 044 isolated from the newborn. *Trudy Inst. Epidem. Mikrobiol. Sanit.*, **36**, 287–294.

ASTM (1983) *Standard Test method for Total Bacterial Count in Water* (Section II, Standard F488). American Society for Testing and Materials, Philadelphia, PA.

Baggerman, W.I. and Koster, T. (1992) A comparison of enrichment and membrane filtration methods for the isolation of *Campylobacter* from fresh and frozen foods. *Food Microbiol.*, **9**, 87–94.

Barber, F.W., Burke C.P. and Fram, H. (1954) The Millipore filter technique in the dairy industry. *J. Milk Food Technol.*, **17**, 109–112.

Bolton, F.J., Hutchinson D.N. and Coates, D. (1984) Blood-free selective medium for isolation of *Campylobacter jejuni* from feces. *J. Clin. Microbiol.*, **19**, 169–171.

Bolton, F.J., Hutchinson D.N. and Parker, G. (1988) Reassessment of selective agars and filtration techniques for isolation of *Campylobacter* species from faeces. *Eur. J. Clin. Microbiol. Infect. Diseases*, **7**, 155–160.

Bourgeois, C., Mafart P. and Thouvenot, D. (1973) Méthode rapide de détection des contaminants dans la biere par marquage radioactif. *Eur. Brew. Conv. Proc. Cong.*, **14**, 219–230.

Bourgeois, C.M., LeParc, O., Abgrall, B. and Cleret, J.-J. (1984) Membrane filtration of milk for counting spores of *Clostridium tyrobutyricum. J. Dairy Sci.*, **67**, 2493–2499.

Brodsky, M.H. (1982) Automated hydrophobic grid-membrane filter technique for obtaining aerobic plate counts and yeast and mold counts in food, in *Rapid Methods and Automation in Microbiology* (ed. Tilton, R.C.). American Society for Microbiology, Washington, DC, pp. 68–71.

Brodsky, M.H., Entis, P., Sharpe A.N. and Jarvis, G.A. (1982a) Enumeration of indicator organisms in foods using the automated hydrophobic grid-membrane filter technique. *J. Food Prot.*, **45**, 292–296.

Brodsky, M.H., Entis, P., Entis, M.P. *et al.* (1982b) Determination of aerobic plate and yeast and mold counts in foods using an automated hydrophobic grid-membrane filter technique. *J. Food Prot.*, **45**, 301–304.

Busta, F.F. and Speck, M.L. (1965) Enumeration of *Bacillus stearothermophilus* by use of membrane filter techniques to eliminate inhibitors present in milk. *Appl. Microbiol.*, **13**, 1043–1044.

Chain, V.S. and Fung, D.Y.C. (1991) Comparison of Redigel, Petrifilm, Spiral Plate System, Isogrid, and aerobic plate count for determining the numbers of aerobic bacteria in selected foods. *J. Food Prot.*, **54**, 208–211.

Charlebois, R., Trudel R. and Messier, S. (1991) Surface contamination of beef carcasses by fecal coliforms. *J. Food Prot.*, **54**, 950–956.

Chu, J.L., Elo, A., Maissin, R. and Decallonne, J.R. (1987) La quantification des populations de bacteries lactiques par la methode HGMF. *Belg. J. Food Chem. Biotechnol.*, **42**, 65–74.

Claydon, T.J. (1975) A membrane-filter technique to test for the significance of sublethally injured bacteria in retail pasteurized milk. *J. Milk Food Technol.*, **38**, 87–88.

Coleman, M.C. and Bender, C.R. (1957) Microbiological examination of liquid sugar using molecular membrane filters. *Food Technol.*, **11**, 398-403.

Cousins, C.M., Pettipher, G.L., McKinnon C.H. and Mansell, R. (1979) A rapid method for counting bacteria in raw milk. *Dairy Ind. Int.*, **44**, 27–39.

Crum, M.G. (1971) Rapid detection of bacterial cells by radioactive methods, in *Developments in Industrial Microbiology* (ed. Murray, E.), Vol. 12, American Institute of Biological Sciences, Washington, DC, pp. 191–196.

Delaney, J.E., McCarthy, J.A. and Grasso, R.J. (1962) Measurement of *Escherichia coli* type 1 by the membrane filter. *Water Sewage Works*, **109**, 289–294.

Dempster, J.F., Reid S.N. and Cody, O. (1973) Sources of contamination of cooked, ready-to-eat cured and uncured meats. *J. Hyg. Camb.*, **71**, 815–823.

DePaola, A., Hopkins L.H. and McPhearson, R.M. (1988) Evaluation of four methods for enumeration of *Vibrio parahaemolyticus*. *Appl. Environ. Microbiol.*, **54**, 617–618.

Devenish, J.A., Ciebin, B.W. and Brodsky, M.H. (1985) Evaluation of Millipore swab-membrane filter kits. *J. Food Prot.*, **48**, 870–874.

Doyle, M.P. and Schoeni, J.L. (1987) Isolation of *Escherichia coli* O157:H7 from retail fresh meats and poultry. *Appl. Environ. Microbiol.*, **53**, 2394–2396.

Duitschaever, C.L. and Ashton, G.C. (1968) Comparison of counts using a Breed-type smear and Millipore membrane methods on fresh and preserved milk samples. *J. Dairy Sci.*, **51**, 665–667.

Early, J.F. and Patterson, J.T. (1982) Rapid detection of salmonellas in foods, feeds and animal slurries by a modified fluorescent-antibody technique. *Soc. Appl. Bacteriol. Tech. Ser.*, **17**, 73–76.

Ehrlich, R. (1955) Technique for microscopic count of microorganisms directly on membrane filters. *J. Bacteriol.*, **70**, 265–268.

Entis, P. (1983) Enumeration of total coliforms, fecal coliforms and *Escherichia coli* in nonfat dry milk and canned custard by hydrophobic grid membrane filter method: collaborative study. *J. A. O. A. C.*, **68**, 897–904.

Entis, P. (1984a) Enumeration of total coliforms, fecal coliforms and *Escherichia coli* in foods by hydrophobic grid membrane filter: supplementary report. *J. A. O. A. C.*, **67**, 811–812.

Entis, P. (1984b) Enumeration of total coliforms, fecal coliforms and *Escherichia coli* in foods by hydrophobic grid membrane filter: collaborative study. *J. A. O. A. C.*, **67**, 812–813.

Entis, P. (1986) Hydrophobic grid membrane filter method for aerobic plate count in foods: collaborative study. *J. A. O. A. C.*, **69**, 671–676.

Entis, P. (1989) Hydrophobic grid-membrane filter MUG method for total coliform and *Escherichia coli* enumeration in foods – collaborative study. *J. A. O. A. C.*, **72**, 936–950.

Entis, P. (1990) Improved hydrophobic grid membrane filter method, using EF-18 agar for detection of *Salmonella* in foods: collaborative study. *J. A. O. A. C.*, **73**, 734–742.

Entis, P. and Boleszczuk, P. (1983) Overnight enumeration of *Vibrio parahaemolyticus* in seafood by hydrophobic grid membrane filtration. *J. Food Prot.*, **46**, 783–786.

Entis, P. and Boleszczuk, P. (1986) Use of Fast Green FCF with tryptic soy agar for aerobic plate count by the hydrophobic grid membrane filter. *J. Food Prot.*, **49**, 278–279.

Entis, P. and Boleszczuk, P. (1990) Direct enumeration of coliforms and *Escherichia coli* by hydrophobic grid-membrane filter in 24 hours using MUG. *J. Food Prot.*, **53**, 948–952.

Ercolani, G.L. (1976) Bacteriological quality assessment of fresh marketed lettuce and fennel. *Appl. Environ. Microbiol.*, **31**, 847–852.

Erickson, J.P. and Jenkins, P. (1991) Comparative *Salmonella* spp. and *Listeria monocytogenes* inactivation rates in four commercial mayonnaise products. *J. Food Prot.*, **54**, 913–916.

Erickson, J.P. and Jenkins, P. (1992) Behaviour of psychrotrophic pathogens *Listeria monocytogenes, Yersinia enterocolitica*, and *Aeromonas hydrophila* in commercially pasteurized eggs held at 2, 6.7 and 12.8°C. *J. Food Prot.*, **55**, 8–12.

Ewing, W.H. (1972) *Differentiation of Enterobacteriaceae by Biochemical Reactions*. US Dept. of Health, Education and Welfare, Atlanta, GA.

Farber, J.M. and Sharpe, A.N. (1984) Improved bacterial recovery by membrane filters in the presence of food debris. *Appl. Environ. Microbiol.*, **48**, 441–443.

Farber, J.M., Sharpe, A.N. and Kalab, M. (1985) Improved recovery of fecal coliforms from the Ottawa River by membrane filters in the presence of food debris. *Can. J. Microbiol.*, **31**, 16–18.

Fick, A. (1855). Ueber diffusion. *Pogg. Ann.*, **94**, 59–86.

Fields, N.D., Oxborrow, G.S., Puleo J.R. and Herring, C.M. (1974) Evaluation of membrane filter field monitors for microbiological air sampling. *Appl. Microbiol.*, **27**, 517–520.

Fifield, C.W., Hoff, J.E. and Proctor, B.E. (1957) The Millipore filter for enumerating coliform organisms in milk. *J. Dairy Sci.*, **40**, 588–589.

Fliss, I., Simard R.E. and Ettriki, A. (1991) Microbiological quality of different fresh meat species in Tunisian slaughterhouses and markets. *J. Food Prot.*, **54**, 773–777.

Frazier, W.C. and Gneiser, D.F. (1968) Short-time membrane filter method for estimation of numbers of bacteria. *J. Milk Food Technol.*, **31**, 177–179.

Fruin, J.T. and Clark, W.S. (1977) Plate count accuracy: analysts and automatic colony counter versus a true count. *J. Food Prot.*, **40**, 552–554.

Godberson, G.W. (1974) New method for direct counting of bacteria in milk. *XIX International Dairy Congress*, December 2–6, *IE*: 529, New Delhi.

Goff, J.H., Claydon T.J. and Iandolo, J.J. (1972) Revival and subsequent isolation of heat-injured bacteria by a membrane filter technique. *Appl. Microbiol.*, **23**, 857–862.

Gould, W.D., Hagedorn, C., Barnelli, T.R. and Żablotowics, R.M. (1985) New selective media for enumeration and recovery of fluorescent pseudomonads from various habitats. *Appl. Environ. Microbiol.*, **49**, 28–32.

Graves, D.C. and Schipper, I.A. (1966) Membrane filtration technique for isolating organisms from raw milk of normal udders. *Appl. Microbiol.*, **14**, 535–539.

Green, R.L., Scales R.W. and Kraus, S.J. (1974) Radioimmunoassay on polycarbonate membranes: a sensitive and simplified method for the detection and quantitation of antibody. *Appl. Microbiol.*, **27**, 475–479.

Haas, G.J. (1956) Use of the membrane filter in the brewing industry. *Wallerstein Lab. Comm.*, **19**, 7–22.

Halden, H.E. (1957) Methods for count of microorganisms in liquid sugar. *J. Am. Soc. Sugar Beet Technol.*, **9**, 393–399.

Halden, H.E., Leethem, D. and Eis, F.G. (1958) Developments in methods for microbiological control in liquid sugar. *J. Am. Soc. Sugar Beet Technol.*, **10**, 138–141.

Halden, H.E., Leethem, D.D. and Eis, F.G. (1960) Determination of microorganisms in sugar products by the Millipore method. *J. Am. Soc. Sugar Beet Technol.*, **11**, 137–142.

Hambraeus, A. and Benediktsdottir, E. (1980) Airborne nonsporeforming anaerobic bacteria. *J. Hyg.*, **84**, 181–189.

Hedberg, M. and Connor, D.A. (1975) Evaluation of Coli-Count Samplers for possible use in standard counting of total and fecal coliforms in recreational waters. *Appl. Microbiol.*, **30**, 881–883.

Hobbie, J.E., Daley R.J. and Jasper, S. (1977) Use of Nuclepore filters for counting bacteria by fluorescence microscopy. *Appl. Environ. Microbiol.*, **33**, 1225–1228.

Holbrook, R., Anderson, J.M. and Baird-Parker, A.C. (1980) Modified direct plate method for counting *Escherichia coli* in foods. *Food Technol. (Australia)*, **32**, 78–83.

Holley, R.A. and Millard, G.E. (1988) Use of MRSD medium and the hydrophobic grid membrane filter technique to differentiate between pediococci and lactobacilli in fermented meat and starter cultures. *Int. J. Food Microbiol.*, **7**, 87–102.

Ingledew, W.M. and Burton, J.D. (1979) Chemical preservation of spent brewers' yeast. *J. Am. Soc. Brewing Chem.*, **37**, 140–144.

Ingledew, J.M. and Burton, J.D. (1980) Membrane filtration: survival of brewing microbes on the membrane during storage at reduced humidities. *J. Am. Soc. Brewing Chem.*, **38**, 125–129.

Ingledew, W.M. and Burton, J.D. (1981) Samplers for dip testing or swab testing in breweries. *J. Am. Soc. Brewing Chem.*, **39**, 39–40.

ISO (1989) *Microbiology – General Guidance for the Detection of Thermoresistant* Campylobacter. (ISO/TC 34/SC9 Working Document No. 203, Oct 1989.) British Standards Institute, London.

Jannasch, H.W. (1958) Studies on planktonic bacteria by means of a direct membrane filter method. *J. Gen. Microbiol.*, **18**, 609–620.

Kaysner, C.A., Abeyta Jr., C., Wekell, M.M. and Colburn, K.G. (1990) Enumeration and differentiation of *Vibrio parahaemolyticus* and *Vibrio vulnificus* by DNA–DNA colony hybridization using the hydrophobic grid membrane filtration (HGMF) technique. 14th International Symposium, International Committee on Food Microbiology and Hygiene. IUMS-ICFMH 14th International Symposium on Gram-negative Pathogens in Food, 14–19 Aug. 1990, Telemark, Norway.

Kenner, B.A., Rockwood, S.W. and Kabler, P.W. (1957) Isolation of members of the genus *Salmonella* by membrane filter procedures. *Appl. Microbiol.*, **5**, 305–307.

Kirkham, W.K. and Hartman, P.A. (1962) Membrane filter method for the detection and enumeration of *Salmonella* in egg albumen. *Poultry Sci.*, **42**, 1082–1088.

Knabel, S.J., Walker, H.W. and Kraft, A.A. (1987) Enumeration of fluorescent pseudomonads on poultry by using the hydrophobic grid membrane filter method. *J. Food Sci.*, **52**, 837–841, 845.

Lin, C.C.S., Fung, D.Y.C. and Entis, P. (1984) Growth of yeast and mold on trypan blue agar in conjunction with the ISO-GRID system. *Can. J. Microbiol.*, **30**, 1405–1407.

Lundholm, I.M. (1982) Comparison of methods for quantitative determinations of airborne bacteria and evaluation of total viable counts. *Appl. Environ. Microbiol.*, **44**, 179–183.

Mackenzie, E.F.W., Taylor, E.W. and Gilbert, W.E. (1948) Recent experiences in the rapid identification of *Bacterium coli* type I. *J. Gen. Microbiol.*, **2**, 197–204.

MacLeod, R.A., Light, M., White, L.A. and Currie, J.F. (1966) Sensitive rapid detection method for viable bacterial cells. *Appl. Microbiol.*, **14**, 979–984.

Mafart, P., Bourgeois, C., Dutheurtre, B. and Moll, M. (1978) Use of [^{14}C]lysine to detect microbial contamination in liquid foods. *Appl. Environ. Microbiol.*, **35**, 1211–1212.

Mafart, P., Cleret, J.J. and Bourgeois, C. (1981) Détection et évaluation des contaminations microbiennes dans les produits alimentaires, par marquage radioactif et spectrométrie à scintillation liquide. *Analusis*, **9**, 32–34.

McDaniels, A.E., Bordner, R.H., Menkedick, J.R. and Weber, C.I. (1987) Comparison of the hydrophobic-grid membrane filter procedure and standard methods for coliform analysis of water. *Appl. Environ. Microbiol.*, **53**, 1003–1009.

McNab, W.B., Forsberg, C.M. and Clarke, R.C. (1991) Application of an automated hydrophobic grid membrane filter Interpreter system at a poultry abattoir. *J. Food Prot.*, **54**, 619–622.

Mehlman, I.J., Simon, N.T., Sanders, A.C. and Olson, J.C. (1974) Problems in the recovery and identification of enteropathogenic *Escherichia coli* from foods. *J. Milk Food Technol.*, **37**, 350–356.

Merrill, E.P. (1963) Measurement of the bacteriological quality of milk by direct microscopic examination of membrane filters. *J. Dairy Sci.*, **46**, 601–602.

Meynell, G.G. and Meynell, E. (1970) *Theory and Practice in Experimental Bacteriology.* Cambridge University Press, Cambridge, pp. 120–121.

Middlekauff, J.E., Shewey, D.R. and Bono, A.W. (1967) Enumeration and development of yeast colonies from beer. *Proc. Am. Soc. Brewing Chem.*, 187–189.

Millard, G.E., McKellar, R.C. and Holley, R.A. (1989) Counting yogurt starters. *Dairy Ind. Int.*, **54**, 37.

Millard, G.E., McKellar, R.C. and Holley, R.A. (1990) Simultaneous enumeration of the characteristic microorganisms in yogurt using the hydrophobic grid membrane filter system. *J. Food Prot.*, **53**, 64–66, 71.

Molchenov, L.F. (1970) Clinical/epidemiological characteristics of illness caused by *Escherichia coli* 0124. *Voenno Meditsinskii Zh.*, **7**, 54–56.

Moroz, R. (1957) Determination of yeast in sugar liquors using membrane filters. *Int. Sugar J.*, **59**, 70–71.

Motes, M.L., McPhearson Jr., R.M. and DePaola Jr., A. (1984) Comparison of three international methods with APHA method for enumeration of *Escherichia coli* in estuarine waters and shellfish. *J. Food Prot.*, **47**, 557–561.

Muller, G. (1947a) Lactose-fuchsin plate for detection of coli in drinking water. *Z. Hyg. Infektionskr.*, **127**, 187–190.

Muller, G. (1947b) Eine trinkwassergebundene Ruhrepidemie. *Z. Bakteriol. Parasitenkd. Infektionskr. Hyg. Abt. I. Orig.*, **152**, 133–135.

Munson, T.E., Schrade, J.P., Bisciello Jr., N.B. *et al.* (1976) Evaluation of an automated fluorescent antibody procedure for detection of *Salmonella* in foods and feeds. *Appl. Environ. Microbiol.*, **31**, 514–521.

Murphy, J.F. and Silliker, J.H. (1955) Special applications of the membrane filter in the food industry. *Bacteriol. Proc.*, A5.

Nakhodkina, V.Z. and Dvornikova, N.B. (1971) Two methods of determination of microbial contamination in sugar. *Sakharnaya Promyshlennost*, **45**, 36–39.

Nobile, J. (1967) Use of membrane filter technique in the microbiological control for the brewing industry. *Appl. Microbiol.*, **15**, 736–737.

Nutting, L.A., Lomot, P.C. and Barber, F.W. (1959) Estimation of coliform bacteria in ice cream by use of the membrane filter. *Appl. Microbiol.*, **7**, 196–199.

Oxborrow, G.S., Fields, N.D. and Puleo, J.R. (1976) Preparation of pure microbiological samples for pyrolysis gas-liquid chromatography studies. *Appl. Environ. Microbiol.*, **32**, 306–309.

Park, C.E., Stankiewicz, Z.K., Lovett, J. and Hunt, J. (1981) Incidence of *Campylobacter jejuni* in fresh eviscerated whole market chickens. *Can. J. Microbiol.*, **27**, 841–842.

Parrington, L.J., Sharpe, A.N. and Peterkin, P.I. (1993) Improved aerobic colony count technique for hydrophobic grid membrane filters. *Appl. Environ. Microbiol.*, **59**, 2784–2789.

Paton, A.M. and Jones, S.M. (1975) The observation and enumeration of micro-organisms in fluids using membrane filtration and incident fluorescent microscopy. *J. Appl. Bacteriol.*, **38**, 199–200.

Peterkin, P.I., Idziak, E.S. and Sharpe, A.N. (1989) Screening DNA probes using the hydrophobic grid membrane filter. *Food Microbiol.*, **6**, 281–284.

Peterkin, P.I., Langford, C.F., Chambers, J.R. and Sharpe, A.N. (1992a) Use of an electronic HGMF Interpreter system for enumeration of nalidixic acid resistant salmonellae in chicken ceca. *J. Rapid Methods Automat. Microbiol.*, **1**, 173–178.

Peterkin, P.I., Idziak, E.S. and Sharpe, A.N. (1992b) Use of a hydrophobic grid-membrane filter DNA probe method to detect *Listeria monocytogenes* in artificially-contaminated foods. *Food Microbiol.*, **9**, 155–160.

Petersen, N.J., Collins, D.E. and Marshall, J.H. (1973) A microbiological assay technique for hands. *Health Lab. Sci.*, **10**, 18–22.

Pettipher, G.L. (1989) The direct epifluorescence filter technique, in *Rapid Methods in Food Microbiology* (eds Adams, M.R. and Hope, C.F.A.). *Progress in Industrial Microbiology, Vol. 26*, Elsevier, London, pp. 19–56.

Pettipher, G.L. and Rodrigues, U.M. (1981) Rapid enumeration of bacteria in heat-treated milk and milk products using a membrane filtration-epifluorescence microscopy technique. *J. Appl. Bacteriol.*, **50**, 157–166.

Pettipher, G.L. and Rodrigues, U.M. (1982) Rapid enumeration of microorganisms in foods by the direct epifluorescent filter technique. *Appl. Environ. Microbiol.*, **44**, 809.

Pettipher, G.L., Mansell, R., McKinnon, C.H. and Cousins, C.M. (1980) Rapid membrane filtration-epifluorescent microscopy technique for direct enumeration of bacteria in raw milk. *Appl. Environ. Microbiol.*, **39**, 423–429.

Picciolo, G.L., Chappelle, E.W., Thomas, R.R. and McGarry, M.S. (1977) Performance characteristics of a new photometer with a moving filter tape for luminescence assay. *Appl. Environ. Microbiol.*, **34**, 720–724.

Portno, A.D. and Molzahn, S.W. (1977) New methods for the detection of viable micro-organisms. *The Brewers Digest*, **March**, 44–50.

Presswood, W.G. (1981) The membrane filter: its history and characteristics, in *Membrane Filtration: Applications, Techniques and Problems* (ed. Dutka, B.J.). Marcel Dekker, New York, pp. 1–17.

Rayman, M.K., Jarvis, G.A., Davidson, C.M. *et al.* (1979) ICMSF methods studies. XIII. An international comparative study of the MPN plating procedure and the Anderson-Baird-Parker direct plating method for the enumeration of *Escherichia coli* biotype 1 in raw meats. *Can. J. Microbiol.*, **25**, 1321–1327.

Reusse, U. (1959) Quantitative determination of *Brucella* in cows' milk by means of membrane filtration. *Bull. WHO*, **21**, 782–784.

Rey, C.R., Halaby, G.A., Lovgren, E.V. and Wright, T.A. (1982) Evaluation of a membrane filter test kit for monitoring bacterial counts in cannery cooling waters. *J. Food Prot.*, **45**, 1087–1090.

Reynolds, N. and Wood, P.C. (1956) Improved techniques for the bacteriological examination of molluscan shellfish. *J. Appl. Bacteriol.*, **19**, 20–25.

Richards, M. (1970) Routine accelerated membrane filter method for examination of ultra-low levels of yeast contaminants in beer. *Wallerstein Lab. Comm.*, **23**, 97–103.

Richards, G.P. (1978) Comparative study of methods for the enumeration of total and fecal coliforms in the Eastern Oyster, *Crassostrea virginica*. *Appl. Environ. Microbiol.*, **36**, 975–978.

Richards, G.P. (1979) Evaluation of Millipore Coli-Count Samplers for monitoring fecal coliforms in the Blue Crab *Callinectes sapidus*. *Appl. Environ. Microbiol.*, **38**, 341–343.

Ryser, E.T. and Richard, J.A. (1992) Detection of bacteriocin activity in bacteria using hydrophobic grid membrane filters. *Lett. Appl. Microbiol.*, **14**, 104–107.

Schiemann, E. (1955) Rationelle Brauerei-Betriebskontrolle mit dem Membranfilter. *Brauwelt*, **95**, 421–427.

Schrot, J.R., Hess, W.C. and Levin, G.V. (1973) Method for radiorespirometric detection of bacteria in pure culture and in blood. *Appl. Microbiol.*, **26**, 867–873.

Scotter, S.L., Humphrey, T.J. and Henley, A. (1993) Methods for the detection of thermotolerant campylobacters in foods: results of an inter-laboratory study. *J. Appl. Bacteriol.*, **74**, 155–163.

Sharpe, A.N. (1978) Some theoretical aspects of microbiological analysis pertinent to mechanization, in *Mechanizing Microbiology* (eds Sharpe, A.N. and Clark, D.S.). Thomas, Illinois, pp. 19–40.

Sharpe, A.N. (1981) Hydrophobic grid-membrane filters the (almost) perfect system, in *Membrane Filtration: Applications, Techniques and Problems* (ed. Dutka, B.J.). Marcel Dekker, New York, pp. 513–535.

Sharpe, A.N. (1987) The potential for automated food microbiology, in *Food Microbiology: New and Emerging Technologies* (ed. Montville, T.J.), Vol. II. CRC Press, Boca Raton, Florida, pp. 1–14.

Sharpe, A.N. and Michaud, G.L. (1974) Hydrophobic grid-membrane filters: new approach to microbiological enumeration. *Appl. Microbiol.*, **28**, 223–225.

Sharpe, A.N. and Michaud, G.L. (1975) A new concept for automatic counting of bacterial colonies. *Arch. Lebensm. Hyg.*, **26**, 10.

Sharpe, A.N. (1982) Hydrophobic grid-membrane microbiology: principles and practice, in *Rapid Methods and Automation in Microbiology and Immunology* (ed. Tilton, R.C.). American Society of Microbiology, Washington, DC, pp. 53–57.

Sharpe, A.N. and Michaud, G.L. (1978) Enumeration of bacteria using hydrophobic grid-membrane filters, in *Mechanizing Microbiology* (eds Sharpe, A.N. and Clark, D.S.). Thomas, Illinois, p. 16.

Sharpe, A.N. and Peterkin, P.I. (1988) *Membrane Filter Food Microbiology* (Innovation in Microbiology Research Studies Series). Research Studies Press, Letchworth, UK.

Sharpe, A.N., Peterkin, P.I. and Malik, N. (1979) Improved detection of coliforms and *Escherichia coli* in foods by a membrane filter method. *Appl. Environ. Microbiol.*, **38**, 431–435.

Sharpe, A.N., Peterkin, P.I. and Rayman, M.K. (1981) Detection of *Escherichia coli* in foods: indole staining methods for cellulosic and polysulfone membrane filters. *Appl. Environ. Microbiol.*, **41**, 1310–1315.

Sharpe, A.N., Rayman, M.K., Bergener, D.M. *et al.* (1983a) Collaborative study of the MPN, Anderson-Baird-Parker direct plating, and hydrophobic grid-membrane filter methods for the enumeration of *Escherichia coli* biotype 1 in foods. *Can. J. Microbiol.*, **29**, 1247–1252.

Sharpe, A.N., Diotte, M.P., Dudas, I. *et al.* (1983b) Colony counting on hydrophobic grid-membrane filters. *Can. J. Microbiol.*, **29**, 797–802.

Sharpe, A.N., Diotte, M.P., Peterkin, P.I. and Dudas, I. (1986) Towards the truly automated counter. *Food Microbiol.*, **3**, 247–270.

Sharpe, A.N., Rayman, M.K., Malik, N. *et al.* (1987). ICMSF Methods Study XVII. An international comparative study of the direct plate and hydrophobic grid-membrane filter methods for enumeration of *Escherichia coli* in foods. *Can. J. Microbiol.*, **33**, 85–92.

Sharpe, A.N., Diotte, M.P., Dudas, I. *et al.* (1989a) Technique for maintaining and screening many microbial cultures. *Food Microbiol.*, **6**, 261–265.

Sharpe, A.N., Parrington, L.J., Diotte, M.P. and Peterkin, P.I. (1989b) Evaluation of indoxyl-β-D-glucuronide and hydrophobic grid-membrane filters for electronic enumeration of *Escherichia coli*. *Food Microbiol.*, **6**, 267–280.

Silliker, J.H., Andrews, H.P. and Murphy, J.F. (1957) A new non-destructive method for the bacteriological sampling of meats. *Food Technol. (Champaign)*, **11**, 317–318.

Sladek, K.J., Suslavich, R.V., Sohn B.I. and Dawson, F.W. (1975) Optimum membrane structures for growth of fecal coliform organisms. *Proceedings of the Symposium on the Recovery of Indicator Organisms Employing Membrane Filters*, US Dept. of Commerce, Nat. Tech. Inf. Ser. PB–273 587, pp. 46–57.

Stapert, E.M., Sokolski, W.T. and Northam, J.I. (1962) The factor of temperature in the better recovery of bacteria from water by filtration. *Can. J. Microbiol.*, **8**, 809–810.

Steele, T.W. and McDermott, S.N. (1984) Technical note: the use of membrane filters applied directly to the surface of agar plates for the isolation of *Campylobacter jejuni* from faeces. *Pathology*, **16**, 263–265.

Thomason, B.M., Hebert, G.A. and Cherry, W.B. (1975) Evaluation of a semiautomated system for direct fluorescent antibody detection of salmonellae. *Appl. Microbiol.*, **30**, 557–564.

Todd, E.C.D., Szabo, R.A., Peterkin, P. *et al.* (1988) Rapid hydrophobic grid-membrane filter-enzyme labeled antibody procedure for identification and enumeration of *Escherichia coli* O157 in foods. *Appl. Environ. Microbiol.*, **54**, 2536–2540.

Todd, E.C.D., Mackenzie, J.M., Parrington, L.J. *et al.* (1991) Evaluation of *Salmonella* antisera for an optimum enzyme-linked antibody detection of *Salmonella* using hydrophobic grid-membrane filters. *Food Microbiol.*, **8**, 311–324.

Whiting, M., Chrichlow, M., Ingledew, W.M. and Ziola, B. (1992) Detection of *Pediococcus* spp. in brewing yeast by a rapid immunoassay. *Appl. Environ. Microbiol.*, **58**, 713–716.

Winter, F.H., York, G.K. and El-Nakhal, H. (1971) Quick counting method for estimating the number of viable microbes on food and food processing equipment. *Appl. Microbiol.*, **22**, 89–92.

Yoovidhya, T. and Fleet, G.H. (1981) An evaluation of the A–1 most probable number and the Anderson and Baird-Parker plate count methods for enumerating *Escherichia coli* in the Sydney Rock Oyster, *Crassostrea commercialis*. *J. Appl. Bacteriol.*, **50**, 519–528.

Zsigmondy, R. and Bachmann, W. (1918) Ueber neue filter. *Z. Anorg. Allgem. Chem.*, **103**, 119–128.

3 Evaluation of commercial kits and instruments for the detection of foodborne bacterial pathogens and toxins

P.D. PATEL and D.W. WILLIAMS

3.1 Introduction

Food poisoning can arise either through the ingestion of food containing bacteria, which become established in the host, resulting in illness (foodborne bacterial infection), or by ingestion of food containing preformed bacterial toxins (foodborne bacterial intoxication). In England and Wales, the most reported incidences of food poisoning in 1992 were attributed to *Campylobacter* (39 000 isolates), followed by *Salmonella* (31 000 isolates) and *E. coli* O157 (470 isolates). This represents a rise of approximately 18, 11 and 30%, respectively, over the figures for 1991 (Kimbell, 1993). The cases of listeriosis actually dropped from 280 in 1989 to 110 in 1992. Other food-poisoning bacteria such as *Clostridium perfringens*, *Bacillus cereus* and *Staphylococcus aureus* were less frequently attributed with incidences. The consequences of food poisoning can be serious for the individuals inflicted and the food producer involved. The annual costs of food poisoning to the UK food industry for 1990 were estimated at more than £350 million (Mintel, 1990).

In today's competitive market, with a wide variety of food types (e.g. short shelf-life, modified-atmosphere packaged and cooked–chilled), together with the current strict food safety legislation, it is vital to streamline and increase the efficiency of modern food production and to ensure the safety of the products. Rapid microbiological methods must be considered in the context of modern QA systems (Patel, 1993). Most of the commercially available rapid methods to date still provide results that are only of retrospective value. Methods and instrumentation are required that will ensure positive release of products and/or remedial action to be taken during food production. To monitor critical control points (CCPs) within hazard analysis critical control point (HACCP) systems, methods are required that will give results in a matter of minutes, or hours, rather than days. Of course, these should also be considered in the context of suitable sampling plans for the product in question.

The objective of this chapter is to review the range of commercial rapid systems that are available for the analysis of foodborne pathogens and bacterial toxins. It should be noted that the cost-per-test data reported for commercial products have been obtained from the retail price list in 1993. Many of the manufacturers do give general discounts for bulk purchase. For products most recently released on the market, the prices quoted have been obtained through verbal communication with the companies concerned.

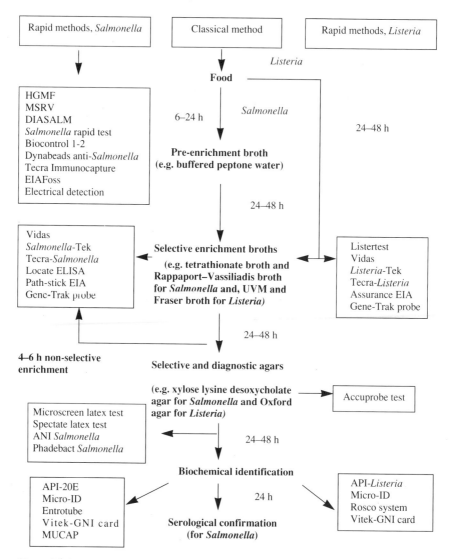

Figure 3.1 Overview of commercial rapid methods for *Salmonella* and *Listeria* showing points of application within classical cultural techniques.

3.2 Detection of foodborne pathogens

The reader is referred to three excellent books that are currently available, which describe the range of pathogens considered in this chapter in terms of their pathogenesis, characteristics of the disease, properties of the toxic components, foods commonly implicated in transmission and procedures for the isolation, detection and identification of the pathogens and/or toxins (Doyle, 1989; Cliver, 1990; Varnam and Evans, 1991).

3.2.1 Salmonella *and* Listeria

Figure 3.1 schematically represents an overview of the range of commercial rapid methods and their points of application within conventional cultural procedures for the isolation and detection of *Salmonella* and *Listeria*. Numerous cultural methods have been proposed for the isolation of these organisms from foods, and some have received accredited status. In foods, *Salmonella* and *Listeria* are often present in low numbers, largely outnumbered by competing organisms, and occasionally in a stressed state. Consequently, cultural isolation procedures tend to include an initial non-selective pre-enrichment stage designed to allow the resuscitation and growth of the target pathogen. The subsequent selective enrichment stage utilises growth media that are inhibitory to competitor cells whilst still maintaining or allowing the growth of the target organism. The selective enrichment process aims to improve the ratio of *Salmonella* or *Listeria* to competitor cells. Colonies are then isolated from the enriched culture by plating on to selective and differential agar media. Any presumptive positive colony is then confirmed biochemically and identified serologically. A range of preselective and selective media used for the enrichment and isolation of *Salmonella* (Andrews, 1985) and *Listeria* (Farber and Peterkin, 1991; Veld *et al.*, 1991) has been reported.

The advantages of cultural techniques for the microbiological examination of foods include high sensitivity (capable of detecting one viable cell) and relatively low cost per sample. The cultural methods are, however, cumbersome, labour-intensive and time-consuming. For *Salmonella* and *Listeria*, 5–7 days may be required for the isolation and identification of the organism, depending on the type of food and the extent of contamination.

A wide range of alternative techniques is now commercially available for the rapid analysis of these pathogens (Figure 3.1). Generally, the sensitivity of a technique dictates at which point during cultural enrichment it can be employed effectively. Further analytical details of the techniques for *Salmonella* and *Listeria* are reviewed below and summarised in Tables 3.1 and 3.2, respectively.

3.2.1.1 Accelerated conventional tests. In the case of *Salmonella*, attempts have been made to shorten the time of cultural enrichment in order to increase the speed of analysis (D'Aoust *et al.*, 1990). In their evaluation, the authors found 109 high-moisture and 18 low-moisture foods to be *Salmonella*-positive following 24 h pre-enrichment followed by a further 24 h selective enrichment procedure. In comparison, a short 6 h selective enrichment following a 24 h pre-enrichment identified approximately 91 and 72% of the contaminated samples, respectively. Andrews (1989) observed that reducing the classical 16–24 h

Table 3.1 Summary of alternative methods available for the detection and identification of *Salmonella*

Test system	Company	Cost/test[a] (£)	Evaluation reference
MSRV	LabM	<1.00	De Smedt and Bolderdijk (1990)
DIASALM	LabM	<1.00	Anon (1993a); Van der Zee (1993)
Salmonella Rapid Test	Unipath	3.39	Holbrook et al. (1989); Blackburn and Patel (1991)
1-2 test	BioControl Systems	5.87	Oggel et al. (1990)
Dynabeads anti-*Salmonella*	Dynal	0.85	Blackburn et al. (1991); refer to chapter 4
Salmonella-Tek ELISA	Organon Teknika	2.60	Blackburn and Patel (1990); St Clair and Klenk (1990)
Teca *Salmonella* ELISA	Tecra Diagnostics	2.92	Limbiri et al. (1990)
Tecra Immunocapture ELISA	Tecra Diagnostics	3.44	Flint and Hartley (1993)
Tecra UNIQUE *Salmonella* ELISA	Tecra Diagnostics	6.70	NA[b]
Locate *Salmonella* ELISA	Rhône-Poulenc	1.60	Bennett et al. (1993b)
Salmonella EIA	Transia	2.60	NA
Path-stick EIA	Lumac	4.73	Van Beurden and Mackintosh (1992)
Assurance EIA	BioControl Systems	5.24	Feldsine et al. (1992)
Automated VIDAS ELISA	bioMérieux	3.57	Blackburn et al. (1994)
Automated EIAFoss ELISA	Foss Electric	5.00	NA
Gene-Trak *Salmonella* probe	Gene-Trak Systems	4.20	Rose et al. (1991)
Automated Bactometer,	bioMérieux	<1.00	Refer to chapter 5
Malthus,	Radiometer	<1.00	
RABIT &	Don Whitley	<1.00	
BacTrac	Sy-Lab	<1.00	
Microscreen latex	Mercia Diagnostics	2.50	Manafi and Sommer (1992)
Spectate colorimetric latex	Rhône-Poulenc Diagnostics	3.40	Clark et al. (1989); D'Aoust et al. (1991)
Phadebact *Salmonella* test	Launch Diagnostics	2.38	NA
ANI-*Salmonella*	Ani Biotech Oy	0.80	NA
API-20E	bioMérieux	2.08	Cox et al. (1987)
Salmonella confirmation strip	LabM	1.40	Dealler et al. (1992)
Vitek-GNI card	bioMérieux	2.80	Knight et al. (1990)
Micro-ID	Organon Teknika	3.15	Cox et al. (1987)
MUCAP	Biolife	0.50	Manafi and Sommer (1992)
Enterotube	Roche Diagnostics	1.40	Cox et al. (1987)

[a] As of November 1993.
[b] NA, not available at the time of preparation of the manuscript.

Table 3.2 Summary of alternative methods available for the detection and identification of *Listeria*

Test system	Company	Cost/test[a] £	Evaluation reference
Listertest	Vicam	6.50	Jackson *et al.* (1992)
Listeria-Tek ELISA	Organon-Teknika	8.13	Haines and Patel (1988); Meier and Terplan (1993)
Tecra *Listeria* ELISA	Tecra Diagnostics	4.02	Noah *et al.* (1991)
Listeria EIA	Transia	5.83	NA[b]
Automated VIDAS ELISA	bioMérieux	4.10	Bailey and Cox (1993)
Assurance EIA	BioControl Systems	6.93	NA
Accuprobe Culture Confirmation Test	GenProbe	3.18	Williams and Patel (1990)
Gene-Trak *Listeria* spp. probe	Gene-Trak Systems	7.40	Url *et al.* (1993)
Gene-Trak L.*monocytogenes* probe	Gene-Trak Systems	8.70	NA
API-*Listeria*	bioMérieux	3.32	Bannerman *et al.* (1991)
Listeria confirmation strip	LabM	1.80	Williams and Patel (1991); Dealler and Rotowa (1991)
Vitek-GPI card	bioMérieux	2.80	Weiss (1989)
Micro-ID	Organon-Teknika	4.20	Williams and Patel (1991); Robison and Cunningham (1991); Higgins and Robison (1993)
Listzym kit	Bioconnections	2.85	Williams and Patel (1991); Kerr *et al.* (1991)

[a]As of November 1993.
[b] NA, not available at the time of preparation of the manuscript.

pre-enrichment period of 6–8 h yielded unacceptably high levels of false-negative results.

The hydrophobic grid-membrane filter technique (HGMF; QA Laboratories Ltd; Entis *et al.*, 1982), described in detail in chapter 2, comprises a 24 h pre-enrichment stage, followed by a shortened selective enrichment (6–8 h). Briefly, a portion of selective enrichment broth is filtered through the HGMF, which is then placed on a selective agar medium and incubated at 42°C for 18–24 h. A time saving of 1 day is achieved compared with the classical techniques. Entis *et al.* (1982) claim that the nature of the filters enables the detection of *Salmonella* in the presence of higher levels of background flora than could be tolerated by conventional plating. In practice, however, delays can arise due to the need to purify colonies from the filter grid cell (which may not necessarily allow the growth of single-cell type), prior to biochemical identification (Blackburn and Patel, 1991).

3.2.1.2 *Motility enrichment of* Salmonella. A range of motility enrichment-based tests is available for *Salmonella*. A major limitation of these tests is the inability to detect non-motile strains (e.g. *S. pullorum* and *S. gallinarum*). These serotypes are, however, rarely encountered in foods (<0.1%; Holbrook *et al.*, 1989). Modified semi-solid Rappaport–Vassiliadis (MSRV) medium (De Smedt *et al.*, 1986) and recently DIASALM (Anon., 1993a) have been used in combination with a short (6 h) pre-enrichment step. The continuous grey–green opaque zone extending out from the point of inoculation on the MSRV agar plate after 18 h is indicative of the presence of *Salmonella*. Further confirmation can be achieved by removing the cells from the leading edge of the zone, plating onto classical selective/differential media, followed by confirmation using API 20E strips and serology. Cells from the leading edge can also be examined by a specific latex agglutination test. In collaborative studies, the MSRV technique was shown to be rapid (presumptive positive results obtained in 48 h) and highly reliable for the detection of *Salmonella* (De Smedt and Bolderdijk, 1990). In these studies, MSRV was compared with a cultural procedure using Rappaport–Vassiliadis and selenite cystine enrichment broths. From a total of 450 test samples, comprising raw meats and processed foods artificially contaminated with *Salmonella*, 347 positive results were obtained using MSRV compared with 320 by the cultural procedure. The MSRV method has been adopted by the International Office of Cocoa, Chocolate and Sugar Confectionery as a standard method for the detection of *Salmonella* (IOCCC, 1990). More recently, O'Donoghue and Winn (1993) suggested that the MSRV method could be used to replace their in-house conventional method for the detection of *Salmonella* in both high- and low-moisture foods. This was based on evaluation using 237 samples, some either artificially or naturally contaminated. Van der Zee (1993) compared the MSRV and DIASALM media for sensitivity towards a range of *Salmonella* strains. They concluded that the DIASALM formulations gave better results than the MSRV formulation.

Two other kits available commercially are the 1-2 Test (BioControl Systems Inc.) and the *Salmonella* Rapid Test (SRT, Unipath Ltd; Figure 3.2). The *Salmonella* 1-2 Test is a disposable plastic device comprising two plastic chambers: a short inoculation chamber containing a selective medium (tetrathionate–brilliant green–serine broth), and a larger motility chamber containing a non-selective semi-solid agar medium. A polyvalent antiserum against *Salmonella* is added to the motility chamber and this gradually diffuses into the semi-solid medium. Motile *Salmonella* can migrate from the inoculation chamber into the motility chamber, where an immunoprecipitation band occurs upon reaction with the antiserum (Oggel *et al.*, 1990). The 1-2 Test module is generally inoculated with a portion of 24 h pre-enrichment broth and incubated for a further 24 h. Compared with classical methods, the results are obtained 24 h earlier. In a brief evaluation of the 1-2 Test (Blackburn and Patel, 1991), salmonellae were detected at levels above 10^2 cfu/ml in the presence of competitors at approximately 10^7 cfu/ml. The test was simple to use and required no specialised equipment. When applied to six food types, including dried whole egg, milk chocolate and raw deboned turkey, there was an overall

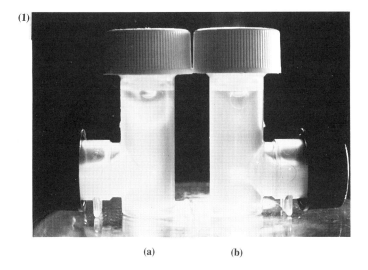

(a)   (b)

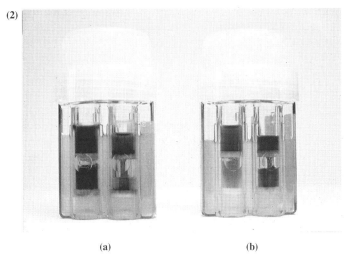

(a)   (b)

Figure 3.2 BioControl 1-2 (1) and *Salmonella* rapid (2, Unipath) tests; (a) and (b) represent presumptive-positive and negative results, respectively.

96.1% agreement between the culture methods (BAM/AOAC) and the 1-2 Test (Flowers and Klatt, 1989). The false-negative rates were 3.6 and 1.7%, respectively.

The SRT is another disposable system, which comprises a single cultural vessel containing a selective medium for *Salmonella*. Within the culture vessel there are two reaction tubes, each containing a lower selective medium and an upper indicator medium, separated by a porous partition. A portion of a 24 h pre-enriched food sample is added to the culture vessel and the system incubated for 24 h. During this period, *Salmonella* actively migrates through the reaction

Table 3.3 Comparison of *Salmonella* rapid test (SRT) with a conventional cultural method (CCM) for the detection of *Salmonella* in foods[a]

Food	No. of samples[b]		No. of *Salmonella* positive samples	
			SRT	CCM
Chocolate	2	(1)	1	1
Cocoa powder	2	(1)	1	1
Cooked chilled beef	2	(1)	1	1
Cooked chilled chicken	4	(2)	2	2
Egg powder	3	(0)	1	1
Salami	2	(1)	1	1
Skimmed-milk powder	9	(1)	1	1
Raw beef	4	(2)	2	2
Raw chicken	2	(1)	1	1
Raw pork	2	(1)	2	2
Raw prawns	2	(1)	1	1
Raw trout	2	(1)	1	1
Raw turkey	2	(1)	1	1
Total	38	(14)	16	16

[a] Data taken from Blackburn and Patel (1991).
[b] Values in parenthesis are the number of samples that were artificially contaminated with *S. enteritidis.*

tubes, where its presence is indicated in the upper portion of the tubes by an identifiable colour change (blackening or red–black coloration) resulting from characteristic biochemical reactions. Presumptive positive results can be further tested using a latex agglutination test supplied with the SRT. In a recent study (Blackburn and Patel, 1991), a total of 36 food samples, including raw poultry and processed low-moisture foods, was analysed using a conventional cultural method and the SRT. A summary of the results is given in Table 3.3. There was 100% agreement between the two methods, with 16 samples identified as *Salmonella*-positive. A total time saving of 1–3 days over the conventional methods was found to be possible using the SRT. Although simple to use, the SRT was cumbersome when a large number of samples needed to be analysed. In addition, interpretation of the colour changes was highly subjective and not clear-cut with certain samples (e.g. artificially contaminated turkey and trout).

3.2.1.3 Immunological tests. The specificity of antibodies (polyclonal and monoclonal) for the corresponding analytes has widely encouraged the development of commercial rapid test systems in microbiological analysis. A range of these is described below.

Immunomagnetic techniques. The immunomagnetic separation technique has been successfully used as a rapid (10 min) alternative to the 24 h cultural selective enrichment step for the detection of *Salmonella* (for details, refer to chapter 4). The particles used in that study are now marketed commercially by the Norwegian company, Dynal.

Immunomagnetic-based techniques are also available from Vicam (USA) for the detection of genus *Listeria* (Listertest Lift) and *L. monocytogenes* specifically (Listertest Mac). Both tests are carried out directly from liquid foods or food homogenates without prior requirement of any form of enrichment (Jackson *et al.*, 1992). Anti-*Listeria* antibody-coupled magnetic particles are mixed with foods and shaken for approximately 2 h to allow any *Listeria* to bind to the particles. *Listeria*-bound particles are then separated and washed prior to assaying. In the Listertest Lift, the *Listeria*-bound particles are plated onto the Listertest plate and incubated overnight. Any presumptive positive *Listeria* colony on the plate is then identified by Western blotting, in which a membrane is overlaid on the colonies to obtain an imprint. After fixation of the cells, *Listeria* is detected by sequential reaction with a primary monoclonal antibody and a secondary antibody–enzyme conjugate. If *Listeria* is present, a coloured spot appears following application of the enzyme substrate.

The Listertest Mac identifies pathogenic *Listeria*. In this test, the *Listeria*-bound particles are placed onto a monolayer of macrophages in a tissue culture flask. The macrophages internalize the bacteria-bound particles. Non-internalized bacteria are killed by antibiotics in the Listertest medium. After infection of macrophages, the tissue culture medium is removed and the monolayer of macrophages is overlaid with Listertest agar. After overnight incubation, if *L. monocytogenes* is present, colonies will appear on the overlay agar. Both tests are quantitative, since no cultural enrichment is required. The sensitivity for most food samples investigated was claimed to be 5–10 cfu/g, with a final result in about 24 h (Jackson *et al.*, 1992). This means that, for samples containing <100 cfu of *Listeria*/ml, either a cultural enrichment or centrifugation steps may be necessary prior to assaying. It is also difficult to envisage a widespread industrial uptake of the Listertest Mac test, owing to the requirement for setting up cumbersome tissue culture techniques.

Enzyme-linked immunosorbent assay (ELISA). The most widely used tests for *Salmonella* and *Listeria* are ELISAs. Figure 3.3 shows examples of a range of commercial immunological tests, whilst Figure 3.1 shows the points of application of these tests.

A typical commercial ELISA protocol for *Salmonella* is outlined in Figure 3.4. The limit of detection for most ELISAs generally ranges from 10^4 to 10^6 cfu/ml, depending upon the species or serotype (Blackburn and Patel, 1990; Curiale *et al.*, 1990a). Consequently, extensive cultural enrichment of a food sample is required prior to applying the ELISA (Figure 3.1). In most cases, a presumptive positive result is obtained in approximately 52 h. The Tecra immunocapture ELISA, which has recently become available, uses a sensitised dipstick to capture *Salmonella* within 20 min from 24 h food pre-enrichment broths (Flint and Hartley, 1993). The captured *Salmonella* is then incubated in M-broth for a further 5 h to increase the level of the target cells for subsequent detection of the heat-solubilised *Salmonella* antigen by using the

(a)
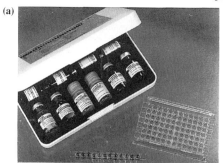
(b)
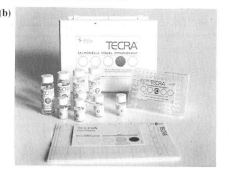

(c)

(d)
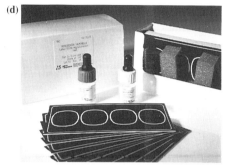

Figure 3.3 Examples of commercial ELISA kits and latex agglutination tests. (a) *Listeria*-Tek. (b) Tecra-*Salmonella*. (c) Spectate. (d) Monoscreen.

Tecra visual immunoassay. When compared with the classical FDA method using a range of foods (e.g. cheese, milk powder, meat and fish), a close agreement was found with the Tecra immunocapture ELISA (Flint and Hartley, 1993). However, the former test gave a result in 96 h and the latter in only 24 h. A modified version of the immunocapture test, the Tecra unique *Salmonella*, is about to be released on the market. The manufacturers claim that it can give a result in less than 22 h. In this test, *Salmonella* captured from an overnight pre-enrichment broth on the dipstick is allowed to multiply in M-broth for a further 4 h, prior to direct detection on the dipstick using an enzyme immunoassay.

In an evaluation of the Tecra *Salmonella* visual immunoassay (Blackburn *et al.*, 1988), five different foodstuffs (including raw and processed foods) were artificially contaminated with low levels (1–10 cfu/25 g) of *Salmonella*. In all cases where the conventional procedure isolated *Salmonella*, a positive result had been obtained using the ELISA. However, the ELISA showed two false-positive results using raw chicken. This was attributed to high numbers of competing flora. Figure 3.5 shows the response of a range of *Salmonella* serotypes and non-*Salmonella* with the *Salmonella*-Tek ELISA (Organon-Teknika; Blackburn and Patel, 1990). The sensitivity of the ELISA varied with the serotype, but was generally between 10^4 and 10^6 cfu/ml. Table 3.4 summarises the results obtained using a range of processed and raw foods. A

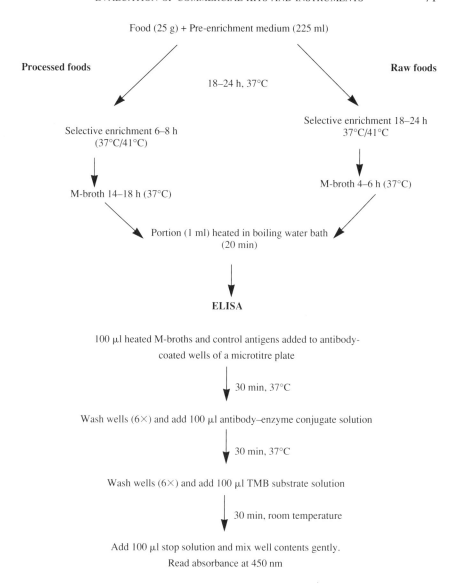

Figure 3.4 *Salmonella*-Tek immunoassay.

100% agreement was found between the *Salmonella*-Tek and the conventional cultural method. A common problem with many ELISA systems for *Salmonella* is cross-reaction with *Citrobacter freundii*. The *Salmonella*-Tek also showed cross-reactions with two strains of *C. freundii* (Blackburn and Patel, 1990), and similar findings have been reported elsewhere (Curiale *et al.*, 1990a).

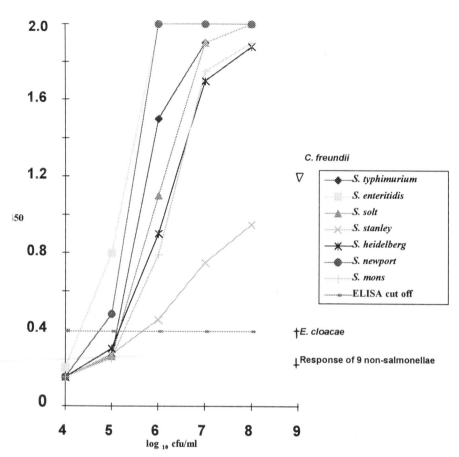

Figure 3.5 Sensitivity and specificity of *Salmonella*-Tek for salmonellae and non-salmonellae cultured in M-broth.

A recent arrival on the market is the Path-stick immunoassay (Table 3.1) for *Salmonella* (Van Beurden and Mackintosh, 1992). The assay has a sensitivity of approximately 10^6 cfu/ml and is based on a dipstick, which will develop two coloured lines on a reaction membrane in the presence of *Salmonella*. In the absence of *Salmonella*, only one line will develop. The assay time is 20 min following 2 days of cultural enrichment. Only one strain of *C. freundii* was found to cross-react in the assay.

In an evaluation of the *Listeria*-Tek and Tecra *Listeria*, comparing their performance against an officially recognised AOAC method (Anon., 1988), it was concluded that there was no statistical difference between the ELISAs and the cultural method (Noah *et al.*, 1991). These ELISAs will only identify *Listeria* to the genus level and further biochemical tests are required to specifi-

(1)

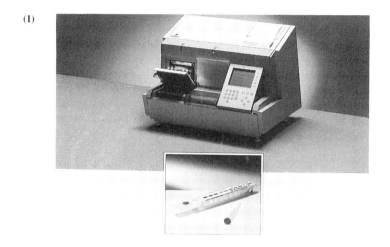

(2)

Figure 3.6 Automated mini-Vidas (1) and EIA Foss (2) for the detection of bacterial pathogens.

cally identify *L. monocytogenes*, considered to be the major human pathogen (Farber and Peterkin, 1991). The manual ELISAs, although practically simple to execute, require a high level of technical skill, particularly in handling small volumes of reagents, preventing reagent carry-over, and ensuring thorough washing at appropriate stages. These steps can be automated to a certain extent using automated washers and reagent dispensers. Recently, however, fully automated ELISAs (Figure 3.6) have become available for the analysis of foodborne pathogens.

Table 3.4 Summary of results showing comparison of *Salmonella*-Tek with the conventional cultural method (CCM)[a]

Food	No. of samplesb		ELISAc	CCM
Skimmed-milk powder	7	(0)	0	0
Egg powder	3	(0)	2	2
Cocoa powder	2	(1)	1	1
Chocolate	2	(1)	1	1
Cooked chilled chicken	2	(1)	1	1
Cooked chilled beef	2	(1)	1	1
Salami	2	(1)	1	1
Raw beef	2	(1)	1	1
Raw pork	2	(1)	2	2
Raw chicken	2	(1)	1	1
Raw turkey	2	(1)	1	1
Raw trout	2	(1)	1	1
Raw prawns	2	(1)	1	1
Total	32	(11)	14	14

[a] Data taken from Blackburn and Patel (1990).
[b] Values in parenthesis are the numbers of samples that were artificially contaminated with *S. enteritidis*.
[c] All ELISA positives were confirmed for the presence of *Salmonella* by subjecting the corresponding selective broths and/or M-broths to conventional analysis.

In the Vidas system (bioMérieux), for each assay, a predispensed disposable reagent strip containing individual reagents (e.g. diluents, antibodies, conjugates and substrates) is required, together with the corresponding solid phase receptacle (SPR). The SPR is essentially a pipette-tip that is coated inside with the primary antibody used to capture the target antigen. The SPR also serves as a pipetting device for sampling and transfer of reagents during the assay. Appropriate protocols specific for an assay, based on a fluorescent end-point, are downloaded from the system computer during start-up of an assay. In their evaluation, Bennett *et al.* (1993a) found that the sensitivity of the Vidas for the genus *Listeria* ranged between 10^5 and 10^6 cfu/ml. The assay results were obtained in approximately 45 min. When tested with a range of foods using four different enrichment protocols, it was concluded that different enrichment protocols were required depending on the food type, with improved results being obtained after 48 h enrichment compared with 24 h. Bailey and Cox (1993) evaluated the Vidas for *Listeria* detection in meat and poultry products. They showed that the USDA and Vidas procedures detected 27 and 20 out of 60 samples, respectively, as *Listeria*-positive. The false-negative results by Vidas were attributed to low sensitivity (10^5–10^6 cfu/ml) of the technique and the suppression of the growth of *Listeria* by competitors in raw chicken samples.

The EIAFoss (Foss Electric) is an automated fluorescent end-point based ELISA, which uses immunomagnetic particles as a solid phase. In its protocol, Foss Electric (UK, Ltd) recommends a two-step enrichment procedure for foods prior to assaying for *Salmonella*. The first step is a pre-enrichment in *Salmonella* enrichment broth, followed by a short post-enrichment in a modified mannose broth, the total period being <24 h. The broths are then boiled before the ELISA procedure, which takes approximately 1.5 h. The assay is claimed to have a sensitivity of approximately 10^5 cfu/ml, and, in an external evaluation (Ref. No. C:\WP51\BDA\D\TESTING.EIA, Foss Electric, UK), the EIAFoss showed better performance than the reference method.

Latex agglutination tests. To date, latex agglutination tests (Table 3.1), utilising latex particles sensitised with the primary antibodies, are available for *Salmonella* but not for *Listeria*. These test are commonly used to confirm suspect colonies isolated on solid agar media. Thus, high levels (approximately 10^7–10^9 cfu) are required to give a positive agglutination reaction (Blackburn and Patel, 1989). The tests are, however, very rapid (3 min), simple in practice and, in the case of the colorimetric Spectate test (Rhône-Poulenc), can give an indication of the *Salmonella* serogroup. Blackburn and Patel (1989) further showed that the Microscreen latex agglutination test for *Salmonella* (Mercia Diagnostics) agreed 100% with the API-20E, which, unlike the latex test, takes 16–24 h to provide results.

3.2.1.4 Nucleic acid probes. The methods described previously are all dependent on the detection of specific biochemical or serological characteristics of the target organism. The genome of the organism is responsible for the expression of these characteristics and it is, therefore, not surprising that rapid test kits have been developed which rely on the interaction of nucleic acid probes with the conserved or marker sequences of target cell nucleic acid.

There are two types of nucleic acid probe tests (colorimetric and chemiluminescent) available for *Salmonella* and *Listeria* (Figures 3.7 and 3.8). Briefly, the Gene-Trak system comprises a dipstick coated with polythymidine residues and two nucleic acid probes that bind to the target cell ribosomal RNA (rRNA); a capture probe containing polyadenosine residues; and a detector probe, labelled with fluorescein. Each probe contains a short sequence of nucleotides complementary to the target sequence. Following hybridisation of the probes with complementary rRNA, extracted from the target pathogen, the dipstick is used to capture the hybrid complex. Finally, the presence of target rRNA is assessed by the addition of an enzyme-labelled anti-fluorescein antibody. Addition of the substrate results in the generation of a coloured product, the extent of coloration being proportional to the target pathogen concentration.

The Gene-Trak assay for the detection of *Salmonella* and *Listeria* is generally applied after 48 h incubation of the food-enrichment media, thereby providing a timesaving of approximately 24–48 h over conventional methods. The sensitivity of the assay for *Salmonella* and *Listeria* is reported to be between 10^5 and 10^7 cfu/ml (Curiale *et al.*, 1990a). In an evaluation of the Gene-Trak test for *Salmonella* (Curiale *et al.*, 1990a), 99% of 294 *Salmonella* strains tested gave positive results. However, three strains belonging to *Salmonella* sub-group V

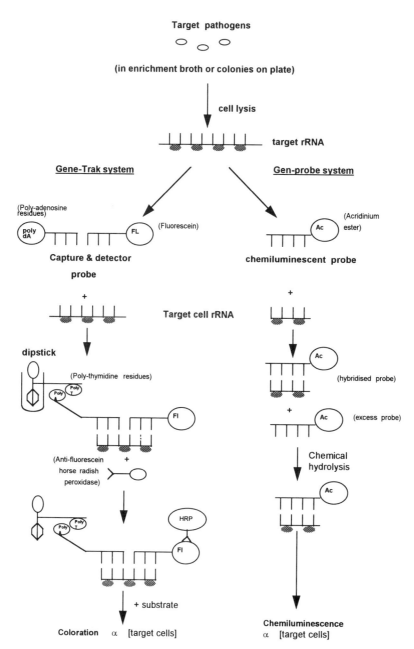

Figure 3.7 Basis of the nucleic acid probe assays for foodborne pathogens (e.g. *Salmonella* and *Listeria*).

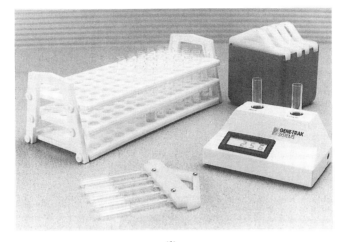

(1)

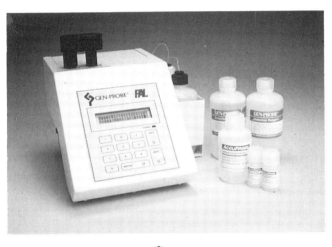

(2)

Figure 3.8 Examples of commercial nucleic acid probe kits for the detection of bacterial pathogens (1) Gene-Trak system for *Salmonella*, (2) GenProbe system for *L.monocytogenes*.

gave false-negative reactions. Out of 100 non-*Salmonella* tested, none was found to give a positive response with this assay. In collaborative studies (Curiale *et al.*, 1990b), the Gene-Trak test for *Salmonella* was compared with the standard cultural method for detection in a range of foods, including non-fat dry milk, milk chocolate, soya isolate, dried whole egg, ground black pepper and raw ground turkey. Twelve samples of each food (10 artificially contaminated with *Salmonella*) were examined by 11 collaborating laboratories. The results obtained indicated that there were no significant differences between the conventional detection procedures and the Gene-Trak test. The Gene-Trak system, although simple to execute, does require a skilled technician to perform the assay. Furthermore, the number of steps involved in the assay makes it particularly labour-intensive and cumbersome when large numbers of samples need to be analysed. Automation of these assays is required to gain widespread support in the industry.

The AccuProbe culture confirmation test (GenProbe Inc.) is a chemiluminescent DNA probe system based on the principle of DNA hybridisation protection (Figure 3.7). The assay is specific for *L. monocytogenes*. Unlike the heterogeneous colorimetric Gene-Trak assays, the Accuprobe test is a homogeneous assay, where the separation of excess reagents is not required. In a recent evaluation, the Accuprobe assay was found to be very simple to use and rapid, providing results within 30 min (Williams and Patel, 1991). The sensitivity of the Accuprobe test for *L. monocytogenes* ranged from 10^5 to 10^6 cfu/ml. The assay was also highly specific, identifying all 21 *L. monocytogenes* cultures tested with no cross-reaction occurring with 36 competitor organisms. The manufacturers recommend that the AccuProbe test should be applied to colonies from solid agar media. Williams and Patel (1991), however, demonstrated the potential of this technique to detect *L. monocytogenes* from *Listeria* selective enrichment broths after 48 h incubation. This represents a time saving of 2–3 days and 6–8 days over most other commercial rapid tests and classical methods, respectively.

3.2.1.5 Automated electrical systems. Electrical systems based on the principles of conductance, impedance and capacitance are the most highly automated systems available for the detection and estimation of microorganisms of importance in the food industry. Thus, chapter 5 is devoted entirely to the subject and covers the hardware and software capabilities of a range of commercial systems and the areas of application of this versatile cultural-based technique. The scope of immunomagnetic techniques in enhancing the speed and reliability of electrical systems is described in chapter 4.

3.2.1.6 Biochemical identification tests. A range of simple and rapid biochemical identification systems is commercially available as alternatives to the conventional tests for both *Salmonella* and *Listeria* (see examples in Tables 3.1 and 3.2, respectively). These tests are applied to colonies on selective/diagnostic media or to subsequently purified colonies on non-selective agars. All incorporate a battery of biochemical and enzymic reactions, which produce identifiable colour changes. The reaction profiles can then be interpreted manually (e.g. API and Micro-ID systems) or by using a range of commercially available automated instruments, including the ATB system (bioMérieux, Freney *et al.*, 1990), Infobac (Unipath), Vitek AMS (bioMérieux) and the Biolog system (Biolog Inc.; Klinger *et al.*, 1992), in order to identify the organism. These tests generally provide an identification to the genus level for *Salmonella* and *Listeria*. In the case of *Listeria*, specific identification of *L. monocytogenes* is possible using the API-*Listeria* test (bioMérieux).

The MUCAP test (Biolife) for *Salmonella* is based on the ability of the organism to produce the enzyme caprylate esterase, which acts on the substrate 4-methylumbelliferyl-caprylate to produce, within 5 min, a blue fluorescent product observed under ultraviolet light. In a recent evaluation (Manafi and Sommer, 1992), this test was found to detect all of the 74 *Salmonella* strains examined, but gave false-positive reactions with eight out of 100 non-*Salmonella* tested.

3.2.2 Campylobacter

Acute gastroenteritis due to *Campylobacter* infection has been identified as one of the most common forms of diarrhoeal disease in humans (Walker *et al.*, 1986). In England and Wales in 1990, an annual incidence of 35 000 *Campylobacter* infections was reported (Skirrow, 1991). *Campylobacter jejuni* accounted for more than 90% of these infections, although less frequent infections can arise from *C. coli* and *C. lari*.

All *Campylobacter* spp. are microaerophilic and require reduced levels of oxygen (5–6%), being inhibited by the normal atmospheric oxygen level of 21%. The conventional cultural procedures for the isolation of *Campylobacter* are carried out under strict gaseous conditions and involve pre-enrichment, selective enrichment, followed by plating on selective diagnostic media and biochemical confirmation. The procedures are lengthy, often taking 4–5 days to obtain presumptive positive results and a further 2–5 days for biochemical confirmation (Skirrow *et al.*, 1982). For further details concerning the range of classical methods and media employed for *Campylobacter* detection, the reader is referred to Bolton *et al.* (1986), Merino *et al.* (1986) and Griffiths and Park (1990).

Scotter *et al.* (1993) reported results of a collaborative evaluation of three cultural methods, two of which were proposed by the International Standards Organisation (ISO), for the detection of thermotolerant *Campylobacter* in minced chicken skin. It was found that the ISO method 2, which involved a non-selective agar in combination with a 0.65 μm membrane filter, was significantly better for detecting campylobacters at levels of 2 or 10 cells per 10 g, but the procedure involved extensive manipulations.

Table 3.5 shows a range of alternative methods that are available for the detection and identification of *Campylobacter* in foods, whilst Figure 3.9 shows a typical cultural enrichment procedure together with points of application of alternative tests.

3.2.2.1 Nucleic acid hybridisation. The DNA probes for the detection of *Campylobacter* include the SNAP (synthetic nucleic acid probe) hybridisation test for *C. jejuni* and *C. coli*, and the AccuProbe culture confirmation system.

The SNAP kit uses a synthetic probe of 15–35 bases in length, with alkaline phosphatase attached to the thymidine base on the probe via a spacer molecule. Colonies isolated on selective/diagnostic plates are subcultured on a non-selective medium and then reconstituted in nutrient broth, supplemented with FBP (containing ferrous sulphate, sodium metabisulphite and sodium pyruvate, Drewett and Patel, 1989). Whole cell DNA is then extracted using lysozyme and sodium dodecyl sulphate treatment. The extracted DNA is immobilised onto nylon filters and then hybridised with the probe. The presence of *Campylobacter* DNA can then be established by the addition of a NBT/BCIP (nitro blue tetra-zolium/5-bromo-4-chloro-3-indolyl phosphate) substrate, which is converted to an insoluble purple precipitate in the presence of the enzyme-labelled probe.

In an evaluation by Olive *et al.* (1990), the SNAP system could detect down to 5 ng of *C. jejuni* DNA, equivalent to approximately 10^5 cfu of bacteria. Drewett and Patel (1989) reported a sensitivity of approximately 10^6 cfu of *Campylobacter*. The assay detected all the 18 *C. jejuni* isolates, but did not show any reaction with 23 other non-*Campylobacter* isolates tested. When applied to foods, the total assay time following 48 h cultural enrichment procedures was approximately 6.5 h; 2.5 h for target cell DNA immobilisation, probe hybridis-ation and washing; and up to 4 h for the detection. The assay was, however, cumbersome and labour-intensive, and not easily adapted to busy laboratories.

The AccuProbe culture confirmation test, principally similar to the test for *Listeria monocytogenes*, allows the detection of *C. jejuni, C. coli* and *C. lari* from colonies grown on primary isolation agar media (e.g. Preston agar) within 30 min. A positive signal can also arise with *C. jejuni* subsp. *doylei* and 'urease-positive thermophilic' *Campylobacter* (information provided to Laboratory Impex by Dr David Wareing, Preston Public Health Laboratory). The manufac-turers claim that positive confirmation of *C. jejuni* may be obtained directly using enrichment broth cultures. However, this will depend on the assay sensi-tivity, specificity and, indeed, the likely productivity of *Campylobacter* in the food-enrichment broth used.

3.2.2.2 Immunological and biochemical tests. Commercial immunological and biochemical tests for *Campylobacter* are limited (Table 3.5), unlike such tests for *Salmonella* and *Listeria*. The sensitivity of the Microscreen, Campyslide and Meritec latex agglutination tests has been reported to be in the range approximately 10^6–10^8 cfu/ml (Hazeleger *et al.*, 1992). However, when applied to 48 h chicken enrichment broths, 69% of the *Campylobacter*-positive enrichment broth cultures were positive with the Microscreen, whilst the Meritec and Campyslide tests detected 63% and 15% of the positive samples, respectively.

Scotter *et al.* (1993) reported in their study using minced chicken skin that the degree of agreement between the traditional biochemical tests and the Microscreen latex test was dependent upon the method of cultural enrichment. In one of their cultural methods, identical results were obtained between the

Table 3.5 Summary of alternative methods available for the detection and identification of *Campylobacter*

Test system	Company	Cost/test[a] (£)	Evaluation reference
Microscreen latex	Mercia Diagnostics	1.78	Hazeleger *et al.* (1992)
Campyslide	Becton Dickinson	3.27	Scotter *et al.* (1993)
Meritec-Campy	Meridian Diagnostics	1.15	Hazeleger *et al.* (1992)
Accuprobe culture confirmation test	GenProbe	3.18	NA[b]
Gene-Trak *Campylobacter* probe	Gene-Trak Systems	5.60	NA
Nucleic acid probe (SNAP)	Molecular Biosystems	NA	Drewett and Patel (1989)
API-Campy	bioMérieux	4.24	Megraud *et al.* (1987)

[a] As of November 1993.
[b] NA, not available at the time of preparation of the manuscript.

biochemical test and the latex agglutination result. However, problems of non-specific agglutination were found when the latex test was applied to colonies taken directly from the selective media. This was attributed to insufficient flagellar antigen production in the selective media.

The API Campy for *Campylobacter* consists of two parts; the first allows the detection of enzymes produced by the organism and the second comprises nine growth tests, and a test for hydrogen sulphide production. Both the API and latex agglutination tests are applied to isolated colonies from selective and non-selective agar plates. The significant difference between these tests is that the latex test gives results in <1 min whilst the API takes 18 h.

3.2.3 Escherichia coli

Although many strains of *E. coli* are harmless, there are currently five major classes of *E. coli* recognised as causative agents of human diarrhoea (Olsvik *et al.*, 1991): vero cytotoxin-producing *E. coli* (VTEC), also called enterohaemorrhagic *E. coli* (EHEC); enterotoxigenic *E. coli* (ETEC); enteropathogenic *E. coli* (EPEC); enteroinvasive *E. coli* (EIEC); and enteroaggregative *E. coli* (EAggEC). Human diarrhoea caused by *E. coli* is attributed to the production of specific enterotoxins and, to date, only known to be important for the VTEC and ETEC groups. Rapid methods for the detection of the enterotoxins are described in Section 3.3.2.1.

The VTEC, of many different serotypes but especially of serotype O157:H7, are now recognised as an increasingly important cause of human diarrhoea in North America and Europe. A recent major outbreak (Foodmaker Inc./Jack In The Box) in the USA involved 500 laboratory-confirmed cases and four deaths

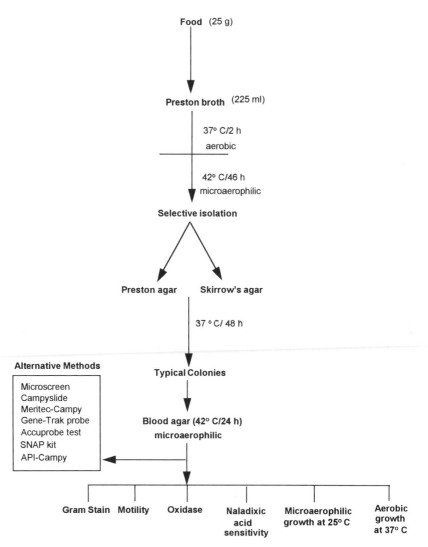

Figure 3.9 Typical cultural isolation and detection procedure for *Campylobacter jejuni/coli* showing points of application of alternative methods.

(Davies *et al.*, 1993). In the UK, an outbreak of *E. coli* O157 has recently been confirmed, involving six cases – four children and two adults (Anon., 1993b). The two adults had haemorrhagic colitis, and three children developed haemolytic uraemic syndrome.

The USDA cultural method (Okrend *et al.*, 1990) for the isolation of *E. coli* O157:H7 from meat involves the following steps: preparation of 1:10 homogenate of the meat in modified EC broth with novobiocin followed by incubation for 24 h at 35°C; serial decimal dilutions of the enrichment broth are made in Butterfield's phosphate diluent and then spread-plated on MacConkey sorbitol agar (MSA) or MSA with 5-bromo-4-chloro-3-indoxyl-β-D-glucuronic acid cyclohexylammonium salt (MSA-BCIG) agar plates. The plates are incubated overnight at 42°C; several of the sorbitol-negative colonies (white) from the MSA plates and sorbitol-negative, BCIG negative colonies (white) from the MSA-BCIG plates are selected. Each colony is spread in the centre of a section of the eosin methylene blue agar (EMB) as well as stabbed in the corresponding section on the phenol red sorbitol agar + 4-methylumbelliferyl-β-D-glucuronide (PRS-MUG) plate. These plates are incubated overnight at 35°C and then examined. The isolates that are sorbitol-negative (no colour change), MUG-negative (no fluorescence under UV light) and show typical dark-purple growth with a green metallic sheen on EMB are then confirmed with tube biochemical tests and an H7 agglutination test. The overall time is approximately 6–7 days to confirm the positives.

For rapid techniques other than those available commercially for the detection of verocytotoxin-producing *E. coli*, the reader is referred to the reviews by Karmali (1988) and Smith and Scotland (1993).

3.2.3.1 Enzyme immunoassays. An increasing number of immunological tests are becoming available for the detection of *E. coli* O157/*E. coli* O157:H7 (Table 3.6).

The EHEC-Tek is a microtitre plate-based ELISA, which is applied after 24 h cultural enrichment of a test sample. A portion of the test broth is heat-treated prior to detection of *E. coli* O157:H7 within 2 h by the ELISA. If presumptive-positive, the result can be confirmed by the classical cultural techniques.

The AMPCOR dipstick and single-step tests are recommended for the presumptive detection of *E. coli* O157:H7 in food and food ingredients. Thus, if positive results are obtained, further confirmation by classical techniques is required. The tests are based on polyclonal anti-*E. coli* O157:H7 antibodies immobilised in a horizontal line on a membrane strip. The tests are applied after 24 and 48 h enrichment of the samples. In the case of the dipstick assay, a portion of the sample is reacted with the specific antibody coupled to an enzyme. The complex (*E. coli* O157:H7 bound antibody–enzyme) is then captured by incubating the dipstick in the sample. Following addition of the substrate, a horizontal blue to black line will form in the middle of the dipstick if *E. coli* O157:H7 is present. The assay gives a result within 20 min and is claimed to have a sensitivity of approximately 10^4 cfu/ml.

The single-step assay procedure is similar to that described for the dipstick assay except that the detection principle is simpler. In this case, a test sample is incubated with the specific polyclonal antibodies, which are bound to gold particles. Thus, in the presence of *E. coli* O157:H7, a complex of *E. coli*-antibody-bound gold is formed. A portion of the sample is added to a plastic device containing anti-*E. coli* O157:H7 antibodies immobilised in a horizontal fashion near the bottom of a membrane strip (test window). Another capture reagent (not specific for *E. coli* O157:H7) is immobilised in the upper part of the membrane (control window). If *E. coli* O157:H7 is present, a red–purple line results in both the test and control windows, whilst in the absence of *E. coli* O157:H7 only the control line appears. The test is claimed to give a result in 20 min and has a sensitivity ranging from 10^4–10^6 cfu/ml.

The Petrifilm Test kit is a membrane-based immunoassay for the detection of *E. coli* O157:H7. Samples to be tested are enriched in a broth for 6–8 h at 37°C with shaking. Serial dilutions (1:10 or 1:100) of the enrichment broth are then applied to the Petrifilm *E. coli* plates, which are incubated for 16–18 h at 42°C. The colonies are then replicated to reactive disc membranes. After washing to remove any unbound material, the discs are treated with an enzyme-labelled anti-*E. coli* O157 antibody. After further washing and addition of the non-chromogenic substrate, the appearance of blue-coloured colonies is indicative of the presence of *E. coli* O157. In a recent evaluation of this test (Okrend *et al.*, 1990), where meat samples were enriched for 6–8 h in modified EC broth with novobiocin, prior to inoculation of the Petrifilm *E. coli* plates, *E. coli* O157 was detected within 26–28 h, although isolation and confirmation of the presumptive positive colonies required an additional 3–4 days. Sernowski and Ingham (1992) found that the HEC O157™ ELISA when evaluated using 74 ground beef

Table 3.6 Summary of alternative methods for the detection of *E. coli* O157 group

Test system	Company	Cost/test[a] (£)	Evaluation reference
EHEC-Tek ELISA for *E. coli* O157:H7	Organon Teknika	5.73	Padye and Doyle (1991)
Petrifilm HEC EIA for *E. coli* O157:H7	3M Health Care	8.46	Okrend *et al.* (1990)
Dip-stick EIA for *E. coli* O157:H7	Ampcor	4.80	NA[b]
One-step EIA for *E. coli* O157:H7	Ampcor	4.00	NA
Path-stick EIA for *E. coli* O157:H7	Lumac	NA	NA
Dynabeads anti-*E.coli* O157	Dynal	1.50	NA
E. coli O157 latex	Unipath	0.20	Bettelheim *et al.* (1993)
Prolex TM *E. coli* O157 latex	Pro-lab Diagnostics	0.40	Rice *et al.* (1992)
Microscreen *E.coli* O157 latex	Mercia Diagnostics	1.05	NA

[a] As of November 1993.
[b] NA, not available at the time of preparation of the manuscript.

samples, had low specificity and showed cross-reactions (e.g. with *Hafnia alvei*). Unlike the ELISAs, based on dipstick and microtitre plate, the Petrifilm assay appears to be cumbersome and labour-intensive.

3.2.3.2 Bead-based immunoassays. The most recent arrival on the market is the Dynal anti-*E. coli* O157 system based on the use of immunomagnetic particles. No independent data are yet available on its performance characteristics. The manufacturers recommend the following steps:

(i)   6 h pre-enrichment of a food sample in BPW-VCC (buffered peptone water with vancomycin, cefixime and cefsulodin);

(ii) treat 1 ml of homogenate with 20 μl of the immunoDynabeads and allow to incubate for 30 min;

(iii) separate and wash the particles, then reconstitute in 50 μl of washing buffer;

(iv) spread the particles on to TC-SMAC (MacConkey medium containing sorbitol, tellurite and cefixime) by swabbing one-third of the plate and then dilute by streaking with a loop;

(v)   read the plate for sorbitol-negative (colourless) colonies after incubation of the plate at 37°C for 18–24 h.

The selected colonies should be confirmed serologically.

Most VTEC of serogroup O157 (either H7 or non-motile) do not ferment sorbitol within 18 h, whereas most *E. coli* strains do. On sorbitol MacConkey agar, VT-producing O157 strains appear as colourless colonies. These colonies may also result from some non-fermenting *E. coli* strains of other serogroups or bacteria of other genera. Thus, confirmation as *E. coli* and as serogroup O157 is essential. There are several commercial latex agglutination tests available for the detection of *E. coli* O157 within minutes (Table 3.6). A common problem associated with these assays is cross-reaction with the organism *E. hermanii*, which is also incapable of fermenting sorbitol and has been isolated from beef and unpasteurised milk (Lior and Borczyk, 1987). Pro-Lab Diagnostics states that antibodies cross-reacting with *E. hermanii* have been removed in the preparation of their latex kit. The latex tests are applied to colonies (non-sorbitol fermenting) isolated directly from sorbitol MacConkey agar or from sub-culture onto a non-selective medium. Any positive reaction should then be confirmed by biochemical testing.

3.2.4 Yersinia enterocolitica

Yersinia enterocolitica is a coccoid-shaped Gram-negative bacterium belonging to the genus *Yersinia* in the Enterobacteriaceae family. It is a recognised foodborne pathogen and several food-associated outbreaks of yersiniosis have been reported. Related species, e.g. *Y. frederiksenii* and *Y. kristensenii*, are generally referred to as non-pathogenic or environmental *Yersinia*. It is the species *Y. enterocolitica* that is significant in terms of food poisoning. There are five biotypes of *Y. enterocolitica*, with biotype 1 comprising a wide range of serotypes. Over 50 serotypes are reported within the species *Y. enterocolitica*, of which 0:1, 2a, 3, 0:3, 0:5, 27, 0:8 and 0:9 are generally considered to be pathogenic to humans (Halligan, 1990). The serotype 0:6,30 has also been associated with disease.

The serotypes 0:3 (biotype 4), 0:8 (biotype 1bis) and 0:9 (biotype 2) are mainly involved in human disease. The serotypes 0:3 and 0:9 are frequently encountered in the UK, whilst the biotype 4 (serotype 0:3) and biotype 3 (serotype 0:5 and 27) are responsible for most human gastrointestinal infections in Europe, Canada and the Republic of South Africa. The serotype 0:8 strains are frequently isolated in the USA (Halligan, 1990). An investigation on the occurrence of *Yersinia* spp. in a range of foods and related environments has been reported elsewhere (Greenwood and Hooper, 1985; Halligan, 1990).

In a recent outbreak (15 cases) in the USA, *Y. enterocolitica* serotype 0:3 was isolated, with the food vehicle involving pork chitterlings (Lee *et al.*, 1990). Although several foods act as vehicles in foodborne yersiniosis, milk is involved most frequently. In the 1985 UK outbreak involving 36 cases, pasteurised milk was again implicated, with the isolation of *Y. enterocolitica* serotypes 0:10K, 0:6 and 30 (Greenwood and Hooper, 1990).

As with the previously described food-poisoning bacteria, this organism is often present in low numbers in foods, and therefore, enrichment is generally required prior to plating on selective media to detect *Yersinia*. An excellent review describing the range of enrichment media, plating media, biochemical differentiation within the genus *Yersinia*, biotype identification scheme and comparative evaluation of pre-enrichment and plating media has been published recently (De Boer, 1992). In addition, other workers have also reported comparative evaluations of media and methods (Schiemann, 1983; Walker and Gilmour, 1986; Greenwood and Hooper, 1989). The overall conclusion was that no single isolation procedure allowed the recovery of all pathogenic strains of *Y. enterocolitica*. With this in mind, the following parallel isolation procedures were recommended (De Boer, 1992).

1. Enrichment in irgasan ticarcillin chlorate (ITC) for 2 days at 24°C; plating on *Salmonella–Shigella* desoxycholate calcium chloride (SSDC) agar and incubation for 2 days at 30°C.
2. Pre-enrichment in tryptone soya broth (TSB) for 1 day at 24°C; enrichment in bile oxalate sorbose (BOS) for 5 days at 24°C; alkali treatment (mixing 0.5 ml enriched broth with 4.5 ml of 0.5% KOH in 0.5% NaCl for 5 s); and plating on cefsulodin irgasan novobiocin (CIN) agar with incubation for 2 days at 24°C.

Identification of *Yersinia* spp. from related genera (e.g. *Hafnia* and *Enterobacter*) requires a further battery of tests including urease and motility, overall taking approximately 3 days. Differentiation of foodborne species within the genus *Yersinia* (e.g. *Y. enterocolitica* from *Y. frederiksenii*) is achieved by a range of biochemical tests, including sugar fermentation tests (De Boer, 1992). Final serotyping and biotyping (Wauters *et al.*, 1991), and virulence testing (Prpic *et al.*, 1985) is necessary to differentiate pathogenic from environmental strains.

3.2.4.1 Alternative commercial tests. A range of alternative methods has been reported in the literature for the detection and identification of *Y. enterocolitica*. These include enzyme immunoassay for pathogenic *Y. enterocolitica* (Kaneko and Maruyama, 1989), DNA hybridisation assay using synthetic oligonucleotide probes against virulent *Yersinia* (Miliotis *et al.*, 1989; Nesbakken *et al.*, 1991) and polymerase chain reaction (PCR)-based assay for *Y. enterocolitica* (Wren and Tabaqchali, 1990). However, only a few commercial systems are currently available. These are considered below.

A HGMF technique is available for the detection of *Y. enterocolitica* (QA Laboratories). However, the method is based on extensive cultural enrichment and appears to be highly cumbersome. The results are generally obtained after 7 days and an additional 3 days are required for the biochemical screening of *Y. enterocolitica*. It is further recommended that an isolate should be confirmed serologically. The HGMF technical details are considered in chapter 2.

Immunological tests. There are two types of latex agglutination test formats available; a slide agglutination test for the detection and serogrouping of *Y. enterocolitica* (Progen) and a test card latex agglutination test for confirmation of *Y. enterocolitica* (ANI Biotech Oy). The Progen test uses monoclonal antibodies and allows accurate identification of *Y. enterocolitica* serogroups 0:3 and 0:9 (£1/test) within 1 min from colonies isolated on enrichment media (Schmidt and Sethi, 1987). The ANI card system allows detection and identification of *Y. enterocolitica* serogroups 0:3, 0:8 and 0:6 (£0.80/test). In this test, antibody-sensitised latex beads are dried onto a test card. In the presence of *Y. enterocolitica*, the blue latex beads will agglutinate within a minute.

Nucleic acid hybridisation and biochemical tests. Principally, the colorimetric assay for *Y. enterocolitica* available from Gene-Trak Systems is similar to the assays for *Salmonella* and *Listeria* described previously. The assay uses synthetic oligonucleotide probes directed against the 16S ribosomal RNA sequences found in all *Y. enterocolitica*. In a preliminary evaluation using environmental and food samples, the assay (£5.60/test) exhibited comparable specificity and sensitivity to the standard FDA/BAM procedure (Chan *et al.*, 1988). However, unlike the classical method, which takes 14–17 days, the Gene-Trak required 2–3 days of cultural enrichment, followed by a 3 h hybridisation assay. Since the assay is specific for all *Y. enterocolitica*, it does not discriminate between virulent and avirulent strains.

Recently, Monafi and Holzhammer (1993) compared three commercially available methods (API 20E, Vitek AMS and Gene-Trak DNA hybridisation) for the identification of *Y. enterocolitica*. Of 83 isolates on CIN agar, the Gene-Trak and Vitek AMS gave 25 and 26 positive results, respectively, compared with 40 (after 24 h incubation) and 33 (after 48 h incubation) for the API 20E. It was stated that the Gene-Trak assay identified *Y. enterocolitica* even in mixed culture, provided the level of *Yersinia* was equal to or above 1×10^6 cfu/ml.

Miniaturised kits based on a range of biochemical tests are valuable for the rapid identification of *Y. enterocolitica* strains. Examples of these include the API 20E (bioMérieux) and Minitek system (BBL), both of which require 18–24 h incubation prior to reading the results. When assessed with 25 typical and atypical *Y. enterocolitica* strains, the API 20E correctly identified all the strains compared with 80% identification by the Minitek system (Restaino *et al.*, 1979). Of the five misidentified strains, four were *C. freundii*. More recently, Sharma *et al.* (1990) evaluated the API 20E for the identification of 183 *Yersinia* isolates. They reported an over 90% correct identification rate for *Y. enterocolitica* and *Y. frederiksenii*, whilst concluding that the results were unacceptable for *Y. intermedia*.

3.3 Detection of bacterial toxins

Food poisoning due to bacterial toxins can occur either by the consumption of preformed toxins (i.e. foodborne intoxication) in foods (e.g. staphylococcal, *Bacillus cereus* and *Clostridium botulinum* food poisoning) or by the consumption of large numbers of an organism that produces toxins (i.e. foodborne infection) within the gastrointestinal tract (e.g. *Clostridium perfringens* food poisoning).

In England and Wales, during the period 1989 to 1991, 252 outbreaks due to bacterial causes (other than *Salmonella*) of food poisoning were recorded (Dr Paul Sockett, personal communication). However, the number of outbreaks recorded declined from 91 and 95 outbreaks in 1989 and 1990, respectively, to 66 in 1991. A total of 3687 cases was associated with the outbreaks reported (*Cl. perfringens*, 3064; *Cl. botulinum*, 27; *Staph. aureus*, 212; and *B. cereus* and other *Bacillus* spp., including *B. subtilis, B. licheniformis, B. pumilis* and *B. brevis*, 384). *Clostridium perfringens, Bacillus* spp. and *Staph. aureus* accounted for 60, 29 and 10% of the outbreaks, respectively. Staphylococcal (36% of outbreaks) and *Bacillus* (22% of outbreaks) food poisoning was more commonly associated with family outbreaks than *Clostridium* (4% of outbreaks) food poisoning. Most of the *Clostridium* outbreaks were related to meals at restaurants and receptions. Staphylococcal food poisoning, with data on several hundred cases in the UK between 1969 and 1990, was recently reviewed by Wieneke *et al.* (1993). For some other major recorded outbreaks of bacterial food poisoning in the USA, Canada and Europe, the reader is referred to Doyle *et al.* (1993).

3.3.1 Staphylococcal enterotoxins

The ingestion of staphylococcal enterotoxins is a common cause of food poisoning. Currently, seven serologically distinct staphylococcal enterotoxins (A, B, C1, C2, C3, D and E) have been identified, although the types A and D are most frequently encountered in staphylococcal food-poisoning outbreaks (Tranter, 1991). The conventional procedures for the extraction and detection of these enterotoxins take between 5 and 11 days (Holbrook and Baird-Parker, 1975), and are based on cumbersome and lengthy extraction procedures prior to detection by double-immunodiffusion in an agarose gel. For details concerning other methods, including animal tests, tissue culture, haemagglutination assays and radial immunodiffusion, the reader is referred to the reviews by Fey (1986) and Bergdoll (1991).

Recently, techniques based on nucleic acid probes with (Wilson et al., 1991) and without (Jaulhac et al., 1992) PCR amplification have been described for the detection of enterotoxigenic Staph. aureus. Commercial systems based on these techniques are, however, not available as yet. A variety of commercial immuno-logical rapid systems is available for the detection of staphylococcal enterotoxins from foods. These techniques can provide results earlier and, generally, more reliably than the conventional techniques.

The range of commercial kits is shown in Table 3.7 together with, where available, references relating to their performance characteristics.

Table 3.7 Summary of alternative methods for the detection of staphylococcal enterotoxins

Test system	Company	Cost/test[a] (£)	Evaluation reference
Tecra ELISA	Tecra Diagnostics	5.58	Bodnaruk and Patel (1990); Park et al. (1993)
Bommeli SET EIA	Fluorochem Ltd	3.05	Haines and Stannard (1987); Park et al. (1992)
Tube ELISA	Transia	8.25	Wieneke (1991)
Dip-stick ELISA	Transia	D[b]	Bodnaruk and Patel (1990); Wieneke (1991)
Plate ELISA	Transia	NA[c]	NA
Ridascreen ELISA	R-Biopharm GmbH	2.85	Park et al. (1993)
Automated VIDAS ELISA	bioMérieux	13.43 (inc. tubes)	NA
RPLA	Unipath	6.16	Haines and Stannard (1987); Rose et al. (1989)
ANI-SET latex test	ANI Biotech Oy	NA	NA

[a] As of November 1993.
[b] D, Discontinued.
[c] NA, not available at the time of preparation of the manuscript.

3.3.1.1. Enzyme immunoassays.   Bodnaruk and Patel (1990) evaluated the Tecra EIA and Transia dipstick ELISA. In both kits, the extraction procedure for the enterotoxins depended on the food type, although for certain food categories (e.g. milks, low-fat spreads) the Tecra procedure was simpler than the Transia protocol. With these foods, extraction of the enterotoxin involved two pH adjustments followed by a chloroform extraction and a centrifugation step (Transia). In the case of the Tecra test, a pH adjustment followed by the addition of a sample additive reagent (supplied with the kit) was required for extraction.

Both kits were able to detect all seven enterotoxins. However, the Tecra assay did not identify the toxin type. The sensitivity of the techniques was approximately 1 ng/ml (Bodnaruk and Patel, 1990). For practical purposes, this relates to detection of the enterotoxins at levels >0.5 μg per 100 g food, assuming a five-fold dilution of a food type. In comparison, the emetic dose for the enterotoxins ranges between 0.1 and 1 μg per 100 g food, depending on the age group and whether an individual is immunocompromised (Notermans and Wernars, 1991).

Figure 3.10 (Bodnaruk and Patel, 1990) shows the results of the Tecra assay when applied to a range of foods spiked with 1 μg enterotoxin per 100 g food. Although there was variability in the ELISA readings, all the spiked samples were found to give values above the cut-off value, while the control samples were all well below that level. Recently, significant false-positive results using the Tecra assay have been observed with sea foods, particularly shellfish (e.g. mussels (85% false-positives) and clams (32% false-positives)) (Park *et al.*, 1993). This was attributed to the presence of a substance that was heat-labile, reacted specifically with normal rabbit or calf serum and did not bind to copper chelate–Sepharose gel. The problems were, therefore, resolved by procedures

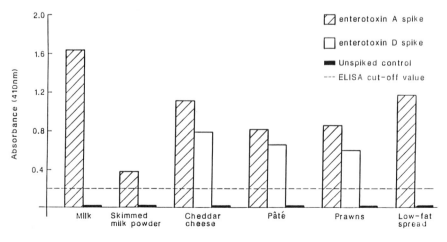

Figure 3.10 Detection of staphylococcal enterotoxin in artificially contaminated foods using Tecra ELISA.

involving one of the following treatments of the food extracts:

(i)   heat-treatment (70°C for 3 min);
(ii)   treatment with rabbit or calf serum for 2 min; and
(iii) metal chelate affinity chromatography (2 h) with copper chelate–
       Sepharose.

3.3.1.2 Latex agglutination tests. In an evaluation of the reversed passive
latex agglutination technique (RPLA) and SET-EIA tests, staphylococcal
enterotoxin types A, B, C and D, artificially inoculated (1 ng/g of sample) into
Cheddar cheese, ham, prawns, chicken liver pâté and skimmed-milk powder,
were correctly detected and identified (Haines and Stannard, 1987). No non-
specific reactions were observed with the SET-EIA. However, some
cross-reactions were found with the cheese and prawn extracts, which showed
agglutination of the control latex particles in the RPLA. Rose *et al.* (1989)
showed that non-specific reactions using the RPLA, particularly in some dairy
products, could be obviated by the addition of 10 mM sodium hexametaphos-
phate to the diluent provided in the assay.

Table 3.8 summarises the relative merits and drawbacks of the Transia and
Tecra ELISAs, as well as the Bommeli ELISA and a latex agglutination test for
the rapid detection of staphylococcal enterotoxins in foods. Wieneke (1991)
compared four commercial kits (Bommeli ELISA, Transia dipstick and tube
ELISAs, and the Unipath RPLA) for the detection of staphylococcal enterotoxins
from outbreaks of food poisoning. Overall, 14 of 18 foods were positive using
the Bommeli ELISA, compared with nine or 10 with the RPLA and Transia

Table 3.8 Comparison of staphylococcal enterotoxin detection methods

	Unipath [a] RPLA	Bommeli [a] ELISA	Transia ELISA [b]	Tecra ELISA [b]
Time for test completion	21 h	26 h	7 h	7 h
Labour intensity				
(1) extraction	High	High	High	Low
(2) test	Low	High	Low	Low
Solid phase	Latex particles	Polystyrene beads	Dip-stick	Microtitre plate
Sensitivity (ng/ml)	1	0.1	0.5[c]	1
Enterotoxins detected	A, B, C, D	A, B, C, D	A, B, C1, C2, C3, D, E	A, B, C1, C2, C3, D, E
Enterotoxin identification	Yes	Yes	Yes	No
Subjectivity	Yes	No	Yes	No
Specialised equipment	No	Yes	No	Yes

[a] Data taken from Haines and Stannard (1987).
[b] Data taken from Bodnaruk and Patel (1990). This dip-stick assay is now available as a tube assay
in which antibodies against the seven enterotoxins are coated on to the surface of a tube (Cortecs
Diagnostics).
[c] Manufacturer's claim.

ELISAs. The Bommeli SET-EIA test is a solid-phase ELISA and comprises polystyrene beads coated with specific antibodies against the enterotoxins A, B, C and D. The RPLA uses latex particles sensitised with antibodies against the same enterotoxins. Neither test both detected the enterotoxin E and provided results in approximately 24 h following extraction of the enterotoxins from foods. The SET-EIA was more sensitive than others at detecting enterotoxin at levels >0.1 ng/ml.

Wieneke and Gilbert (1987) compared the performance characteristics of four techniques (two commercial, i.e. the Bommeli ELISA and Unipath latex aggluti-nation, and a gel diffusion and a plate ELISA) for the detection of staphylococcal enterotoxins in foods from outbreaks of food poisoning. The least sensitive of the tests was the gel diffusion. The RPLA was easy to perform, but the sensitivity of the ELISA methods was needed to detect enterotoxins in some of the foods.

3.3.2 Enterotoxin detection kits for enteric pathogens

Table 3.9 shows the range of commercial kits available for the detection of the enterotoxins produced by enteric pathogens. The table also shows references, where available, that provide information concerning the performance character-istics of the tests.

Table 3.9 Commercial tests for detection of enterotoxins produced by *Cl. perfringens, B. cereus* and *E. coli*.

Test system	Company	Cost/test[a] (£)	Evaluation reference
PET-RPLA (*Cl. perfringens* enterotoxin)	Unipath	3.50	Berry *et al.*, (1986)
BCET-RPLA (*B. cereus* diarrhoeal type)	Unipath	3.78	Buchanan and Schultz (1992); Notermans and Tatini (1993)
Tecra ELISA (*B. cereus* diarrhoeal type)	Tecra Diagnostics	5.58	Notermans and Tatini (1993)
VET-RPLA (*E. coli* LT)	Unipath	3.50	Yam *et al.* (1992)
E. coli ST EIA	Unipath	1.98	Scotland *et al.* (1989)
Phadebact ETEC-LT	Launch Diagnostics	2.38	Speirs *et al.* (1991)
Nucleic acid probes for STA1, STA2 and LTI	Bresatec	NA[b]	Scotland *et al.* (1989); Medon *et al.* (1988)

[a] As of November 1993.
[b] NA, not available at the time of preparation of the manuscript.

3.3.2.1 Enterotoxigenic Escherichia coli (ETEC).

The ETEC is commonly associated with travellers' diarrhoea (rather than with foodborne disease), partic-ularly for individuals returning from countries with lower hygienic standards, where ETEC is a major cause of diarrhoea in infants and young children (Doyle and Padhye, 1989).

ETEC is capable of producing one of two types of toxin, or perhaps even both heat-stable and heat-labile enterotoxins, depending on their plasmid content (Doyle and Cliver, 1990). Two forms of heat-stable enterotoxins have been described, termed as STI (or ST_A) and STII (or ST_B), the former being implicated in human disease. Similarly, two types of heat-labile enterotoxins are recognised (LTI and LTII). The toxin LTI is closely related to the cholera toxin and is important in human disease.

A range of *in vivo* tests (e.g. ligated gut and rabbit skin) for the detection of *E. coli* LT and ST, and, for the latter, several *in vitro* tests (e.g. Chinese hamster ovary cell and radioimmunoassay) have been reviewed by Linggood (1982). A recent approach to the preparation of non-radiolabelled probes and detection of LT and ST has been described by Danbara (1990). Routine testing using animals and tissue cultures is not suitable for many laboratories without the necessary facilities, and therefore simpler alternative testing procedures have been developed. A range of techniques is now commercially available for the detection of *E. coli* enterotoxins (Table 3.9). These comprise three immuno-logical tests and nucleic acid hybridisation probes. The immunological tests include a latex agglutination test for the *E. coli* heat-labile toxin (RPLA), a principally different coagglutination test for the heat-labile toxin (Phadebact ETEC-LT) and an ELISA for the *E. coli* heat-stable toxin (ST-EIA). The coagglutination test involves the use of non-viable *Staph. aureus* cells naturally containing cell-surface protein A. Antibodies against the LT are coated onto the cells simply by mixing; the protein A binds to the Fc portion of the antibody, allowing the antigenic Fab sites to be freely exposed on the surface of *Staph. aureus*. The sensitised *Staph. aureus* will agglutinate in the presence of the LT toxin within 10 min.

In a recent evaluation of the VET-RPLA test and Phadebact ETEC-LT kit (Spiers *et al.*, 1991), 48 toxigenic strains were tested for LTI and the results compared with those from two cell culture assays (Y-1 and Vero). All the assays identified 23 cultures as LTI-positive. The VET-RPLA and the cell culture assays were more sensitive (32 to 2000-fold) than the Phadebact test, but required approximately 24–36 h to obtain results. The Phadebact test was rapid, giving results within 10 min of sample application. The authors concluded that the VET-RPLA was a good substitute in laboratories without cell culture facilities, whilst the Phadebact test might be the method of choice as a simpler alternative. Chapman and Daly (1989) found that the culture medium adversely affected the detection of heat-labile toxin by the coagglutination test. Of the four media tested, only the modified blood agar gave a positive result with all strains of *E. coli* known to produce the toxin. Eighty-six strains negative by the Y-1 cell culture assay were also negative by the coagglutination assay. It was concluded that the Phadebact kit was a simple, sensitive and economical method for detecting the heat-labile enterotoxin.

The ST-EIA test is a competitive enzyme immunoassay that uses microtitre wells coated with synthetic *E. coli* heat-stable toxin. The test involves incubating enzyme-labelled monoclonal antibody against STI and a test sample in a microtitre plate. The presence of STI in the test sample results in a reduction in the level of conjugate binding to the microtitre well. The manufacturer's claimed sensitivity for heat-stable enterotoxin is 10 ng/ml, with the total assay time of approximately 3–4 h (including sample preparation). Scotland *et al.* (1989) evaluated the ST-EIA and the SNAP hybridisation probe in comparison with the infant mouse test for the detection of the heat-stable enterotoxin ST_A. A hundred strains of *E. coli* carrying the ST_{A1} and ST_{A2} genes were found to be positive in all the three tests. An additional 100 strains were negative in the ELISA and the mouse test. Unlike the ELISA and the mouse tests, the probe test also detected two of the strains carrying ST_{A1} genes.

3.3.2.2 Bacillus cereus. Two distinct types of gastroenteritis are associated with *B. cereus* food poisoning. A longer-onset 'diarrhoeal syndrome' involves puddings, proteinaceous foods and vegetable sauces, whilst the acute-onset 'emetic syndrome' is associated largely with cooked rice. The characteristic symptoms of these illnesses are believed to be caused by the diarrhoeal entero-toxin and the emetic toxin (Kramer *et al.*, 1982; Kramer and Gilbert, 1989). Details concerning properties and pathogenesis due to these enterotoxins, together with foodborne disease caused by other *Bacillus* spp., have been summarised by Lund (1990).

According to Kramer and Gilbert (1989), the emetic toxin (preformed in the food) is largely responsible for *B. cereus* food poisoning in the UK. On the other hand, the diarrhoeal enterotoxin is both preformed in the food and formed in the intestine. In England and Wales, between 1986 and 1988, there was a greater than six-fold increase in the number of reported cases (65 in 1986 and 418 in 1988) of food poisoning due to *Bacillus* spp. (Lund, 1990). Most of these were attributed to *B. cereus*. For details concerning the cultural isolation and identifi-cation methods, and serological typing schemes for *Bacillus* spp. in foods, together with biological methods (e.g. ligated rabbit ileal loop and monkey feeding) for the detection of the enterotoxin, the reader is referred to Kramer *et al.* (1982) and Kramer and Gilbert (1989). Recently, Jackson (1991) reported the development of a 96-well membrane-based fluorescent immunoassay for the specific detection of the diarrhoeal enterotoxin at levels >50 ng/ml. Immunological methods have not been developed for the emetic toxin, because of difficulties encountered in the purification and characterisation of the toxin. Instead, a range of tissue culture techniques has been used for the detection of the emetic toxin (Jackson, 1991).

There is both an ELISA and an RPLA test available commercially for the detection of the diarrhoeal enterotoxin (Table 3.9). Griffiths (1990) used the RPLA to show that 85% of the total (83 strains) psychrotrophic *Bacillus* spp. produced diarrhoeogenic toxin during growth at 25°C. The majority of these strains were *B. cereus*. In addition, it was further reported that four psychrotrophic bacilli also produced the toxin during growth in milk at temperatures ranging from 6 to 21°C. The RPLA assay showed sensitivity of approximately 2 ng/ml toxin.

Van Netten *et al.* (1990) used the RPLA to determine the enterotoxin production by psychrotrophic *B. cereus* during investigations of three outbreaks of *B. cereus* food poisoning in Spain and the Netherlands. They showed that the growth and enterotoxin production could be prevented at temperatures below 4°C and pH values not exceeding 5.0. Although controls were included in the study to validate that no false-negative and false-positive reactions were obtained, the authors commented that the specificity of the RPLA needed to be substantiated.

The RPLA was compared with two biological assays, based on cytotonic response in HEp-2 and Chinese hamster ovary cells, for the detection of the diarrhoeal toxin produced by 12 *B. cereus* isolates in broth cultures (Buchanan and Schultz, 1992). While there was an overall agreement between the biological assays and the RPLA, there were distinct differences for specific strains. The RPLA did not detect two of the nine diarrhoeogenic strains that were positive by the biological assays. Conversely, an equal number of false-positives was noted when compared with the biological assays. Heat-treated extracts gave negative response in the biological assays, whilst the RPLA gave positive response, indicating that the antigenic sites were not denatured. It was concluded that the results of RPLA assays should be further confirmed using an appropriate biological assay.

The Tecra ELISA kit for the detection of the diarrhoeal toxin in foods, food ingredients and related products has only recently been launched on the market. The sample preparation protocols are simple for certain food types (e.g. milk and milk powder) and relatively labour-intensive for others (e.g. cheese and dehydrated food ingredients). The assay is a classic microtitre-plate-based sandwich ELISA, which takes approximately 4 h to complete. In a recent comparison of the Tecra ELISA and the latex agglutination test, Notermans and Tatini (1993) found no correlation between the two tests when examined for the production of diarrhoeal toxin from several strains of *B. cereus*. In this study, it was observed that, out of three strains that had been implicated in foodborne outbreaks, all yielded positive results by the Tecra assay, but only one by the latex agglutination test. It was claimed that, in trials conducted by the manufacturer, the Tecra assay had a sensitivity of 1 ng/ml of extrac using artificially contaminated foods, with no false-positives or negatives recorded.

3.3.2.3 Clostridium perfringens.   Five serotypes (A–E) of *Cl. perfringens* are recognised, based on several major extracellular toxins produced. Of these serotypes, only A and C cause significant disease in man. *Cl. perfringens* food poisoning is caused by the release of heat-labile enterotoxin (mol. wt 36 000 ± 4000) when large numbers of vegetative cells are ingested, and which multiply and sporulate in the intestine.

For conventional methods for the detection, enumeration and identification of *Cl. perfringens* in foods, the reader is referred to Mead *et al.* (1982), whilst methods for typing *Cl. perfringens* have been described by Stringer *et al.* (1982). A wide range of classical methods, based on biological (e.g. tissue culture) and serological (e.g. immunodiffusion) detection has been used for the detection of the *Cl. perfringens* enterotoxin. These have been reviewed previously (Giugliano *et al.*, 1983). Several ELISAs and, recently, DNA probes to detect the enterotoxin gene (Doyle *et al.*, 1993) have been reported. However, to the authors' knowledge, none of these is available commercially as yet.

Direct detection of the enterotoxin in faeces has proved to be another potentially useful epidemiological criterion. The commercial reversed passive latex agglutination test (RPLA) (Table 3.9) and an in-house ELISA (Bartholomew *et al.*, 1985) for *Cl. perfringens* enterotoxin have been evaluated by Berry *et al.* (1986). In this evaluation, 131 faecal specimens from 24 different suspected *Cl. perfringens* food-poisoning cases were examined. Of these, 94% of the specimens gave the same results using both tests. Discrepancies between the two methods were encountered with eight samples and were attributed to very low toxin concentrations (<7 ng/g faeces). One of these samples was classified negative by the RPLA, but positive by the ELISA. The authors concluded that failure to detect this low toxin-positive sample was unlikely to be of any practical significance, since most outbreaks involved toxin concentrations between 1 and 100 μg/g (Bartholomew *et al.*, 1985).

3.4 Future perspectives

There is no doubt that the range of rapid or alternative detection techniques for foodborne pathogens and microbial toxins will increase significantly in the future. The range of assay formats, from manual and disposable systems to highly automated instruments, will also continue to increase, to cater for the needs of small, medium and large food manufacturers. These techniques are also likely to be actively adopted by other authorities, including public analysts and environmental health officers. Overall, the modern technology will help ensure product quality and safety, in compliance with the strict UK food safety legislation.

In addition to the commercial systems already described, there are a number of other detection systems that are near-market; two of these are briefly considered. Genetic modification of bacteriophages form the basis of the lux gene (Ulitzer and Kuhn, 1987) and bacterial ice-nucleation diagnostic (BIND) technologies (Wolber and Green, 1990). These involve the introduction of either lux genes (from naturally luminescent bacteria, e.g. *Vibrio fisheri*) or ice-nucleation genes (from certain psychrotrophic *Pseudomonas* spp.) into the genome of a bacteriophage by recombinant DNA procedures. The bacteriophage used is specific for the target bacterial cell (e.g. *Salmonella*).

In the case of the lux gene test, the genes are transferred to the host cell and encode for light-producing proteins, whilst ice-nucleation proteins are expressed by the target bacteria, in the case of the BIND test. The target cells can then be detected either by measuring luminescence (lux gene test) or by monitoring the freezing (in combination with an indicator dye) of super-cooled test samples (BIND test). A sensitivity of 10^2 cfu/ml has been reported using the lux gene system for *Salmonella* (Stewart *et al.*, 1989). However, although this level of sensitivity is lower than reported for other rapid test systems described previously, an enrichment procedure is still necessary. Using this technology, however, results may be obtained within a single working day (Stewart and Williams, 1992). The major limitation of these systems at present is the lack of suitable bacteriophages that can infect all strains of the target genus (e.g. *Salmonella*), without the infection of closely related genera. There is also the problem of potential interference from background luminescence or ice-nucleation proteins produced by certain food-spoilage bacteria. For further details concerning the lux gene system, the reader is referred to chapter 7.

The PCR in combination with molecular probes is described in chapters 4, 6 and 10. The scope of the techniques based on genetic markers and nucleic acid fingerprinting as applied to food analytical areas has recently been described by Patel (1994). This powerful technology has already been applied to the detection of low numbers of foodborne pathogens (e.g. *Listeria*). A recent product on the market, EnviroAMP (Perkin-Elmer Cetus), is capable of detecting 100 cells of *Legionella* per ml in tap waters by PCR (Anon., 1992). It is envisaged that future tests for foodborne pathogens will incorporate simple procedures based on the separation technology (see chapters 4 and 10) in combination with nucleic acid amplification (e.g. PCR) and non-radiolabelled detection (Bailey *et al.*, 1977).

References

Andrews, W.H. (1985) A review of cultural methods and their relation to rapid methods for the detection of *Salmonella* in foods. *Food Technol.*, **39**, 77–82.

Andrews, W.H. (1989) Methods for recovering injured "classical" enteric pathogenic bacteria (*Salmonella, Shigella* and enteropathogenic *Escherichia coli*) from foods, in *Injured Index and Pathogenic Bacteria: Occurrence and Detection in Foods, Water and Feeds* (ed. Ray, B.). CRC Press Inc., Boca Raton, FL.

Anon. (1988) *Listeria* isolation; Revised method of analysis, in *Bacteriological Analytical Manual*. Ch. 29. *Listeria* Federal Register 53, Washington, DC.

Anon. (1992) *EnviroAMP PCR Based Kit for Environmental Testing*. Preliminary information. Perkin-Elmer Cetus Instrument News.

Anon. (1993a) Diagnostic *Salmonella*-selective semisolid medium (DIASALM). *Int. J. Food Microbiol.*, **17**, 230–233.

Anon. (1993b) *Escherichia coli* O157 in Sheffield. *CDR*, **3** (23) 1.

Bailey, J.E., Fazel-Makjlessi, J., McQuitty, D.N. *et al.* (1977) Characterization of bacterial growth by means of flow microfluorometry. *Science*, **198**, 1175–1176.

Bailey, J.S. and Cox, N.A. (1993) Automated ELISA detection of *Listeria* from meat and poultry products using the Vidas system. *Dairy Food & Environ. Sanit.*, **March**, 182.

Bartholomew, B.A., Stringer, M.F., Watson, G.N. and Gilbert, R.J. (1985) Development and application of an enzyme linked immunosorbent assay for *Clostridium perfringens* type A enterotoxin.

J. Clin. Pathol., **38**, 222–228.

Bannerman, E., Yersin, M.N., Catimel, B. *et al.* (1991) Evaluation of a new identification system for *Listeria*. Presented at the 5th European Congress of Clinical Microbiology and Infectious Diseases, 9–11 Sept., Oslo, Norway, Poster No. 1266.

Bennett, A.R., Bobbitt, J.A. and Betts, R.P. (1993a) Evaluation of the Vidas *Listeria* system for the detection of *Listeria* in foods, in *SAB 62nd Annual Meeting and Summer Conference*, University of Nottingham, 13–16 July, 1993, p. xvi.

Bennett, A.R., Bobbitt, J.A. and Betts, R.P. (1993b) An evaluation of the Locate *Salmonella* screening test, in *SAB 62nd Annual Meeting and Summer Conference*, University of Nottingham, 13–16 July, 1993, p. xix.

Bergdoll, M.S. (1991) *Staphylococcus aureus*. *J. A. O. A. C.*, **74**, 706–710.

Berry, P.R., Stringer, M.F. and Uemura, T. (1986) Comparison of latex agglutination and ELISA for the detection of *Clostridium perfringens* type A enterotoxin in faeces. *Lett. Appl. Microbiol.*, **2**, 101–102.

Bettelheim, K.A., Evangelidis, H., Pearce, J.L. *et al.* (1993) Isolation of a *Citrobacter freundii* strain which carries the *Escherichia coli* O157 antigen. *J. Clin. Microbiol.*, **31**, 760–761.

Blackburn, C. de W. and Patel, P.D. (1989) *Brief Evaluation of the Microscreen* Salmonella *Latex Slide Agglutination Test for the Detection of* Salmonella *in Foods* (Leatherhead FRA Tech. Note No. 86). Leatherhead Food Research Association, Leatherhead, Surrey.

Blackburn, C. de W. and Patel, P.D. (1990) *Evaluation of* Salmonella-*Tek ELISA for the Detection of* Salmonella *in Foods* (Leatherhead FRA Tech. Note No. 89). Leatherhead Food Research Association, Leatherhead, Surrey.

Blackburn, C. de W. and Patel, P.D. (1991) *Brief Evaluation of the* Salmonella *Rapid Test (Unipath Ltd) and Hydrophobic Grid Membrane Filter Technique for the Detection of Salmonellae in Foods* (Leatherhead FRA Tech. Note No. 91). Leatherhead Food Research Association, Leatherhead, Surrey.

Blackburn, C. de W., Stannard, C.J. and Gibbs, P.A. (1988) *Evaluation of the Tecra* Salmonella *Visual Immunoassay for the Detection of* Salmonella *in Foods* (Leatherhead FRA Tech. Note No. 65). Leatherhead Food Research Association, Leatherhead, Surrey.

Blackburn, C. de W., Patel, P.D. and Gibbs, P.A. (1991) Separation and detection of *Salmonella* using immunomagnetic particles. *Biofouling*, **5**, 143–156.

Blackburn, C. de W., Curtis, L.M., Humpheson, L. and Petitt, S.B. (1994) *Evaluation of the Vitek Immunodiagnostic Assay System (VIDAS) for the Detection of* Salmonella *in Foods* (Leatherhead FRA Tech. Note No. 109). Leatherhead Research Association, Leatherhead, Surrey.

Bodnaruk, P.W. and Patel, P.D. (1990) *Evaluation of Tecra and Transia Enzyme Immunoassay for the Detection of Staphylococcal Enterotoxins in Foods* (Leatherhead FRA Tech. Note No. 90). Leatherhead Food Research Association, Leatherhead, Surrey.

Bolton, F.J., Hutchinson, D.N. and Coates, D. (1986) Comparison of three selective agars for the isolation of campylobacters. *Eur. J. Clin. Microbiol.*, **5**, 466–468.

Buchanan, R.L. and Schultz, F.J. (1992) Evaluation of the Oxoid BCET-RPLA kit for the detection of *Bacillus cereus* diarrhoeal enterotoxin as compared to cell culture cytotonicity. *J. Food Prot.*, **55**, 440–443.

Chan, S., Pitman, T., Shah, J., King, W., Lane, D. and Lawrie, J. (1988) Non-radioactive DNA assay for detection and identification of foodborne *Yersinia*, in *Abstracts of the 88th Annual Meeting of the American Society for Microbiology*. Washington, DC, p. 280.

Chapman, P.A. and Daly, C.M. (1989) Comparison of Y1 mouse adrenal cell and coagglutination assays for detection of *Escherichia coli* heat-labile enterotoxin. *J. Clin. Pathol.*, **42**, 755–758.

Clark, C., Candlish, A.G. and Steell, W. (1989) Detection of *Salmonella* in foods using a novel coloured latex test. *Food Agric. Immunol.*, **1**, 3–9.

Cliver, D.O. (1990) *Foodborne Diseases*. Academic Press, New York.

Cox, N.A., Fung, D.Y.C., Bailey, J.S. *et al.* (1987) Miniaturised kits, immunoassays and DNA hybridisation for recognition and identification of foodborne bacteria. *Dairy Food Sanit.*, **7**, 628–631.

Curiale, M.S., McIver, D., Weathersby, S. and Planer, C. (1990a) Detection of salmonellae and other Enterobacteriaceae by commercial deoxyribonucleic acid hybridization and enzyme immunoassay kits. *J. Food Prot.*, **53**, 1037–1046.

Curiale, M.S., Klatt, M.J. and Mozola, M.A. (1990b) Colorimetric deoxyribonucleic acid hybridisation assay for rapid screening of *Salmonella* in foods: collaborative study. *J. A. O. A. C.*, **73**, 248–256.

Danbara, H. (1990) Identification of enterotoxigenic *Escherichia coli* by colony hybridisation using biotinylated LTIh, STIa and STIb enterotoxin probes, in *Gene Probes for Bacteria* (eds Macario, A.J.L. and De Macario, E.C.). Academic Press, New York, pp. 167–177.

D'Aoust, J.Y., Sewell, A. and Jean, A. (1990) Limited sensitivity of short (6 h) selective enrichment for detection of foodborne *Salmonella*. *J. Food Prot.*, **53**, 562–566.

D'Aoust, J.Y., Sewell, A.M. and Greco, P. (1991) Commercial latex agglutination kits for the detection of foodborne *Salmonella*. *J. Food Prot.*, **54**, 725–730.

Davies, M., Osaka, C., Gordon, D. *et al.* (1993) Update: Multistate outbreak of *Escherichia coli* O157:H7 infections from hamburgers – Western United States, 1992–1993. *Morb. Mortal. Weekly Rep.*, **42**, 258–263.

Dealler, S.F. and Rotowa, N.A. (1991) Rapid screening of colonies from *Listeria* selective agar. *J. Hosp. Infect.*, **17**, 147–150.

Dealler, S.F., Collins, J. and James, A.L. (1992) A rapid heat-resistant technique for presumptive identification of *Salmonella* on desoxycholate–citrate agar. *Eur. J. Clin. Microbiol. Infect. Diseases*, **11**.

De Boer, E. (1992) Isolation of *Yersinia enterocolitica* from foods. *Int. J. Food Microbiol.*, **17**, 75–84.

De Smedt, J.M. and Bolderdijk, R. (1990) Collaborative study of the International Office of Cocoa, Chocolate, and Sugar Confectionery on the use of motility enrichment for *Salmonella* detection in cocoa and chocolate. *J. Food Prot.*, **53**, 659–664.

De Smedt, J. M., Bolderdijk, R., Rappold, H. and Lautenschlaeger, D. (1986) Rapid *Salmonella* detection in foods by motility enrichment on modified semisolid Rappaport–Vassiliadis medium. *J. Food Prot.*, **49**, 510–514.

Doyle, M.P. (1989) *Foodborne Bacterial Pathogens*. Marcel Dekker, New York.

Doyle, M.P. and Cliver, D.O. (1990) *Escherichia coli*, in *Foodborne Diseases* (ed. Cliver, D.O.). Academic Press, New York, pp. 209–215.

Doyle, M.P. and Padhye (1989) *Escherichia coli*, in *Foodborne Bacterial Pathogens* (ed. Doyle, M.P.). Marcel Dekker, New York, pp. 236–281.

Doyle, M.E., Steinhart, C.E. and Cochrane, B.A. (1993) *Food Safety 1993*. Food Research Institute, University of Wisconsin–Madison, Marcel Dekker, New York.

Drewett, I.M. and Patel, P.D. (1989) *Rapid Identification of* Campylobacter jejuni *using the DuPont SNAP Hybridisation Kit* (Leatherhead FRA Tech. Note No. 85). Leatherhead Food Research Association, Leatherhead, Surrey.

Entis, P., Brodsky, M.K., Sharpe, A.N. and Jarvis, G.A. (1982) Rapid detection of *Salmonella* spp. in food by use of the ISO-GRID hydrophobic grid membrane filter. *Appl. Environ. Microbiol.*, **43**, 261–268.

Farber, J.M. and Peterkin, P.I. (1991) *Listeria monocytogenes*, a food-borne pathogen. *Microbiol. Rev.*, **55**, 476–511.

Feldsine, P.T., Falbo-Nelson, M.T. and Hustead, D.L. (1992) Polyclonal enzyme immunoassay method for detection of motile and non-motile *Salmonella* in foods: Collaborative study. *J. A. O. A. C.*, **75**, 1032–1044.

Fey, H. (1986). Staphylococcal enterotoxins, in *Antigen Detection to Diagnose Bacterial Infections, Vol. II: Applications* (ed. Kohles, R.B.). CRC Press, Boca Raton, Florida.

Flint, S.H. and Hartley, N.J. (1993) Evaluation of the TECRA immunocapture ELISA for the detection of *Salmonella typhimurium* in foods. *Lett. Appl. Microbiol.*, **17**, 4–7.

Flowers, R.S. and Klatt, M.J. (1989) Immunodiffusion screening method for detection of motile *Salmonella* in foods. Collaborative study. *J. A. O. A. C.*, **72**, 303–311.

Freney, J., Herve, C., Desmonceaux, M. *et al.* (1990) Description and evaluation of the semiautomated 4-hour ATB 32E method for identification of members of the family Enterobacteriaceae. *J. Clin. Microbiol.*, **29**, 138–141.

Giugliano, L.G., Stringer, M.F. and Draser, B.S. (1983) Detection of *Clostridium perfringens* enterotoxin by tissue culture and double-gel diffusion methods. *J. Med. Microbiol.*, **16**, 233–237.

Greenwood, M.H. and Hooper, W.L. (1985) *Yersinia* spp. in foods and related environments. *Food Microbiol.*, **2**, 263–269.

Greenwood, M.H. and Hooper, W.L. (1989) Improved methods for the isolation of *Yersinia* species from milk and foods. *Food Microbiol.*, **6**, 99–104.

Greenwood, M.H. and Hooper, W.L. (1990) The source of *Yersinia* spp. in pasteurised milk: an investigation at a dairy. *Epidemiol. Infect.*, **104**, 351–360.

Griffiths, M.W. (1990) Toxin production by psychrotrophic *Bacillus* spp. present in milk. *J. Food Prot.*, **53**, 790–792.

Griffiths, P.L. and Park, R.W.A. (1990) Campylobacters associated with human diarrhoeal disease. *J. Appl. Bacteriol.*, **69**, 281–301.

Haines, S.D. and Patel, P.D. (1988) *Evaluation of* Listeria-*Tek Kit for the Rapid Detection of* Listeria *spp. in Foods* (Leatherhead FRA Tech. Note No. 76). Leatherhead Food Research Association, Leatherhead, Surrey.

Haines, S.D. and Stannard, C.J. (1987) *Brief Evaluation of Two Kits for the Detection of Staphylococcal Enterotoxins in Foods* (Leatherhead FRA Tech. Note No. 54). Leatherhead Food Research Association, Leatherhead, Surrey.

Halligan, A.C. (1990) *The Emerging Pathogens –* Yersinia, Aeromonas *and Verotoxigenic* E. coli *(VTEC) – A Literature Survey* (Leatherhead FRA Food Focus, No. 11). Leatherhead Food Research Association, Leatherhead, Surrey.

Hazeleger, W.C., Beumer, R.R. and Rombouts, F.M. (1992) The use of latex agglutination tests for determining *Campylobacter* species. *Lett. Appl. Microbiol.*, **14**, 181–185.

Higgins, D.L. and Robison, B.J. (1993) Comparison of Micro-ID *Listeria* method with conventional biochemical methods for identification of *Listeria* isolated from food and environmental samples. Collaborative study. *J. A. O. A. C.*, **76**, 831–838.

Holbrook, R. and Baird-Parker, A.C. (1975) Serological methods for the assay of staphylococcal enterotoxins, in *Some Methods for Microbiological Analysis* (eds Board, R.G. and Lovelock, D.W.). SAB Tech. Ser. No. 8, Academic Press, London, pp. 108–126.

Holbrook, R., Anderson, J.M., Baird-Parker, A.C. *et al.* (1989) Rapid detection of *Salmonella* in foods – a convenient two-day procedure. *Lett. Appl. Microbiol.*, **8**, 139–142.

IOCCC (International Office of Cocoa, Chocolate and Sugar Confectionery) (1990) Microbiological examination of chocolate and other cocoa products. *IOCCC Analyt. Meth.*, 118–199.

Jackson, S.G. (1991) *Bacillus cereus*. *J. A. O. A. C.*, **74**, 704–706.

Jackson, B.J., Chen-Wu, J.L., Hansen, T.J. *et al.* (1992) Quantitative *Listeria* testing for foods within 24 hours without enrichment. 1992 IFT Annual Meeting, New Orleans, Abstract No. 398.

Jaulhac, B., Bes, M., Bornstein, N. *et al.* (1992) Synthetic DNA probes for detection of genes for enterotoxins A, B, C, D, E and TSST-1 in staphylococcal strains. *J. Appl. Bacteriol.*, **72**, 386–392.

Kaneko, S. and Maruyama, T. (1989) Evaluation of enzyme immunoassay for the detection of pathogenic *Yersinia enterocolitica* and *Yersinia pseudotuberculosis* strains. *J. Clin. Microbiol.*, **27**, 748–751.

Karmali, M.A. (1988) Laboratory diagnosis of verocytotoxin-producing *Escherichia coli* infections. *Oxoid Culture*, **9** (2).

Kerr, K.G., Rotowa, N.A., Hawkey, P.M. and Lacey, R.N. (1991) Evaluation of the Rosco system for the identification of *Listeria* species. *J. Med. Microbiol.*, **35**, 193–196.

Kimbell, H. (1993) Food poisoning – what are we going to do about it and whose responsibility is it anyway? Consumer Panel Meeting, 7 October, 1993, London, distributed by MAFF Consumer Panel Secretariat.

Klinger, J.M., Stowe, R.P., Obenhuber, D.C. *et al.* (1992) Evaluation of the Biolog automated microbial identification system. *Appl. Environ. Microbiol.*, **58**, 2089–2092.

Knight, M.T., Wood, D.W., Black, J.F. *et al.* (1990) Gram-negative identification card for identification of *Salmonella, Escherichia coli* and other Enterobacteriaceae isolated from foods. Collaborative study. *J. A. O. A. C.*, **73**, 729–733.

Kramer, J.M. and Gilbert, R.J. (1989) *Bacillus cereus* and other *Bacillus* species, in *Foodborne Bacterial Pathogens* (ed. Doyle, M.P.). Marcel Dekker, New York, pp. 21–70.

Kramer, J.M., Turnbull, P.C.B., Munshi, G. and Gilbert, R.J. (1982) Identification and characterisation of *Bacillus cereus* and other *Bacillus* species associated with foods and food poisoning, in *Isolation and Identification Methods for Food-Poisoning Organisms*, SAB Tech. Ser. No. 17 (eds Corry, J.E.L., Roberts, D. and Skinner, F.A.). Academic Press, London, pp. 261–286.

Lee, L.A., Gerber, A.R., Conswav, D.R. *et al.* (1990) *Yersinia enterocolitica* 0:3 infections in infants and children associated with the household preparation of chitterlings. *New Eng. J. Med.*, **322** (14) 984–987.

Limbiri, M., Mavridou, A., Richardson, S.C. and Papadakis, J.A. (1990) Comparison of the Tecra *Salmonella* immunoassay with the conventional cultural method. *Lett. Appl. Microbiol.*, **11**, 182–184.

Linggood, M.A. (1982) *Escherichia coli*: Detection of enterotoxins, in *Isolation and Identification Methods for Food-Poisoning Organisms*, SAB Tech. Ser. No. 17 (eds Corry, J.E.L., Roberts, D. and Skinner, F.A.). Academic Press, London, pp. 227–237.

Lior, H. and Borczyk, A.A. (1987) False positive identification of *Escherichia coli* O157. *Lancet*, **8528**, 333.

Lund, B. (1990) Foodborne illness. Foodborne disease due to *Bacillus* and *Clostridium* species. *Lancet*, **336**, 982–986.

Manafi, M. and Sommer, R. (1992) Comparison of three rapid screening methods for *Salmonella* spp.: 'MUCAP Test, MicroScreen Latex and Rambach Agar'. *Lett. Appl. Microbiol.*, **14**, 163–166.

Mead, G.C., Adams, B.W., Roberts, T.A. and Smart, J.L. (1982) Isolation and enumeration of *Clostridium perfringens*, in *Isolation and Identification Methods for Food Poisoning Organisms*, SAB Tech. Ser. No. 17 (eds Corry, J.E.L., Roberts, D. and Skinner, F.A.). Academic Press, London, pp. 99–108.

Medon, P.P., Lanser, J.A., Monckton, P.R. *et al.* (1988) Identification of enterotoxigenic *Escherichia coli* isolates with enzyme-labelled synthetic oligonucleotide probes. *J. Clin. Microbiol.*, **26**, 2173–2176.

Megraud, F., Belbouri, A., Monget, D. and Gayral, P. (1987) A micromethod to identify *Campylobacter* spp. Preliminary results, in *Proceedings of the 4th International Workshop on* Campylobacter *Infections*. 16–18 June, Goteborg, Sweden.

Meier, K. and Terplan, G. (1993) Investigation of cheese and other foodstuff samples with the *Listeria*-Tek ELISA. *Lett. Appl. Microbiol.*, **17**, 97–101.

Merino, F.J., Aguillo, A., Villasante, P.A. *et al.* (1986) Comparative efficacy of seven selective media for isolating *Campylobacter jejuni*. *J. Clin. Microbiol.*, **24**, 451–452.

Miliotis, M.D., Galen, J.E., Kaper, J.B. and Morris, J.G. (1989) Development and testing of a synthetic oligonucleotide probe for the detection of pathogenic *Yersinia* strains. *J. Clin. Microbiol.*, **27**, 1667–1670.

Mintel (1990) *Food Safety* (Mintel Special Report). Mintel Publications, London.

Monafi, M. and Holzhammer, E. (1993) Comparison of the API 20E, Vitek-, and Gene-Trak® systems for identification of *Y. enterocolitica*, in *Seventh International Congress on Rapid methods and Automation in Microbiology and Immunology*. London, Abstr. P7/7, p. 72.

Nesbakken, T., Kapperud, G., Dommarsnes, K. *et al.* (1991) Comparative study of a DNA hybridisation method and two isolation procedures for detection of *Yersinia enterocolitica* 0:3 in naturally contaminated pork products. *Appl. Environ. Microbiol.*, **57**, 389–394.

Noah, C.W., Ramos, N.C. and Gipson, M.V. (1991) Efficiency of two commercial ELISA kits compared with the BAM culture method for detecting *Listeria* in naturally contaminated foods. *J. A. O. A. C.*, **74**, 19–821.

Notermans, S. and Tatini, S. (1993) Characterisation of *Bacillus cereus* in relation to toxin production. *Neth. Milk Dairy J.*, **47**, 71–77.

Notermans, S. and Wernars, K. (1991) Immunological methods for detection of foodborne pathogens and their toxins. *Int. J. Food Microbiol.*, **12**, 91–102.

O'Donoghue, D. and Winn, E. (1993) Comparison of the MSRV method with an in-house conventional method for the detection of *Salmonella* in various high and low moisture foods. *Lett. Appl. Microbiol.*, **17**, 174–177.

Oggel, J.J., Nundy, D.C. and Chesley, J.R. (1990) Modified 1-2 test system as a rapid method for detection of *Salmonella* in foods and feeds. *J. Food Prot.*, **53**, 656–658.

Okrend, A.J.G., Rose, B.E. and Matner, R. (1990) An improved screening method for the detection and isolation of *Escherichia coli* O157:H7 from meat, incorporating the 3M Petrifilm Test Kit-HEC- for hemorrhagic *Escherichia coli* O157:H7. *J. Food. Prot.*, **53**, 963–940.

Olive, D.M., Johny, M. and Sethi, S.K. (1990) Use of an alkaline phosphatase-labelled synthetic oligonucleotide probe for detection of *Campylobacter jejuni* and *Campylobacter coli*. *J. Clin. Microbiol.*, **28**, 1565–1569.

Olsvik, Ø., Wasteson, Y., Lund, A. and Hornes, E. (1991) Pathogenic *Escherichia coli* found in foods. *Int. J. Food Microbiol.*, **12**, 103–114.

Padye, N.V. and Doyle, M.P. (1991) Rapid procedure for detecting enterohemorrhagic *E. coli* O157:H7 in food. *Appl. Environ. Microbiol.*, **57**, 2693–2698.

Park, C.E., Akhtar, M. and Rayman, M.K. (1992) Nonspecific reactions of a commercial enzyme-linked immunosorbent assay kit (TECRA) for detection of staphylococcal enterotoxins in foods. *Appl. Environ. Microbiol.*, **58**, 2509–2512.

Park, C.E., Akhtar, M. and Rayman, M.K. (1993) Simple solutions to false-positive staphylococcal enterotoxin assays with seafood tested with an enzyme-linked immunosorbent assay kit (TECRA). *Appl. Environ. Microbiol.*, **59**, 2210–2213.

Patel, P.D. (1993) Real-time microbiological methods in quality assurance. *Food Technol. Int. Eur.*, 201–204.

Patel, P.D. (1994) The use of DNA fingerprinting in food analysis. *Food Technol. Int. Eur.*, in press.

Prpic, J.K., Robins-Browne, R.M. and Davey, R.B. (1985) *In vitro* assessment of virulence in *Yersinia enterocolitica* and related species. *J. Clin. Microbiol.*, **22**, 105–110.

Restaino, L., Grauman, G.S., McCall, W.A. and Hill, W.M. (1979) Evaluation of the Minitek and API 20E systems for identification of *Yersinia enterocolitica*. *J. Food Prot.*, **42**, 120–123.

Rice, E.W., Sowers, E.G., Johnson, C.H. (1992) Serological cross-reactions between *Escherichia coli* O157 and other species of the genus *Escherichia*. *J. Clin. Microbiol.*, **30**, 1315–1316.

Robison, B.J. and Cunningham, C.P. (1991) A research note: accuracy of Micro-ID *Listeria* for identification of members of the genus *Listeria*. *J. Food Prot.*, **54**., 798–801.

Rose, S.A., Banks, P. and Stringer, M.F. (1989) Detection of staphylococcal enterotoxins in dairy products by the reversed passive latex agglutination (SET-RPLA) kit. *Int. J. Food Microbiol.*, **8**, 65–72.

Rose, B.E., Llabres, C.M. and Bennett, B. (1991) A research note: evaluation of a colorimetric DNA hybridisation test for detection of salmonellae in meat and poultry products. *J. Food Prot.*, **54**, 127–131.

Schiemann, D.A. (1983) Comparison of enrichment and plating media for recovery of virulent strains of *Yersinia enterocolitica* from inoculated beef stew. *J. Food Prot.*, **46**, 957–964.

Schmidt, M. and Sethi, K.K. (1987) Production and characterisation of monoclonal antibodies specific for pathogenic serogroups 0:3 and 0:9 of *Yersinia enterocolitica*. *Z. Bakteriol. Hyg. 1 Abt. Orig. A*, **265**, 113–123.

Scotland, S.M., Willshaw, G.A., Said, B. *et al.* (1989) Identification of *Escherichia coli* that produces heat-stable enterotoxin STA by a commercially available enzyme-linked immunoassay and comparison of the assay with infant mouse and DNA probe tests. *J. Clin. Microbiol.*, **27**, 1697–1699.

Scotter, S.L., Humphrey, T.J. and Henley, A. (1993) Methods for the detection of thermotolerant campylobacters in foods. Results of an inter–laboratory study. *J. Appl. Bacteriol.*, **74**, 155–164.

Sernowski, I.P. and Ingham, S.C. (1992) Low specificity of the IIEC O137™ ELISA in screening ground beef for *Escherichia coli* O157:H7. *J. Food Prot.*, **55**, 545–547.

Sharma, N.K., Doyle, P.W., Gerbasi, S.A. and Jessop, J.H. (1990) Identification of *Yersinia* species by the API 20E. *J. Clin. Microbiol.*, **28**, 1443–1444.

Skirrow, M.B. (1991) *Campylobacter*, in *Foodborne Illness* (eds Waites, W.M. and Arbuthnott, J.P.). Edward Arnold, London.

Skirrow, M.B., Benjamin, J., Razi, M.H.H. and Waterman, S. (1982) Isolation, cultivation and identification of *Campylobacter jejuni* and *C. coli*, in *Isolation and Identification Methods for Food-Poisoning Organisms*, SAB Tech. Ser. No. 17 (eds Corry, J.E.L., Roberts, D. and Skinner, F.A.). Academic Press, London, pp. 313–329.

Smith, H.R. and Scotland, S.M. (1993) Isolation and identification methods for *Escherichia coli* O157 and other verocytotoxin producing strains. *J. Clin. Pathol.*, **46**, 10–17.

Spiers, J., Stavric, S. and Buchanon, B. (1991) Assessment of two commercial agglutination kits for detecting *Escherichia coli* heat-labile enterotoxin. *Can. J. Microbiol.*, **37**, 877–880.

St. Clair, V.J. and Klenk, M.M. (1990) Performance of three methods for the rapid identification of *Salmonella* in naturally contaminated foods and feeds. *J. Food. Prot.*, **53**, 961–964.

Stewart, G.S.A.B. and Williams, P. (1992) Review article lux genes and the applications of bacterial bioluminescence. *J. Gen. Microbiol.*, **138**, 1289–1300.

Stewart, G.S.A.B., Smith, A.T. and Denyer, S.P. (1989) Genetic engineering of bioluminescent bacteria. *Food Sci. Technol. Today*, **3**, 19–22.

Stringer, M.F., Watson, G.N. and Gilbert, R.J. (1982) *Clostridium perfringens* Type A: serological typing and methods for the detection of enterotoxin, in *Isolation and Identification Methods for Food-Poisoning Organisms*, SAB Tech. Ser. No. 17 (eds Corry, J.E.L., Roberts, D. and Skinner, F.A.). Academic Press, London, pp. 111–132.

Tranter, S.H. (1991) Foodborne staphylococcal illness, in *A Lancet Review. Foodborne Illness* (eds Waites, W.M. and Arbuthnott, J.P.). Edward Arnold, London.

Ulitzer, S. and Kuhn, J. (1987) Introduction of lux genes into bacteria, a new approach for specific determination of bacteria and their antibiotic susceptibility, in *Bioluminescence and Chemiluminescence. New Perspectives* (eds Schlomerich, J., Andreesen, R., Kapp, A. *et al.*). Wiley, Bristol, pp. 463–473.

Url, B., Heitzer, A. and Brandl, E. (1993) Determination of *Listeria* in dairy and environmental samples: comparison of a cultural method and a colorimetric nucleic acid hybridisation assay. *J. Food Prot.*, **56**, 581–585.

Van Beurden, R. and Mackintosh, R. (1992) A study of the performance of a new, rapid dip-stick immunoassay for the detection of salmonellae in food, in *International Food Technology Exposition and Conference*, 15–18 November, The Hague, The Netherlands, Abstr. 156, p. 136.

Van der Zee, H. (1993) Comparison of self prepared and commercially available brands of diagnostic semi-solid *Salmonella* agars, in *Seventh International Congress on Rapid Methods and Automation in Microbiology and Immunology*, 12–15 Sept., London, Abstr. P8/3, p. 73.

Van Netten, P., Van de Moosdijk, A., Van Hoensel, P. *et al.* (1990) Psychrotrophic strains of *Bacillus cereus* producing enterotoxin. *J. Appl. Microbiol.*, **69**, 73–79.

Varnam, A.H. and Evans, M.G. (1991) *Foodborne Pathogens: An Illustrated Text*. Wolfe Publishing Ltd, Aylesbury.

Veld, P.H. In't. and Boer, E. de (1991) Recovery of *Listeria monocytogenes* on selective agar media in a collaborative study using reference samples. *Int. J. Food Microbiol.*, **13**, 295–300.

Walker, R.I., Caldwell, B., Lee, C. *et al.* (1986) Pathophysiology of *Campylobacter enteritis. Microbial. Rev.*, **50**, 81–94.

Walker, S.J. and Gilmour, A. (1986) A comparison of media and methods for the recovery of *Yersinia enterocolitica* and *Yersinia enterocolitica*-like bacteria from milk containing simulated raw milk microfloras. *J. Appl. Bacteriol.*, **60**, 175–183.

Wauters, G., Aleksic, S., Charlier, J. and Schulze, G. (1991) Somatic and flagellar antigens of *Yersinia enterocolitica* and related species. *Contr. Microbiol. Immunol.*, **12**, 239–243.

Weiss, L.H. (1989) Comparison of the automicrobic system to a conventional method for the biochemical identification of *Listeria* spp., in *Proceedings of the 103rd AOAC Annual Meeting and Exposition*, St. Louis. Abstr. 259.

Wieneke, A.A. (1991) Comparison of four kits for the detection of staphylococcal enterotoxin in foods from outbreaks of food-poisoning. *Int. J. Food Microbiol.*, **14**, 305–312.

Wieneke, A.A. and Gilbert, R.J. (1987) Comparison of four methods for the detection of staphylococcal enterotoxin in foods from outbreaks of food-poisoning. *Int. J. Food Microbiol.*, **4**, 135–143.

Wieneke, A.A., Roberts, D. and Gilbert, R.J. (1993) Staphylococcal food poisoning in the United Kingdom, 1969–90. *Epidemiol. Infect.*, **110**, 519–531.

Williams, D. and Patel, P.D. (1991) *A Brief Evaluation of Five Methods for the Rapid Detection of Listeria spp.* (Leatherhead FRA Tech. Note No. 99). Leatherhead Food Research Association, Leatherhead, Surrey.

Wilson, I.G., Cooper, J.E. and Gilmour, A. (1991) Detection of enterotoxigenic *Staphylococcus aureus* in dried skimmed milk: use of the polymerase chain reaction for amplification and detection of staphylococcal enterotoxin genes entB and entC1 and the thermonuclease gene *nuc. Appl. Environ. Microbiol.*, **57**, 1793–1798.

Wolber, P.K. and Green, R.L. (1990) New rapid method for the detection of *Salmonella* in foods. *Trends Food Sci. Technol.*, **1**, 80–82.

Wren, B.R. and Tabaqchali, S. (1990) Detection of pathogenic *Y. enterocolitica* by the polymerase chain reaction. *Lancet*, **336**, 693.

Yam, W.C., Lung, M.L. and Ng, M.H. (1992) Evaluation and optimisation of a latex agglutination assay for detection of cholera toxin and *Escherichia coli* heat-labile toxin. *J. Clin. Microbiol.*, **30**, 2518–2520.

4 Microbiological applications of immunomagnetic techniques

P.D. PATEL

4.1 Introduction

In today's competitive food industry with diverse range of products (e.g. short shelf-life foods, modified atmosphere packaged products and minimally processed products), and governed by strict food legislation, microbiological safety has become a key issue. Legally required to demonstrate 'due diligence' under the 1990 UK Food Safety Act, food manufacturers are demanding analytical techniques that are simple to use, cost-effective, robust, reliable and can provide results in real-time. The majority of current microbiological techniques, particularly for the analysis of foodborne pathogens, give results that are only of retrospective value and do not allow proactive or reactive measures to be implemented during modern food production.

For pathogen analysis, a range of classical and commercially available rapid (or more aptly, alternative) methods has been considered in chapter 3. It is clear that the methods based on immunology, molecular biology, electrical conductance, etc., rely on time consuming cultural enrichment steps in order to increase the productivity of the target pathogen to a minimum threshold level, prior to detection. An alternative to the cultural enrichment procedures is to use rapid biophysical and immunological techniques for the 'enrichment' of the desired microorganisms. For details concerning the use of solid phase separation techniques (e.g. ion exchangers), the reader is referred to chapter 7 and a review by Patel and Wood (1985).

The use of magnetic particles for immunoaffinity purification or 'enrichment' has been demonstrated in various microbiological applications, including the detection of anti-staphylococcal enterotoxin from culture supernatants (Patel *et al.*, 1984), staphylococcal enterotoxins from foods (Patel, 1985; Stannard *et al.*, 1987; Patel and Gibbs, 1988), *Salmonella* from mixed cultures (Patel, 1984), isolation of K88+ *Escherichia coli* from faeces (Lund *et al.*, 1988), isolation of protein-A-bearing strains of *Staphylococcus aureus* from milk (Johne and Jarp, 1988), localisation of exopolysaccharide in *Staph. aureus* (Johne *et al.*, 1989), *Clostridium perfringens* enterotoxin type A in faecal and food samples (Cudjoe *et al.*, 1991), *E.coli* O157 (Olsvik *et al.*, 1991; Fratamico *et al.*, 1992), *Vibrio parahaemolyticus* in faecal and food samples (Tomoyasu, 1992), *Listeria monocytogenes* from a range of food matrices (Skjerve *et al.* 1990), *Salmonella* serogroup C1 in blood and stool samples (Luk and Lindberg, 1991) and sulphate-reducing bacteria from oil-field waters (Christensen *et al.*, 1992). An excellent volume comprising proceedings of the first John Ugelstad Conference, held in Oxford, on applications of magnetic particles in cellular and molecular biology (mammalian and microbial systems) has recently been published (Kemshead, 1991).

The objectives of this chapter are to consider the development and evaluation of immunomagnetic techniques in food microbiological analysis, and to review the progress to date indicating the extent of applications of this exciting technology.

4.2 ImmunoDynabeads for separation and concentration of *Salmonella*

This study describes the development and use of antibody-linked streptavidin-coated Dynabeads (M280; 2.8μm) for the rapid separation and concentration of salmonellae from broth cultures and from a range of raw and processed food enrichment broths. Optimisation of several steps in the assay procedures has also been considered, together with a range of end-point detection systems following the enrichment of salmonellae. The *Salmonella* immunoDynabeads are now available commercially (Dynal, UK) as rapid alternatives to the conventional growth-based selective enrichment procedures.

4.2.1 Development of Salmonella immunoDynabead particles

4.2.1.1 Preparation of biotinylated antibodies. BacTrace CSA-1 antibodies (Dynatech Laboratories Ltd) against *Salmonella* were reconstituted (0.33 mg/ml) as directed by the manufacturers. The antibody solution was allowed to react for 24 h with NHS-biotin (Sigma) in dimethyl formamide (100 μl, 25 mg/ml). The mixture was then dialysed extensively to remove the excess reagents, and the biotinylated antibody used in subsequent studies.

An enzyme-linked immunosorbent assay (ELISA) approach, involving titration of the biotinylated antibodies (adsorbed to microtitre wells) against dilutions of streptavidin-alkaline phosphatase conjugate, was used to confirm the preparation of biotinylated antibodies.

4.2.1.2. Preparation and reactivity of immunoDynabeads. The streptavidin particles (10min mg/ml) were allowed to react with the biotinylated antibody solution (1 ml) for 24 h at 4°C. The resulting immunoDynabeads were then washed four times with 0.15M phosphate-buffered saline (PBS) using the Dynal MPC-6 magnetic particle concentrator after each wash.

The reactivity of immunoDynabeads with salmonellae and non-salmonellae was assessed as follows. In general, microbial cultures (1–5 ml) at appropriate dilutions were incubated with immunoDynabeads (10–50 μl) for 10 min under gentle agitation. The immunoDynabeads were then separated and washed (four times) prior to final resuspension in 50-100 μl PBS. The bound cells were then enumerated by spreading the particles on solid media (e.g. xylose lysine and brilliant green agar plates).

An illustration of the interaction of *Salmonella typhimurium* with immunoDynabeads is shown in Figure 4.1, and the specificity of reactivity of the immunoDynabeads (M280) is shown in Figure 4.2

Figure 4.1 Interaction of *S.typhimurium* with immunomagnetic Dynabeads M450 (approximate magnification × 10 000).

4.2.1.3 Optimisation of Salmonella *immunoDynabeads.*   Figure 4.3 shows the interaction of salmonellae with Dynabeads M280 (anti-rabbit IgG or strepta-vidin) coupled to antisera obtained from four commercial sources. Except for the KPL preparation (raised in goat), the remaining antisera (raised in rabbits) were bound to the antirabbit IgG-Dynabeads. The biotinylated KPL antibody used with the streptavidin Dynabeads showed significant immunological binding with the three *Salmonella* serotypes.

The effects of varying quantity of immunoDynabeads on the isolation of *S. enteritidis* from 1, 5 and 10 ml culture volumes are shown Figure 4.4. Maximum binding occurred between 0.2 and 0.5mg bead quantity with a gradual increase up to 1 mg beads. More cells were bound to the immuno-Dynabeads (0.5 mg) using a volume of 5 ml than when using a 1 ml volume. At the same bead quantity, there was little difference in the number of cells bound using 5 or 10 ml of cell suspensions.

4.2.2 Evaluation of Salmonella immunoDynabeads in food enrichment broths

The classical method for the detection of salmonellae in foods is schematically shown in Figure 4.5, together with points of application of the immunoDynabeads. Figure 4.6 shows the procedure for immunomagnetic isolation of salmonellae. In the present study, the processed foods were each

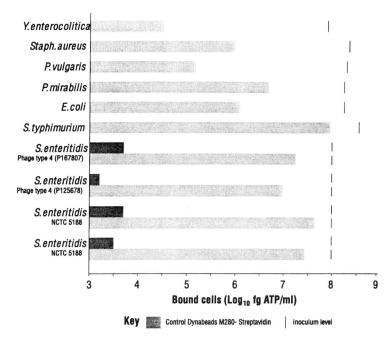

Figure 4.2 Interaction of foodborne bacteria with Dynabeads M280-streptavidin coated with BacTrace CSA-1 antibody.

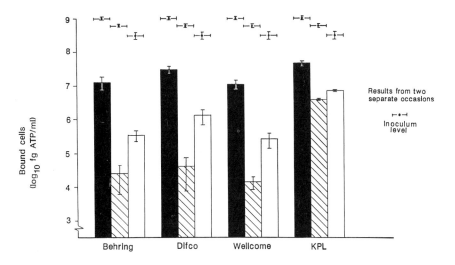

Figure 4.3 Binding of *S.typhimurium* (■), *S.enteritidis* (NCTC 5188) (▨), and *S.newport* (□) to M280 Dynabeads (anti-rabbit IgG or streptavidin) coated with a range of antibodies against *Salmonella*.

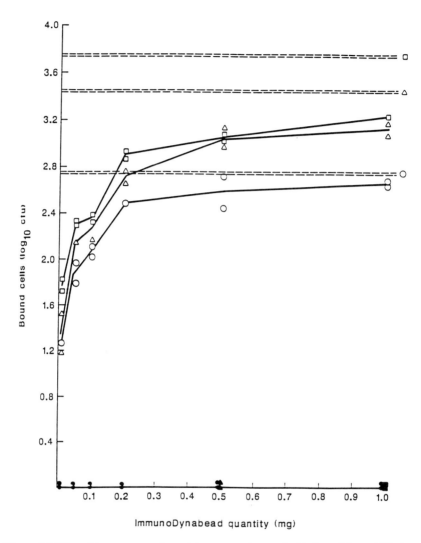

Figure 4.4 Effect of immunomagnetic particle concentration on the binding of *S.enteritidis* from 1 ml (○), 5 ml (△), and 10 ml (□) volumes of buffered peptone water (= inoculum levels ●, ▲, ■, control particles).

inoculated with a reference *S. typhimurium* II 505 strain (courtesy of Paul int' Veld, RIVM). Each raw food was inoculated separately with *S. typhimurium* and *S. enteritidis* (10–100cfu per 25g food). The immunoDynabeads were applied at 6 and 18 h of pre-enrichment and after selective enrichment in the Rappaport-Vassiliadis and selenite cystine broths. As controls, the untreated broths were also streaked on to XL agar. The extent of *Salmonella* growth was estimated as shown in Figure 4.7.

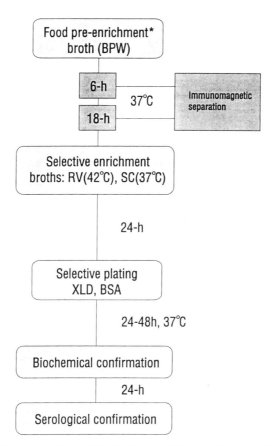

Figure 4.5 Conventional cultural method for the detection of *Salmonella* in foods showing points of application of immunomagnetic separation technique.
BPW, buffered peptone water; RV, Rappaport-Vassiliadis; SC, selenite cystine; XLD, xylose lysine desoxycholate; BSA, bismuth sulphite agar.
*Except for skimmed-milk powder (distilled water + 0.002% brilliant green) and cocoa powder (UHT skimmed-milk + 0.002% brilliant green), BPW was used as a pre-enrichment broth for the foods in this study.

4.2.2.1 Detection of Salmonella *in processed foods.* Table 4.1 shows a summary of results for the detection of *Salmonella* in processed foods. No salmonellae were isolated from the 6 h pre-enrichment broths. After 24 h pre-enrichment, *Salmonella* was isolated from four foods by both immunomagnetic separation and control streaking. For cocoa and skimmed-milk powders containing competitor flora, there was an increase in the isolation of salmonellae with concomitant reduction in the number of competitor flora after immunomagnetic separation from 24 h pre-enrichment broths (Table 4.1, Figure 4.8). After selective enrichment, *Salmonella* was isolated from all the foods. In most cases, there was distinct enrichment of salmonellae after immunomagnetic separation. All *Salmonella* isolates were confirmed serologically as *S. typhimurium*.

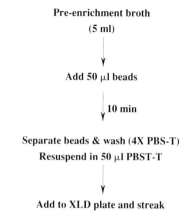

Pre-enrichment broth
(5 ml)

Add 50 μl beads

10 min

Separate beads & wash (4X PBS-T)
Resuspend in 50 μl PBST-T

Add to XLD plate and streak

PBS-T; Phosphate-buffered saline containing 0.05%
Tween 20, XLD; Xyloser lysine desoxycholate

Figure 4.6   *Salmonella* immunoDynabead protocol.

Table 4.1 Detection of *Salmonella* in processed foods using immunoDynabeads

Food	Pre-enrichment (24 h)		Selective enrichment (RV[a] 24 h)	
	C[b]	MS[c]	C	MS
Skimmed-milk powder	3+	3+[d]	+	2+
Cocoa powder	2+	2+[d]	2+	3+
Egg powder	2+	2+	2+	3+
Cooked–chilled chicken	2+	3+	2+	2+

[a] RV, Rappaport–Vassiladis broth.
[b] C, control streak.
[c] MS, magnetic separation.
[d] Reduction in competitor colonies.

4.2.2.2 Detection of Salmonella *in raw foods.* A summary of results for *Salmonella* in raw foods is shown in Table 4.2. After 6 h pre-enrichment, salmonellae were isolated from five out of six contaminated samples using immunomagnetic separation, compared with three by control streaking. For the latter, restreaking was necessary to isolate the *Salmonella*. After selective enrichment in the RV medium, there was a significant enrichment of salmonellae following immunomagnetic separation. Moreover, a natural contaminant confirmed as *S. enteritidis* was isolated from the control chicken sample by magnetic separation, but not by the conventional method.

Table 4.2 Detection of *Salmonella* in raw foods using immunoDynabeads

Food	Pre-enrichment				Selective enrichment RV[a]	
	6 h		24 h			
	C[b]	MS[c]	C	MS	C	MS
Prawns						
Con.[d]	−	−	−	−	−	−
A[e]	+[g]	2+	−	2+[g]	2+	3+
B[f]	−	2+[g]	−	−	2+	3+
Pork						
Con.	−	−	−	−	−	−
A	−	−	−	−	2+	3+
B	2+[g]	2+	3+	3+[g]	2+	3+
Chicken						
Con.	−	−	−	−	−	2+
A	−	2+	3+	3+	2+	3+
B	2+[g]	2+	3+	3+[g]	2+	3+

[a] RV, Rappaport–Vassiliadis broth.
[b] C, control streak.
[c] MS, magnetic separation.
[d] Con., uninoculated.
[e] A, inoculum level 29–74 cfu *S. typhimurium*/25 g.
[f] B, inoculum level 29–74 cfu *S. enteritidis*/25 g.
[g] Reduction in competitor colonies.

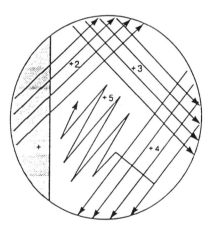

Figure 4.7 Standard streaking method to estimate growth of *Salmonella* (+ to 5+ indicates sector in which most growth/isolated colonies were found).

In this evaluation, the recovery of immunoDynabeads was lower in the case of cocoa powder and aggregation of the particles was evident with the 24 h pre-enrichment broths containing raw foods. In general, however, enrichment of salmonellae was achieved with the immunomagnetic particles when compared with the control samples not treated with the particles. Overall, the immunomag-

Figure 4.8 Xylose lysine desoxycholate plates showing isolation of *Salmonella* (black colonies) from cooked–chilled chicken pre-enrichment broth with (right-hand figure) and without (left-hand figure) immunomagnetic separation.

netic separation technique was shown to improve the sensitivity of the conventional method, and was capable of detecting salmonellae in foods 1-2 days earlier than the conventional method.

4.2.2.3 Related studies reported in the literature.   Evaluation of the commercially available *Salmonella* immunoDynabeads, in comparison with classical cultural methods, has recently been reported (Mansfield and Forsythe, 1993). When applied to a total of 120 foods, comprising high- and low-moisture foods, immunoseparation gave 63 positive recoveries of *Salmonella* compared with 62 for selenite cystine and Rappaport-Vassiliadis broths and 61 for Muller-Kaufmann tetrathionate broth. Immunoseparation, however, gave a result after a 10min enrichment procedure compared with a 24 h (and even 48 h in some cases) enrichment period with the approved cultural selective methods. other commercial sources of magnetic particles have been used to develop alternative immunomagnetic techniques for the purpose of separation and concentration of salmonellae. For example, McClements and Patel (1993) reported the use of BioMag-protein G and streptavidin particles with *Salmonella* and *Listeria*.

Skjerve and Olsvik (1991) and Vermunt *et al.* (1992) covalently coupled the KPL antibodies directly to the Dynabeads M280 (tosyl-activated). The former group showed successful isolation of *S. saint paul* at levels >500 per 25 g from pre-enrichment broths of most foods. However, low recovery of the immunoDynabeads was observed with yoghurt containing high levels (>10^7 cfu/ml) of competing flora, and chicken liver. Vermunt *et al.* (1992) found significant cross-reactions due to *Aeromonas hydrophila* (strain G5) and *Enterobacter agglomerans* with the immunoDynabeads, and attributed this to non-specific binding to the polystyrene beads. Other strains, including *Ent. agglomerans* (strains G7 and G8) and *Klebsiella pneumoniae* either did not cross-react or showed little binding to the *Salmonella*-specific immunoDynabeads. Although, the technique showed enrichment of *Salmonella* from a sample of minced beef using immunoDynabeads, Vermunt *et al.* (1992) suggested that further optimisation of the immunoDynabeads was essential.

In the case of *Listeria monocytogenes*, the corresponding monoclonal antibodies were bound to the Dynabeads M280, sheep anti-mouse IgG, prior to application of the immunoDynabeads to vacuum-packed ham and Brie cheese (Skjerve *et al.*, 1990) The recovery of *L. monocytogenes* ranged from below 2×10^2 in ham grown in UVM-TSB and cheese in UVM, to not less than 2×10^4 *L. monocytogenes* per ml in cheese grown in TSB. Thus, the sensitivity of the separation technique for these food types was not sufficient to allow detection of *L. monocytogenes* within 24 h. Recently, Skjerve *et al.* (1993) reported the development of an immunomagnetic particle-based system for the rapid separation of *Listeria* from fresh poultry carcasses, raw and smoked salmon and samples from production localities. However, it was observed that the separation was 'hard to accomplish' if the background flora reached a level of about 10^8 cfu/g.

4.2.3 SalmonellaimmunoDynabeads and alternative detection systems

The basic immunoDynabead-enrichment procedure for salmonellae can be of significant value in conjunction with a number of end-point detection/amplification techniques, including ELISAs, bioluminescence, automated electrical detection and polymerase chain reaction (PCR). The immunoDynabeads in combination with the PCR technique are considered in detail in chapter 10.

4.2.3.1 Detection of S. enteritidis *from eggs.*   *Salmonella enteritidis* was added at levels ranging from 10<4 to 10<9 cfu/g of egg yolk prior to attempting isolation using the immunoDynabeads M280 coated with the KPL antibodies (Blackburn *et al.*, 1991). Figure 4.9 shows that separated *Salmonella* could be detected down to 1×10^5 cfu/g within 20 min, using the bioluminescent measurement of intracellular ATP. The immunological binding was approximately 1 $\log_{10}$ cycle greater than the non-specific binding at contamination levels of 7.1 and 8.1 $\log_{10}$ cfu/g. After immunomagnetic separation, the non-microbial ATP from the egg yolk was reduced from about 7.5 $\log_{10}$ fg ATP/g to below the limit of sensitivity (<4.2 $\log_{10}$ fg ATP/g of the bioluminescence assay. In practice, the bioluminescence assay is not specific for salmonellae and an estimation of ATP in a given test sample will indicate the levels of general microbial contamination, unless the flora is exclusively salmonellae.

An ELISA for the detection of a range of heat-solubilised salmonellae antigens absorbed to immunoDynabead M280 coated with the KPL antibodies has recently been reported (Figure 4.10; McClements and Patel, 1993). The

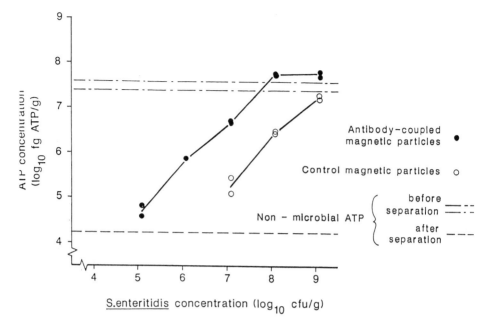

Figure 4.9 Detection of *S.enteritidis* (NCTC 5188) in egg yolk using an immunomagnetic particle/bioluminescence technique.

detection limit ranged between 10^5 and 10^6 cfu/ml. When applied to heat-treated extracts of artificially contaminated egg yolks, *S. enteritidis* could be separated and detected by ELISA down to about 10^6 cfu/g within 3.5 h (Blackburn *et al.*, 1991).

Overall, the use of more sensitive luminometers available on the market (Stanley, 1992), and brief incubation of eggs at appropriate temperatures may enable detection of low levels (approximately $>10^3$ cfu/g) of contamination within a few hours using the techniques described previously.

4.2.3.2 Detection of Salmonella *using immunoDynabeads in combination with conductance.* A number of instrumental techniques, based on electrical measurements of microbial metabolism, are currently available for the analysis of a range of food-spoilage and food-poisoning microorganisms of importance in foods. A review of these techniques in relation to measurement of conductance, capacitance and impedance properties of media supporting growth of microorganisms of importance in foods is presented in chapter 5.

Most conventional microbiological media for the growth and detection of specific groups of organisms, e.g. pathogens, are not suitable for use with the automated electrical systems, so a variety of conductance media that are more selective has been developed (Pugh *et al.*, 1988; Blackburn, 1991). For the detection of foodborne pathogens, the current methods available for use in the

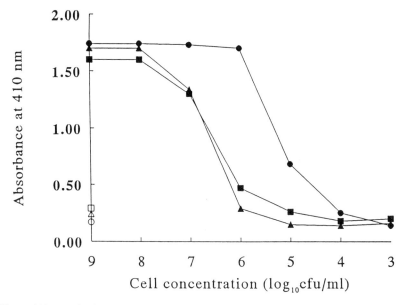

Figure 4.10 Detection by enzyme immunoassay of salmonellae bound to BacTrace anti-*Salmonella*-coated Dynabeads.

	BacTrace anti-*Salmonella* coated particles	Control particles (no antibody)
S.enteritidis	●	○
S.typhimurium	■	□
S.heidelberg	▲	△

electrical systems still rely on extensive time-consuming cultural enrichment steps. For salmonellae, the presumptive positive detection may take approximately 48 h, whilst methods are still under development for reliable detection of *Listeria, Campylobacter, Yersinia*, etc. Clearly, an immunomagnetic 'selective' enrichment step applied during the course of cultural pre-enrichment may enable rapid detection of the pathogens in electrical systems in two ways: (1) separation of the target pathogen from 'background' interference could give a better baseline and distinct electrical curves; and (ii) the concentration of the pathogens from larger volumes of enrichment media might increase the sensitivity and allow rapid detection of the pathogens.

Bird *et al.* (1989) investigated the use of immunomagnetic materials (Bio-Enzabeads and anti-rabbit IgG-Dynabeads) in combination with conductance measurement, for the detection of salmonellae. They found that the Bio-Enzabeads (*c.* 3 mm diameter) coated with monoclonal antibodies against *Salmonella* were inefficient in immobilising salmonellae. In fact, the use of Bio-enzabeads resulted in longer detection times (up to 11.4 h) than the normal conductance method. This was attributed to non-uniform coating of the primary antibody on the surface of the beads, in addition to possible loss in viability of cells adsorbed to the surface. The use of Dynabeads coated with polyvalent o antibodies also did not improve the detection times compared with the normal conductance method. Uncoated control beads were found to give the same detection times as antibody-coated beads, indicating significant non-specific binding.

The authors have recently shown the potential of immunoDynabeads, coated with the KPL antibodies, for enhancing the selectivity and subsequent detection of salmonellae using the Bactometer M-128 (McClements *et al.*, 1990). Some of the results obtained in previously unreported studies are briefly considered below. In a mixed culture containing *S. enteritidis* at a level of approximately 10^5 cfu/ml and a cocktail of three competitors at a level of approximately 10^9 cfu/ml, the standard conductance procedure failed to give a detection curve in the Easter and Gibson medium within 24 h (Figure 4.11). However, following application of the immunoDynabeads to the mixed culture, and inoculation of the particles into the conductance medium, a signal similar to that typical of pure culture of *S. enteritidis* was obtained in 15 h.

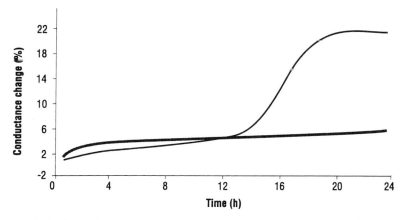

Figure 4.11 Effect of immunomagnetic separation on the conductimetric detection in Easter and Gibson medium of *S.enteritidis* (10^5 cfu/ml) in the presence of three other Enterobacteriaceae (10^9 cfu/ml). Immunomagnetic separation, (———) Direct inoculation (▬▬).

The immunoDynabeads also demonstrated significant value in enhancing the conductance technique as applied to *Salmonella* analysis in raw and processed foods. Figures 4.12 and 4.13 show the effect of the immunomagnetic technique on the conductance curves obtained using a 6 h chicken pre-enrichment broth and 24 h pork pre-enrichment broth, respectively. *Salmonella enteritidis* was inoculated at a level of approximately 10 cfu per 225 ml of the pre-enrichment broths. Figure 4.12 shows that the detection time for *S. enteritidis* in an ornithine medium was reduced by 1.6 h compared with the control sample not treated with

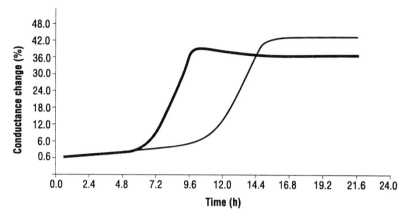

Figure 4.12 Immunomagnetic separation of *Salmonella* from 6 h chicken pre-enrichment broth artificially contaminated with *S.enteritidis* pt. 4 125678. Curve from control well (direct inoculation of 1:10 dilution of pre-enrichment medium) (————), curve from well containing immunomagnetic particles (▬▬▬).

the particles. The standard conductance technique failed to detect *Salmonella* from the 24 h pork pre-enrichment broth after a further 24 h incubation in the Bactometer containing Easter and Gibson medium (Figure 4.13). Following immunomagnetic separation, however, the ratio of *Salmonella* to competitors was improved sufficiently for a detection to be registered in 20 h.

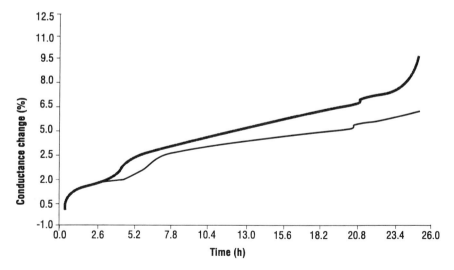

Figure 4.13 Immunomagnetic separation of *Salmonella* from 24 h pork pre-enrichment broth artificially contaminated with *S.enteritidis* pt. 4 125678. Curve from control well (direct inoculation of 1:10 dilution of pre-enrichment medium) (————), curve from well containing immunomagnetic particles (▬▬▬).

The result of the application of the immunomagnetic technique in combination with conductance measurement to skimmed milk powder pre-enrichment broth (225 ml), inoculated with a gelatin capsule containing artificially contaminated (between 5 and 10 cells of *S. typhimurium*) spray-dried milk (Beckers *et al.*, 1985) is shown in Figure 4.14. The consequence of the immunomagnetic technique was a reduction in the detection time by 1.8 h compared with the untreated sample. In addition, an increase in the detection time by 7 h for a false-positive result was observed following the immunomagnetic separation from the uninoculated sample compared with the sample not treated with the particles.

Parmar *et al.* (1992) recently evaluated the commercial *Salmonella*-specific immunoDynabeads, in combination with conductance measurements using the RABIT system (Don Whitley), for the detection of *Salmonella* in skimmed-milk powder pre-enrichment broths. The concluded that the *Salmonella* detection was enhanced by reducing the number and types of competing bacteria present and concentrating the number of salmonellae in the final conductance assay.

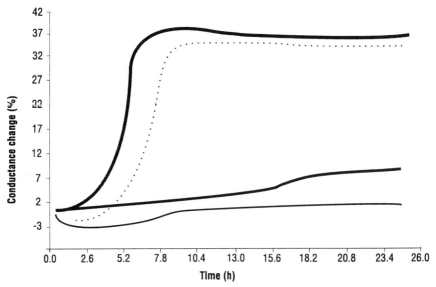

Figure 4.14 Immunomagnetic separation of *S.typhimurium* NCTC 74 from 24 h skimmed-milk powder pre-enrichment broth. Curve from control well (direct inoculation of 1:10 dilution of pre-enrichment medium) (.....), curve from well containing immunomagnetic particles (▬▬), curve from control well inoculated with uncontaminated broth (——), curve from uncontaminated broth after immunomagnetic separation (▬▬).

4.3 Potential of novel colloidal magnetic fluids in microbial analysis

Ferrofluids are colloidal magnetic fluids, of dimensions approximately 50 nm, and available in various ligand-coupled forms including protein A, anti-mouse

IgG, streptavidin and biotin. The advantages of these fluids compared with the particulate magnetic systems are that constant shaking is not required, the reaction kinetics are rapid and there appears to be no need for removal of unreacted primary antibody prior to application of the ferrofluids. Some of the author's work relating to the use of ferrofluids in food microbiological applications has recently been reported (Williams and Patel, 1992; McClements and patel, 1993; Patel *et al.*, 1993).

4.3.1 Antibody-based systems for Salmonella and Listeria

Figure 4.15 shows the specificity for the separation of *S. enteritidis* using the KPL antibodies in combination with protein A magnetite. Greater than 99% recovery of *S. enteritidis* was obtained at all dilutions of ferrofluid. In the case of *Enterobacter*, only 4% of the cells were recovered. Figure 4.16 shows the specific coating of ferrofluid on the surface of *S. enteritidis*.

The results of the application of KPL antibody and protein A magnetite to a mixed culture containing *S. enteritidis* (4.5 $\log_{10}$ cfu.ml) and *E. cloacae* (6.9 $\log_{10}$ cfu/ml) are shown in Figure 4.17. The ratio of *Enterobacter* to *Salmonella* was reduced from 226:1 (before separation) to 1.6:1 (after separation). The inclusion of a washing stage further reduced the ratio to 1:1.1, although the recovery of *Salmonella* fell from >99% to approximately 70%.

Preliminary results showing the application of ferrofluidic system to minced beef are presented in Figure 4.18. The presence of minced beef in buffered peptone water (BPW), but not in PBS, had an adverse effect on the separation of *Salmonella*. When the minced beef was homogenised in BPW (1:10 dilution), only 9% of salmonellae were recovered. This value is increased to >99% when no food material was present in BPW (Figure 4.18). When the minced beef was homogenised and diluted in PBS-gelatin, *Salmonella* recovery was >75%, irrespective of the presence of food material.

4.3.2 Lectin-based systems for food-spoilage yeasts

The classical mycological examination of foods, together with more rapid alternative techniques that are emerging at an increasing frequency, is considered in detail in Chapter 8. The speed, efficiency and reliability of modern rapid methods may be enhanced significantly by inclusion of a physicochemical separation step to 'enrich' the fungal cells or antigens from the 'background' interference due to non-target organisms and the food matrix.

Lectins, defined as proteins or glycoproteins that bind to specific saccharide residues, have previously been examined for their potential use in the detection and identification of food-spoilage and food-poisoning organisms (Haines and Patel, 1989; Williams and Patel, 1990). Concanavalin A (Con A) was found to react with a range of food-spoilage yeasts and, to a limited extent, moulds. Streptavidin-ferrofluid, in combination with biotinylated Con A, were examined

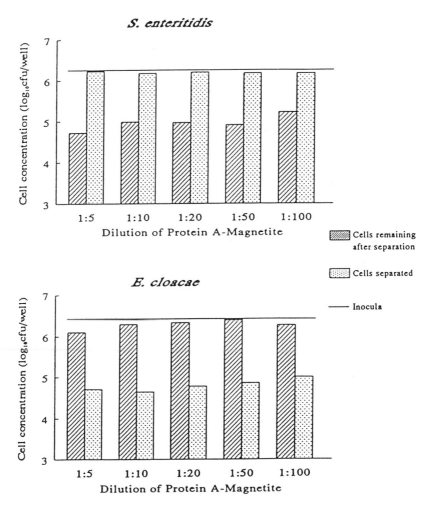

Figure 4.15 Separation of *S.enteritidis* using BacTrace anti-*Salmonella* antibody and protein A-magnetite.

for their value in the separation and concentration of a range of food-spoilage yeasts.

The results from broth cultures and spiked apple juice are summarised in Figures 4.19 and 4.20, respectively. In the case of yeasts in broth cultures, more than 95% of the cells were recovered in each case. Similar results were obtained for apple juice contaminated with the yeasts (Figure 4.20).

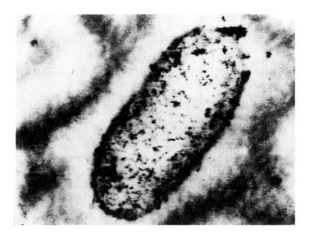

Figure 4.16 Electron micrograph (× 10 000 magnification) showing protein A-magnetite ferrofluid bound to the surface of a cell of *S.enteritidis*.

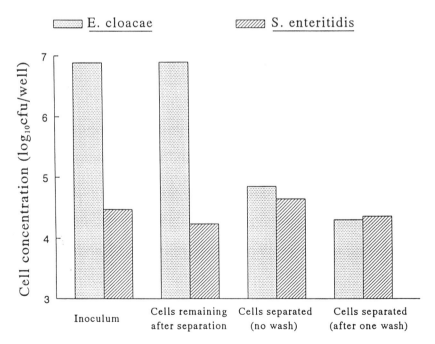

Figure 4.17 Enrichment of *S.enteritidis* from a mixed culture containing *E.cloacae* using BacTrace anti-*Salmonella* antibody and protein A-magnetite.

PBS+gelatin

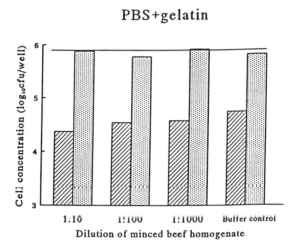

Buffered Peptone Water

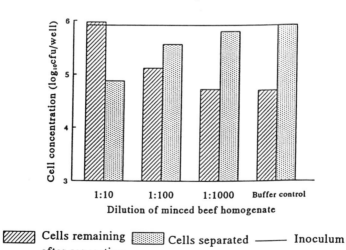

Figure 4.18 Effect of minced beef suspended in two buffer systems, on the separation of *S.enteritidis* using BacTrace anti-*Salmonella* antibody and protein A-magnetite.

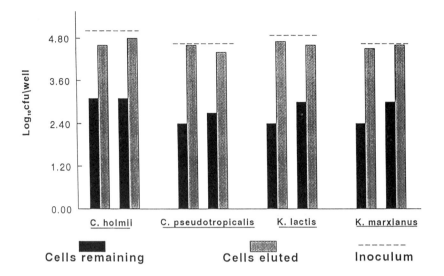

Figure 4.19 Separation of yeasts using Con A lectin and streptavidin-ferrofluid.

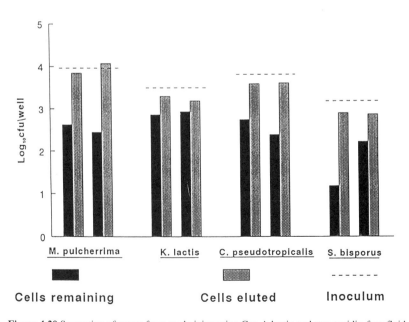

Figure 4.20 Separation of yeasts from apple juice using Con A lectin and streptavidin-ferrofluid.

4.4 Immunomagnetic detection of bacterial toxins

Two types of bacterial foodborne disease are recognised (Patel, 1989). Foodborne bacterial infection is caused by ingestion of food containing viable bacteria that then grow and establish themselves in the host, resulting in illness. This category includes *Cl. perfringens, Salmonella, Listeria* and *Campylobacter jejuni*. Foodborne bacterial intoxication is caused by the ingestion of food containing preformed bacterial toxin resulting from bacterial growth in the food and includes *Staph. aureus, Cl. botulinum* and *B. cereus*. Chapter 3 provides a review of the range of commercial rapid methods that are currently available for the detection of staphylococcal enterotoxins and *Cl. perfringens* enterotoxin. This section is directed specifically to the application of magnetic particle technology to analysis of these toxins.

4.4.1 Staphylococcal enterotoxins (SE)

The author's studies utilised glutaraldehyde-activated Magnogel AcA 44 (Figure 4.21), i.e. polyacrylamide-agarose beads (approximately 140 μm in diameter) containing iron oxide, to which antibodies against the enterotoxins for either sandwich ELISA or inhibition ELISA were covalently coupled. A range of applications, including enzyme immunoassays for the estimation of anti-SE B antibodies (Patel *et al.*, 1984) and SE (Patel and Gibbs, 1984, 1988; Patel, 1985), has been reported previously.

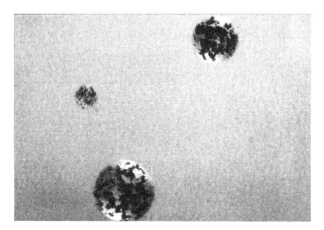

Figure 4.21 Magnogel AcA 44 (range approximately 40–10 μm).

The procedure involved in the application of the immunomagnetic technique to foods was as follows. Briefly, food samples spiked with 1 or 10 μg SE per 100 g were homogenised with equal volumes of distilled water and then centrifuged. The supernatants containing the SE were further clarified by extraction with chloroform and centrifuged. The aqueous layers were then assayed directly, or after further concentration by ultrafiltration, by using the immunomagnetic techniques as solid phase in ELISAs.

Figures 4.22 and 4.23 show the calibration curves for standard enterotoxin B estimated by direct magnetic enzyme immunoassay (MEIA) and inhibition MEIA, respectively, together with response of the SE extracted from cheese and ham. In both cases, the enterotoxin B was detected down to 1 μg per 100 g food. It has been shown that protein A, present in cell walls of certain strains of *Staph. aureus*, can interfere with ELISAs for the SEs (Notermans, 1985). The interference of protein A from food matrices can be completely eliminated by using non-specific antibodies to bind up the protein A prior to assaying SE (Patel and Gibbs, 1988).

It is the author's belief that, with the current availability of a wide variety of particulate and colloidal magnetic systems with numerous functional chemistries, it should be possible to develop simple and reliable assays for the estimation of SEs in foods within 1 h.

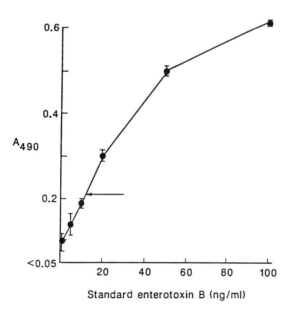

Figure 4.22 Estimation of enterotoxin B in cheese by direct MEIA ◄——, ent. B in cheese; control (i.e. no ent. B or cheese with toxin E) readings <0.05.

4.4.2 *Clostridium perfringens enterotoxin*

In 1992, there were approximately 1000-3000 reported cases of food poisoning in the UK due to *Cl. perfringens* (Sisson *et al.*, 1992). confirmation of an

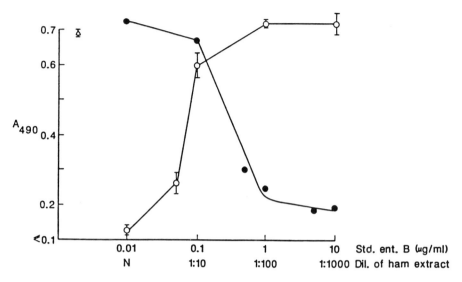

Figure 4.23 Estimation of enterotoxin B in ham by inhibition MEIA. Std. ent. B (●), ent. in ham (○), control (no ent. B) (△).

outbreak requires the isolation of *Cl. perfringens* from the implicated food and the detection of spores and the enterotoxin in faeces (Berry *et al.*, 1987).

Immunological analysis offers simple and rapid alternatives to the classical tissue culture techniques for the detection of the enterotoxin (Metha *et al.*, 1989). However, latex-agglutination-based immunological techniques can result in non-specific agglutination reactions, particularly at lower dilutions of extracts; thus caution needs to be exercised when interpreting results (Rose *et al.*, 1989). Detection of *Cl. perfringens* enterotoxin gene by the polymerase chain reaction has recently been used to differentiate enterotoxigenic strains from the non-enterotoxigenic strains (Saito *et al.*, 1992). The technique can also be applied directly to faecal samples. Although simple and sensitive, the PCR-based technique requires specialist knowledge and technical skills.

Cudjoe *et al.* (1991) reported the use of an immunomagnetic particle system for the rapid separation and concentration of the *Cl. perfringens* type A entero-toxin from faecal and food samples. The concentrated toxin on the particles was then detected by an enzyme immunoassay procedure. The sensitivity of this test was 2.5 ng/ml, compared with 0.1–5 ng/ml for other reported assays (Cudjoe *et al.*, 1991). However, the immunomagnetic ELISA gave results in 4 h, whilst the others, including the commercial reverse passive latex agglutination test (Oxoid), required 24 h or longer.

4.5 Future perspectives

Clearly, the magnetic-particle-based techniques, in combination with protein-binding reagents (e.g. monoclonal/polyclonal antibodies and lectins), are increasingly being introduced in microbial (and non-microbial) analysis as simple systems for the rapid separation and concentration of antigens prior to detection. Several of these techniques are now commercially available, e.g. *Salmonella* immunoDynabeads (Dynal), and Listerlist and Listertest systems for *Listeria* (Vicam). The latter are considered in chapter 3.

Undoubtedly, the range of commercial immunomagnetic systems is likely to increase further with applications directed to other pathogens (e.g. *E.coli* 0157:H7, *Yersinia* and *Aeromonas*) and microbial toxins (e.g. the enterotoxins of *Staph. aureus, Cl. perfringens, B. cereus* and *E. coli*). Since the basic magnetic separation system is universal in application, it is expected that the range of subsequent end-point detection systems will also increase (e.g. ELISAs with chemiluminescence and bioluminescence end-points, nucleic acid hybridisation and electrical conductance). Manual techniques are likely to increase in the near future, whilst robust automated magnetic-particle-based systems should follow as a result of increased confidence in this new technology.

Combination of immunomagnetic separation and molecular amplification techniques (e.g. PCR and Q-beta; Kramer *et al.*, 1991) is a logical progression for detecting very low levels of microorganisms, eventually leading to real-time techniques that are urgently required by the industry. This type of approach is considered in detail in chapter 10. The combination of immunomagnetic separation and PCR has recently been reported for the detection of approximately 0.1 cfu of *S. enteritidis* per g of chicken meat (Fruit *et al.*, 1993a) and 1 cfu of *L. monocytogenes* per g of cheese (Fruit *et al.*, 1993b). In both cases, monoclonal antibodies against the corresponding antigens were immobilised to Magnisort M chromium particles coated with the appropriate secondary antibodies. Similar techniques have also been reported for the detection of *Shigella* spp. (Islam and Lindberg, 1992), and *Y. enterocolitica* in foods (Vardund *et al.*, 1992; Kapperud *et al.*, 1993).

The stable ferrofluidic systems described previously offer even less manipulative, but more rapid and automatable procedures for the separation and concentration of microorganisms, compared with the particle-based systems. These systems are currently being investigated further in the author's laboratory, and it is envisaged that the number of applications of the ferrofluidic systems is also likely to increase. It is conceivable that the scope of magnetic systems will be extended to other industries, including the agricultural, clinical and pharmaceutical industries.

Acknowledgements

The basic development and optimisation of the immunomagnetic separation techniques were supported by the UK Ministry of Agriculture, Fisheries and Food, which is gratefully acknowledged. The author's thanks also go to the Dynal company for supporting *Salmonella* work in food systems. The technical skills of Clive Blackburn and Matthew McClements were invaluable in these studies.

References

Beckers, H.J., Van Leusden, F.M., Meijssen, M.J.M. and Kampelmacher, E.H. (1985) Reference material for the evaluation of a standard method for the detection of *Salmonella* in foods and feeding stuffs. *J. appl. Bacteriol*, **59**, 507-512.

Berry, P.R., Weineke, A.A., Rodhouse, J.C. and Gilbert, R.J. (1987) Use of commercial kits for the detection of *Clostridium perfringens* and *Staphylococcus aureus* enterotoxins, in *Immunological Techniques in microbiology* SAB Tech. Ser. No. 24 (eds Grange, J.M., Fox, A. and Morgan, N.L.). Blackwell Scientific, Oxford, pp. 245-254.

Bird, J.A., Easter, M.C., Hadfield, S.G. *et al.* (1989) Rapid *Salmonella* detection by a combination of conductance and immunological techniques, in *Rapid microbiological methods for foods, beverages and pharmaceuticals*, SAB Tech. Ser. No. (eds Stannard, C.J., Petitt, S.B. and Skinner, F.A.). Blackwell Scientific, Oxford, pp. 165-183

Blackburn, C.de W. (1991) Detection of *Salmonella* in foods using impedance. *Eur. Food Drink Rev.*, **Winter**, 35-40.

Blackburn, C.de W., Patel, P.D. and Gibbs, P.A. (1991) Separation and detection of salmonellae using immunomagnetic particles. *biofouling*, 143-156.

Christensen, B., Torsvik, T. and Lien, T. (1992) Immunomagnetically captured thermophilic sulphate-reducing bacteria from North Sea oil-field waters. *Appl. environ. Microbiol.*, 58, 1244-1248.

Cudjoe, K.S., Thorsen, L.I., Sorensen, T. *et al.* (1991) Detection of *Clostridium perfringens* type A enterotoxin in faecal and food samples using immunomagnetic separation (IMS)-ELISA. *Int. J. Food Microbiol.*, **12**, 313-322.

Fratamico, P.M., Schultz, F.J. and Buchanan, R.L. (1992) Rapid isolation of *Escherichia coli* 0157:H7 from enrichment cultures of foods using an immunomagnetic separation method. *Food Microbiol.*, **9**, 105-113.

Fruit, Ad. C., Widjojoatmodjjo, M.N., Box, A.T.A. *et al.* (1993a) Rapid detection of salmonellae in poultry with the magnetic immuno-polymerase chain reaction assay. *Appl. Environ. Microbiol.*, **59**, 132-346.

Fruit, Ad. C., Torensma, R., Visser, M.J.C. *et al.* (1993b) Detection of *Listeria monocytogenes* in cheese with the magnetic immuno-polymerase chain reaction assay. *Appl. Environ. Microbiol.*, **59**, 1289-1293.

Haines, S.D. and Patel, P.D. (1989) *A fluorescent Lectin Assay for the Rapid Detection of Moulds in Foods* (Leatherhead FRA Res. Rep. No. 656 (Available to members only). Leatherhead Food Research Association, Leatherhead, Surrey.

Islam, D. and Lindberg, A.A. (1992) Detection of *Shigella dysenteriae* type-1 and *Shigella flexneri* in faeces by immunomagnetic isolation and polymerase chain reaction. *J. Clin. Microbiol.*, **30**, 2801-2806.

Johne, B. and Jarp, J. (1988) A rapid assay for protein-A in *Staphylococcus aureus* strains using immunomagnetic monosized polymer particles. *APMIS*, **96**, 43-49.

Johne, B., Jarp, J. and Haaheim, L.R. (1989) *Staphylococcus aureus* exopolysaccharide *in vivo* demonstrated by immunomagnetic separation and electron microscopy. *J. Clin. Microbiol.*, **27**, 1631-1635.

Kapperud, G., Vardund, T., Skjerve, E. *et al.* (1993) Detection of pathogenic *Yersinia enterocolitica* in foods and water by immunomagnetic separation, nested polymerase chain reaction and colorimetric detection of amplified DAN. *Appl. Environ. Microbiol.*, **59**, 2938-2944.

Kemshead, J.T. (1991) *Magnetic Separation Techniques Applied to Cellular and Molecular Biology*. Wordsmiths' Conference Publications, Cromwell Press, Broughton Gifford, Wiltshire.

Kramer, R.F., Tyagi, S., Guerra, C.E. *et al.* (1991) Q-beta amplification assays, in *Rapid Methods and Automation in Microbiology and Immunology* (eds Vaheri, A., Tilton, R.C. and Balows, A. Springer-Verlag, Berlin, pp. 17-23.

Luk, J.M.C. and Lindberg, A.A. (1991) Rapid and sensitive detection of *Salmonella* (0:6,7) by immunomagnetic monoclonal antibody-based assays. *J. Immunol. Meth.*, **137**, 1-8.

Lund, A., Hellemann, A.L. and Vartdal, F. (1988) Rapid isolation of K88+ *Escherichia coli* by using imunomagnetic particles. *J. Clin. Microbiol.*, **26**, 2572-2575.

Mansfield, L.P. and Forsythe, S.J. (1993) Immunomagnetic separation as an alternative to enrichment broths for *Salmonella* detection. *Lett. appl. Microbiol.*, **16**, 122-125.

McClements, M.R., Blackburn, C. de W. and Patel, P.D. (1990) *Immunomagnetic Separation of Salmonella spp. and Detection by Conductance Measurements*. (Leatherhead FRA Res. Rep. No. 678 (available to members only). Leatherhead Food Research Association, Leatherhead, Surrey.

McClements, M.R. and Patel, P.D. (1993) *Magnetic Separation as an Aid to Rapid Microbiological Analysis* (Leatherhead FRA Res. Rep. No. 706 (available to members only). Leatherhead Food Research Association, Leatherhead, Surrey.

Mehta, R., Narayan, K.G. and Notermans, S. (1989) Dot-enzyme linked immunosorbent assay for detection of *Clostridium perfringens* type enterotoxin A. *Int. J. Food Microbiol.*, **9**, 45-50.

Notermans, S. (1985) Enzyme-linked immunosorbent assay of staphylococcal enterotoxins in foods, in *Rapid Methods and Automation in Microbiology and Immunology* (ed. Habermehl, K.O.). Springer-Verlag, Berlin, pp. 649-655.

Olsvik, Ø., Wasteson, Y., Lund, A. and Hornes, E. (1991) Pathogenic *Escherichia coli* found in food. *Int. J. Food Microbiol.*, **12**, 103-113.

Parmar, N., Easter, M.C. and Forsythe, S.J. (1992) The detection of *Salmonella enteritidis* and *S. typhimurium* using immunomagnetic separation and conductance microbiology. *Lett. Appl. Microbiol.*, **15**, 175-178.

Patel, P.D. (1984) The potential of chromatographic techniques for the manipulation of viable microorganisms. *PhD Thesis*, University of Surrey, Surrey.

Patel, P.D. (1985) Application of enzyme immunoassay techniques for the estimation of staphylo-coccal enterotoxins in foods, in *Immunoassays in Food Analysis* (eds Morris, B.A. and Clifford, M.N.). Elsevier Applied Science, London, pp. 141-155.

Patel, P.D. (1989) Bacterial toxins: modern analytical methods and quality control of food, in *Food Safety: New Methods for Research and Control* (eds de Pauli, E., Galli, C.L. and Restani, P.). Masson, Milan, pp. 61-66.

Patel, P.D. and Gibbs, P.A. (1984) Development of an inhibition magnetic enzyme immunoassay (MEIA) technique for determination of staphylococcal enterotoxin B. *Biochem. Soc. Trans.*, **12**, 264-265.

Patel, P.D. and Wood, J.M. (1985) The potential of chromatographic techniques for the manipulation of viable microorganisms, in *Rapid Methods and Automation in Microbiology and Immunology* (ed. Habermehl, K.O.). Springer-Verlag, Berlin, pp. 665-679.

Patel, P.D., Wood, J.M. and Gibbs, P.A. (1984) Development of a magnetic enzyme immunoassay (MEIA) technique for determination of anti-(staphylococcal enterotoxin) immunoglobulin G-type antibodies. *Biochem. Soc. Trans.*, **12**, 266-268.

Patel, P.D. and Gibbs, P.A. (1988) The prevention of protein A interference in immunoassays for *Staphylococcus aureus* enterotoxins, in Immunoassays for Veterinary and Food Analysis-1 (eds Morris, B.A., Clifford, M.N. and Jackman, R.). Elsevier Applied Science, London, pp. 333–337.

Patel, P.D., Williams, D.W. and Haines, S.D. (1993) Rapid separation and detection of food-spoilage yeasts and moulds by means of lectins. In *New Techniques in Food and Beverage Microbiology*, SAB Tech. Ser. No. 31 (eds Kroll, R.G., Gilmour, A. and Sussman, M.). Blackwell Scientific, London, pp. 31-43.

Pugh, S.J., Griffiths, J.L., Arnott, M.L. & Gutteridge, C.S. (1988) A complete protocol using conductance for rapid detection of salmonellas in confectionery materials. *Lett. Appl. Microbiol.*, **7**, 23-27.

Rose, S.A., Patel, N.P., Scott, A.O. and Stringer, M.F. (1989) Immunoassay kits for the detection of toxins associated with foodborne illness, in *Rapid Microbiological Methods for Foods, Beverages and Pharmaceuticals*, SAB Tech. Ser. No. 25. (eds Stannard, C.J., Petitt, S.B. and Skinner, F.A.). Blackwell Scientific, Oxford, pp. 265-283.

Saito, M., Matsumoto, M. and Funabashi, M. (1992) Detection of *Clostridium perfringens* gene by the polymerase chain reaction amplification procedure. Int. J. Food Microbiol., 17, 47–55.

Sisson, P.R., Kramer, J.M., Brett, M.M. *et al.* (1992). Application of pyrolysis mass spectrometry to the investigation of outbreaks of food poisoning and non-gastrointestinal infection associated with *Bacillus* species and *Clostridium perfringens. Int. J. Food Microbiol.*, **17**, 57-66.

Skjerve, E. and Olsvik, Ø. (1991) Immunomagnetic separation of *Salmonella* from foods. *Int. J. Food Microbiol.*, **14**, 11-17.

Skjerve, E., Rorvik, L.M. and Olsvik, Ø. (1990) Detection of *Listeria monocytogenes* in foods by immunomagnetic separation. *Appl. Environ. Microbiol.*, **56**, 3478-3481.

Skjerve, E., Rorvik, L.M., Heidenreich, B. and Cudjoe, K.S. (1993). Immunomagnetic separation of *Listeria* from foods, in Listeria *1992: The Eleventh International Symposium on Problems of Listeriosis*. ISOPOL XI, Eigtved's Pakhus-Copenhagen, 11-14 May, 1992, Book of Abstracts, pp. 21-22.

Stanley, P.E. (1992) A survey of more than 90 commercially available luminometers and imaging devices for low-light measurements of chemiluminescence and bioluminescence, including instruments for manual, automated and specialised operation, for HPLC, LC, GLC and microtitre plates. Part 1: descriptions. *J. Biolumin. Chemilumin.*, **7**, 77-108.

Stannard, C.J., Patel, P.D., Haines, S.D. & Gibbs, P.A. (1987) Magnetic enzyme immunoassay (MEIA) for staphylococcal enterotoxin B, in *Immunological Techniques in Microbiology*, SAB Tech. Ser. No. 24. (eds Grange, J.M., Fox, A. and Morgan, N.L.). Blackwell Scientific, Oxford, pp. 59-73.

Tomoyasu, T. (1992) Development of the immunomagnetic enrichment method selective for *Vibrio parahaemolyticus* serotype-k and its application for food-poisoning study. *Appl. Environ. Microbiol.*, **58**, 2679-2682.

Vardund, T., Kapperud, G., Skjerve, E. *et al.* (1992). Detection of *Yersinia enterocolitica* in food by immunomagnetic separation and polymerase chain reaction, in *Foodborne Infections and Intoxications:* (Proceedings of the 3rd World Congress). Berlin, pp. 543-545.

Vermunt, A.F.M., Franken, A.A.J.M. and Beumer, R.R. (1992) Isolation of salmonellas by immunomagnetic separation. *J. Appl. Bacteriol.*, **72**, 112-118.

Williams, D. and Patel, P.D. (1990) *Detection of Food Pathogens by Novel Lectin Binding Assays* (Leatherhead FRA Res. Rep. No. 674 (available to members only). Leatherhead Food Research Association, Leatherhead, Surrey.

Williams, D. and Patel, P.D. (1992) *Detection and Separation of Microorganisms using Lectins* (Leatherhead FRA Res. Rep. Np. 698 (available to members only). Leatherhead Food Research Association, Leatherhead, Surrey.

5 Automated electrical techniques in microbiological analysis

F.J. BOLTON and D.M. GIBSON

5.1 Introduction

Of all the techniques for rapid and automated microbiology described in the 1970s, those based on direct electrical measurements of media have been the most successful. The approach has been exploited commercially and now there are a variety of instruments available and in use in laboratories throughout the world.

Stewart (1899) has been regarded as the originator of the techniques, but careful reading of his papers leads to the conclusion that he was not measuring in the same fashion as modern instruments. For example, the changes in AC conductivity he reported occurred over many days on cultures of bacteria whose growth would be complete in about 2 days. His system was basically two balanced electrode cells, and he measured the electrical changes or out of balance due to growth detected through inert platinum electrodes. The cells were kept at the same temperature under running water because he noted that there was a large temperature effect (coefficient) on his measurements and that it was easier to hold two cells at the same temperature rather than one cell at a precise constant temperature.

What was being measured? Stewart (1899) compared his results with those of the freezing point of the solution. These are the colligative properties of matter, based on the number of particles in solution, the molar properties, rather than the weight or mass of each particle. As microbes grow in media, they metabolise various components and produce end products, perhaps carbon dioxide, water, ammonia, etc., or partially degraded substrates such as lactic acid from glucose, amines from amino acids, sugar acids from polysaccharides, acetate from lipids, etc. Generally, the metabolism of large molecules results in the production of small molecules in higher numbers in the medium; anabolism synthesises larger molecules within the microbes and not in the medium. Small molecules produced by metabolism are generally ions or charged particles, and they are better charge carriers than their parent molecule. Thus, metabolism and growth results in an increase in the number of charge-carrying particles and in better charge carriers. These are the characteristics exploited by the instruments.

5.2 What do the instruments measure?

As the changes in media described above result in changes in their electrical properties, it seems reasonable to try to measure these directly. Measuring the

DC conductivity of the solution seems an attractive option, but the sensitivity achievable is low and, during the measurements, uncontrolled electrolysis of the medium would occur, thus changing the characteristics of the solution and breaking one of the tenets of measurement, that the act of taking a reading should not alter the test solution. As multiple readings are needed to monitor the changes caused by growth, too many artifacts would be introduced.

AC conductivity measurements have therefore been used. These are variously referred to as conductance, impedance, admittance measurements. With AC, ions or charge carriers only oscillate in the electrical field and there are no changes induced in the medium. Schematically, a culture can be represented as equivalent to a resistance and a capacitance in parallel. However, by placing electrodes in the solution to carry the AC and to make the measurements, a further resistance and capacitance are introduced. The problem facing the instrument designer is to know which components change and which to measure. Firstly, with AC, there is a choice of frequency. Most instruments operate in the range of 2–10 kHz. At the lower end, capacitance is predominant in any measurement, and at the top, conductance. Conductance is mainly a solution effect, capacitance is mainly at or near the electrodes, and impedance a combination, the contributions depending on the precise electronic circuitry. During growth, metabolism affects the conductance more than the capacitance, and cell mass mainly capacitance. This is very much a simplification, but serves to indicate the relative importance of the different electrical components to monitoring microbial cultures.

5.3 When are measurements made?

Electrical assays have been dependent on the development of computers to record the data and to present them in a digested form to the instrument operator. The advantage of electrical measurements is that they are instant, require no reagents and no operator involvement, can be made on multiple samples sequentially, and readings can be taken at any time interval, perhaps every 5–10 min, which means that many data points are available during each doubling of the microbial counts. These yield vast amounts of data, which require storage and analysis. Most potential users do not require information on a detailed analysis of the whole growth curve, but in order to determine say the detection time when change is initiated, all the information is present and has to be sorted. To do this on multiple simultaneous assays without computers would be cumbersome and labour intensive.

5.4 Are assays faster than those by conventional tests?

Traditional microbiology tests are based on visual examination of plates or tubes, usually after a predetermined time interval of 1-3 days. Because they rely for their signal on the increase in cell numbers to yield a colony on a plate (perhaps containing 10^4 cells from a single cell) or a turbid solution (of perhaps $>10^6$ ml), it is generally not worthwhile examining cultures earlier. Automation with say colorimeters and spectrophotometers has not resulted in significant assay time reductions, as human eyesight is quite accurate, although there are instruments under development addressing this problem. Electrical assay instruments are more sensitive, at best responding to 10^3/ml, more generally to 10^5-10^6/ml, and, with continuous monitoring, positive results are displayed on the computer screen as soon as they are available. Of course, being computerised, they can also be transferred to, say, a quality controller for immediate action. Some assays take the same time as conventional tests. This is especially so when the metabolism of the microbes does not produce good charge carriers, e.g. alcohols, but there are instances of ingenious medium formulation which encourage good responses such that assays take only one-fifth of the time of traditional assays, as well as the development of 'indirect' assays (selection 5.10) which overcome such difficulties.

The total conductance or impedance change depends on the medium used and the organisms growing. For example, in their original work, Richards *et al.* (1978) used a strain of *Escherichia coli* which yielded a change of *c*.100 µS. This strain has since been shown to be atypical, due to its time in culture collections, and fresh *E.coli* isolates give 200–300 µS under the same conditions. The changes in many selective media (section 5.9) are much greater. The total change depends on the medium, the sample and the microbes present, and can vary, possibly due to the unpredictability of the behaviour of microbes in adverse growth conditions. However, in a qualitative rather than an absolute sense, total changes are used to indicate that media are giving the expected responses, thereby eliminating the possibility of false-positive results due to small spurious changes in conductance or impedance. Smith *et al.* (1989) found that the point of inflection of the differential curve (actually the value of the second differential) was statistically the best parameter for comparison. It is also the most robust point on the growth curve and, although if used would increase assay times slightly, is more reproducible than conventional detection times.

5.5 Current instrumentation and systems

There have been numerous attempts to commercialise conductance/impedance instruments and at present there are four manufacturers:

- Bactometer   bioMérieux UK Ltd, Grafton House, Grafton Way, Basingstoke, Hants RG22 6HY
- BacTrac   Sy-lab, Vertriesbsges.mbH, Hans-Buchmuller-G.5, A-3002 Purkersdorf, Austria

- Malthus   Malthus Instruments Ltd, The Manor, Manor Royal, Crawley, West Sussex RH10 2PY
- RABIT   Don Whitley Scientific Ltd, 14 Otley Road, Shipley, West Yorkshire BD17 7SE

All are highly automated systems able to measure multiple samples, and present analysed reports automatically. The main characteristics of these instruments, based on recent information from the manufacturers, are shown in Table 5.1.

Table 5.1 Hardware and software characteristics of impedance analysers

Components/ Functions per System	Bactometer	BacTrac	Malthus	RABIT
Maximum system size (Number of tests)	512	240	1200	512
Temperature range (°C)	18–55	4–60	5–56	15–46
Number of temperature combinations	8	6	5	16
Types of incubator units	1	2	3	1
Test volume (ml)	2	5–100	2–100	2–12
Random access for assays	No	Yes	Yes	Yes
Reusable cells	No	Yes	Yes	Yes
Disposable cells	Yes	No	Yes	No
Measurement	Impedance/ capacitance	Impedance/ capacitance	Conductance	Admittance
Indirect conductimetry	No	No	Yes	Yes
Computer	Dedicated	Dedicated	Non-dedicated	Dedicated

5.5.1 Bactometer

The Bactometer (Figure 5.1) is based on the work of Cady (1975). It differs in the means of achieving similar objectives. First, as originally used by Stewart (1899), it has twin cells for each sample, one of which serves as the control and the other as test cell. These are arranged in modules of eight which after inoculation are placed in an air incubator, whose temperature is controlled in the range ambient to 55°C. Each pair of cells, containing stainless-steel electrodes, is initially balanced. Readings are taken regularly and the 'out of balance' signal calculated. The Bactometer operates in a choice of modes, i.e. conductance, impedance and capacitance, and any combination can be selected. The resulting curves are similar to those shown in Figure 5.2 but, as they are a ratio, they are dimensionless and expressed as o change from the initial values. This ratio is expressed as follows:

$$Z = \frac{Z_r}{Z_r + Z_s}$$

Figure 5.1 The Bactometer.

where Z is the impedance value, Z_r is the reference impedance and Z_s is the sample impedance.

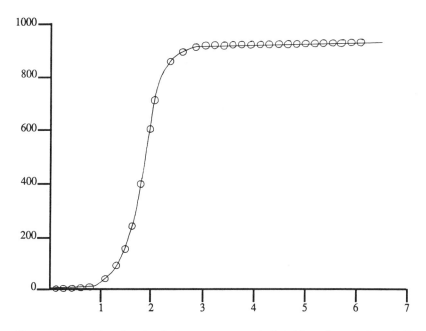

Figure 5.2 Typical impedance/conductance response. x axis, time (h); y axis, conductance (μS).

5.5.2 BacTrac

The BacTrac (Figure 5.3) has a different type of electrode arrangement. It employs four electrodes, one pair for measuring solution changes, the other for electrode changes. The manufacturers claim that these are separate and can be used to determine conductance changes (*M* value) and a capacitance type of effect (*E* value), respectively. The electrodes are made of stainless-steel and are reusable. Individual incubators can take 40 samples at a time and up to six incubators can be linked in one system. The instrument presents the results in terms of impedance change automatically.

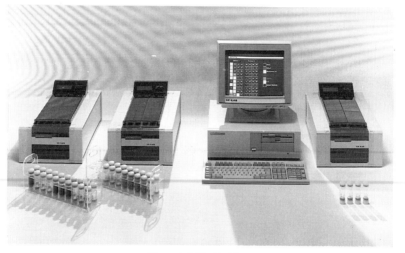

Figure 5.3 The BacTrac.

5.5.3 Malthus

The Malthus instruments (Figure 5.4) are products licensed under the patent granted to Aberdeen University and the Torry Research Station, Aberdeen. They are based on the work of Richards *et al.* (1978). Samples are held in tubes of various configurations which bear a pair of essentially biologically inert electrodes. In the original work, these were made of platinum wire. First commercial versions were of platinum pasted on to ceramic substrates (thick film technology). Both types could be used many times, the latter being virtually indestructible and withstanding repeated autoclaving. Besides these, there are now single-use disposable electrode cell systems with titanium electrodes. The cells containing media and inoculum are placed in water incubators whose temperature is precisely controlled to ±0.005°C in the range 5–55°C. This is necessary for high quality work, as the temperature coefficient of conductance/impedance is 1.6%/Ω per °C and the total change due to growth

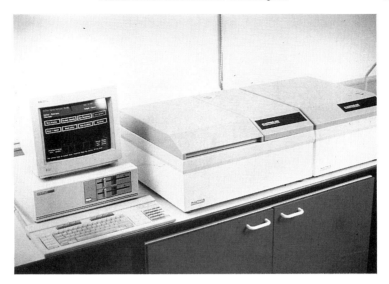

Figure 5.4 The Malthus.

may only be 2% of the starting resistance. There are four types of incubator unit which can take either 20 large volume cell samples or 60, 120, 240 of 2–10 ml samples. The instrument takes 'resistance' readings in Ω on each cell automatically at predetermined time intervals of 6–60 min. The computer program converts these to conductance (in siemens) and then stores the change in conductance. An early reading, usually the fourth, is taken as zero to allow time for temperature equilibration of the sample to the water temperature. A typical plot of the change in conductance with incubation times is shown in Figure 5.2. This resembles a microbial growth curve and, to show their relationships, Richards *et al.* (1978) showed that the change in microbial number parallelled the change in conductance over most of the growth curve. At low numbers, the instrument sensitivity did not permit this to be obvious. Further discussion of the outputs from the various instruments and their interpretation is given in section 5.6.

5.5.4 RABIT

The RABIT (Rapid Automated Bacterial Impedance Technique) instruments (Figure 5.5) are licensed by the same patent. However, instead of using water incubators, the RABIT has metal block incubators, each taking up to 32 individual samples. The electrode cells have stainless steel electrodes protruding through the base of their glass tubes. The tubes fit snugly into holes in the blocks. Readings are taken regularly as with the Malthus and processed similarly.

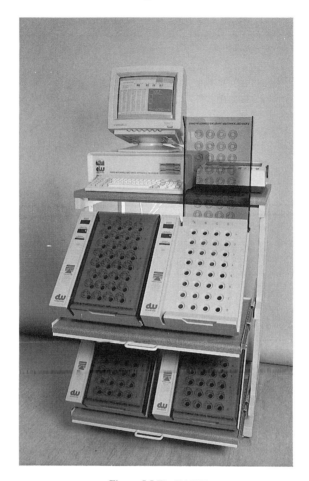

Figure 5.5 The RABIT.

5.6 Instrument outputs

5.6.1 Data

From Figure 5.2 it is clear that the instruments provide a form of data akin to conventional growth curves. The curves show a lag time, a period of accelerating growth, a period of exponential growth, followed by a reduction in the growth rate and finally the stationary phase. The greatest use has been made of the time the culture takes from inoculation to reach the acceleration phase. This is generally referred to as the detection time, the time since inoculation when there is a clear change from the baseline or lag phase values. There is, for most organisms and foods, a good relationship between detection and microbial load and calibration curves such as that shown in Figure 5.6 are usually obtained. A poor relationship in the quantitative sense, usually indicates some irregular behaviour such as that due to antimicrobial constituents in the test material. Calibration curves should be constructed for each assay or class of food, etc., by the analysis of multiple samples, but as commercial samples are often of very uniform microbial load, it is difficult to obtain sufficient spread of values (ideally covering 4 log cycles) except by diluting or concentrating naturally contaminated samples.

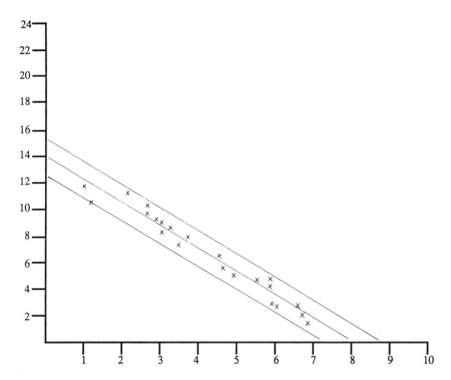

Figure 5.6 Relationship between change of conductance and change in bacterial number. x axis, log no. standard plate count; y axis, detection time (h). This calibration curve was produced using 30 data points and gave a correlation coefficient (r) of 0.985.

As is found with conventional methods for measuring growth curves, there is a period of maximum growth rate, generally half way up the exponential growth phase. Jason (1983) carried out extensive analysis and modelling of such curves. He showed that the simple curves of conductance change with time could be mathematically differentiated and obtained bell-shaped plots of the rate of change of conductance versus the change of conductance. He further showed that the specific growth rate of the culture could be calculated from the coordinate of the peak of that curve.

A fuller treatment of this kind of analysis is given by Gibson and Jason (1993).

5.6.2 Computers and software

All the systems are managed by dedicated software. The computers may be specifically designed for the system and dedicated or may be high specification computers which can be used 'offline' for other purposes. The key functions of the computer and software are listed below.

(i)   To enable input of sample details, i.e. laboratory number, product number and type of food product.
(ii) To assign the test to be performed, i.e. coliforms, *Salmonella*, etc., and the monitoring criteria for each test.
(iii) To assess the data against predetermined criteria. Most systems use software routines which determine the time to detection (TTD) or the detection time (DT). Each system has a default setting which is unique for the analyser. Most systems also permit the user to define and introduce specific detection routines for use with an individual medium or test. In addition, the Malthus system has software routines based on the rate of change and magnitude of response which are used for analysing changes after the DT. These are known as confirmation routines and give an excellent prediction of the type of organism which produced the response in a selective medium. The major advantage is that it removes subjectivity from the interpretation of the results.
(iv) To present data graphically, in text form or in abbreviated or long report format. Most systems also allow data to be transformed to other computer databases i.e. ASCII or Lotus so that the information can be presented in a customised format.
(v)   To store test details and data in libraries and archive information.

5.7 Spoilage assays

5.7.1 Total viable flora

Many assay protocols for determining total viable counts or their equivalents in foods and beverages have been described. Generally, instrument responses, usually in terms of detection times, have been plotted against the microbial load determined by conventional methods. With conductance measurements, the changes found are due to the metabolic or biochemical activities of the organisms present, which result in changes in the number and potential of the ion carriers present. These correlate well with the conventional count, but correlation coefficients of >0.9 between conventional and electrical assays are not expected due to the errors in conventional counts (Gibson, 1989). For example, a colony forming unit on a solid medium is assumed to arise from a single cell, but can be due to a clump of cells with perhaps up to 1000 cells present. In a liquid electrical assay, all these cells can contribute to the change recorded. Debris from the homogenate can mask or interfere with conventional counts. Gross debris in homogenates, for example, fatty lumps from beefburgers, are often deliberately discarded in removing aliquot samples for conventional assays to ease pipetting etc., but can be included in electrical assays provided they do not coat the electrodes. This can bias the assay. Microbes in foods are often in a microaerophilic respiration mode due to restricted access to oxygen. Exposing such organisms to air in plating can be damaging. Injured microbes are known to recover better and faster in liquid media with low oxygen levels. Care must be exercised in comparing novel and conventional assay data. There have been some presentations at scientific symposia reporting high correlation coefficients between the results from electrical assays done at, say 25°C, and conventional assays incubated at 37°C. Such correlations are spurious, as it is rare for the same flora to develop in extracts of foods at these temperatures, and should be disregarded.

As mentioned, detection times and conventional data are computed to yield a calibration curve. Ideally, such a curve should cover a wide range of microbial qualities, which are not always found in commercial samples. Some researchers have extended the quality range by storing the foods for longer than usual or by storing the foods at an elevated temperature. This latter practice requires careful evaluation, as the flora and its metabolic activity can change with storage temperature. Usually, metabolism becomes less efficient as the optimum temperature for growth is approached, so that relatively more ion carriers are produced. Thus, the practice tends to be fail-safe, indicating that the microbial quality is apparently poorer than it actually is.

The medium used for total counts is often a liquid version of that used for conventional assays. Gibson (1989) argued that the composition of the assay medium could be made closer to that of the food being analysed by either making the medium from the food, or by using relatively large volumes of the food homogenate as inoculum, which is diluted sufficiently to eliminate natural antimicrobials. Thus, the results obtained may be more meaningful with regard to the spoilage of the food being assayed. In traditional plating, perhaps only 20 mg of the food are plated and actually assayed. In electrical assays with say 10 ml volumes, 2 g can be assayed. Increasing the assayed sample size by 100 should improve the statistical aspects of relating the results of the sample to the population from which it has been taken.

The spoilage, and therefore quality, of some foods is often due to the development of a limited flora or even a single species. This is more common when the environmental conditions met by the flora are near the limit for microbial growth. An example is the spoilage of fruit juice by *Zygosaccharomyces bailii*. Because of its restricted metabolism, it can be difficult to find or devise a medium which yields electrical change. Various approaches have been made and some unusual compounds have been incorporated into media selected to be nutritionally poor in carbon or nitrogen. These include tartrates (Connelly *et al.*, 1988), organic buffers such as succinate (Owens *et al.*, 1985), electron acceptors such as trimethylamine oxide (Easter *et al.* 1982), etc. With media based on a single substrate yielding the electrical signal, there is the possibility of a food containing an atypical microbe which has lost the ability to produce the charac- teristic reaction. However, if the substrate is relevant to the spoilage, this is of no consequence. One way of overcoming this is to capitalise on a characteristic common to all forms of metabolism, the production of carbon dioxide, the other common metabolite being water. Carbon dioxide is measured in the so-called 'indirect' technique (section 5.10).

5.8 Detection and enumeration of indicator organisms

Indicator organisms have been used as a basis for determining the hygienic quality of foods and dairy products for many years. Traditional methods for detection and enumeration of indicator organisms are widely used for the testing of foods, dairy products, water and for environmental monitoring. Enterobacteriaceae, coliforms, *Escherichia coli* and enterococci are all used as indicator organisms, but their role is not always clear (Petitt, 1989). Tests for Enterobacteriaceae may be of value as indicators of failures during processing or even of post treatment contamination. Detection of the organisms in raw foods, i.e. meats and vegetables, may be more difficult to interpret. Indicator organisms are also used to highlight the potential presence of enteric pathogens, e.g. *Salmonella* spp.

The basis for using indicators is that (i) they are found in large numbers in raw sewage, and (ii) the greater the numbers of these organisms the greater is the likelihood of the presence of enteric pathogens. Hence, any method used for this purpose should give some assessment of the numbers of indicator organisms present. Coliforms and *E.coli* testing is accepted for monitoring drinking-water supplies and electrometric methods have been developed (Irving *et al.*, 1989). Similarly, the quality of shellfish is also based on the numbers of *E.coli* or faecal coliforms (EC Directive 491/91). Coliforms have traditionally been used by the dairy industry to indicate the hygienic quality of milk and other milk based products, but there is now a growing trend towards testing for *E.coli*. Whilst coliform testing of foods is well established, the use of Enterobacteriaceae as a more suitable indicator has been suggested (Mossel, 1985). There is an interna- tional trend to use *E.coli* as the definitive indicator organism in other food products.

Several media and methods have been developed for use with various impedance analysers. Most of these detect changes directly in the growth medium, but newer methods utilising indirect conductimetry have recently been described.

5.8.1 Enterobacteriaceae

Three of the instrument manufacturers produce special media for the detection of Enterobacteriaceae. The exact formulations of these media are confidential, but they are likely to include some peptone, glucose, a buffer, eg. Tris, and a detergent as a selective agent, which may be either bile salts or sodium lauryl sulphate. These media are optimised for each system and whilst direct comparisons cannot be made, an indication of performance is usually available following evaluations with internationally recognised methods. Petitt (1989) has reviewed Enterobacteriaceae testing and evaluated several media formulations. He developed an Enterobacteriaceae medium based on Brain heart Infusion broth. This was superceded by use of Malthus Columbia broth with added Triton X-100 and a selective supplement of vancomycin and novobiocin. This medium, incubated at 37°C, was reported to be superior to both a Malthus Enterobacteriaceae medium and a Bactometer Enterobacteriaceae medium (Petitt, 1989). In this study, regression analysis comparing the logarithm counts of Enterobacteriaceae on violet-red bile glucose agar (VRBGA) and detection times in the new medium showed reasonable results. The correlation coefficient was only 0.74 with results from tests on vegetable products, but was 0.88 with meat products.

Recently, Malthus Instruments Limited has produced a modified Enterobacteriaceae medium in disposable cells. It has been evaluated using solid foods homogenised in 0.1% peptone water to give a 1:10 dilution of the food product, and with undiluted liquid food and beverage. One ml of the 1:10 dilution was inoculated into 2 ml of the conductance medium and tests were incubated overnight in a Malthus 2000 analyser at 36°C. The samples were simultaneously tested by the ISO (1985) procedure. Details of the sample types and results are shown in Table 5.2. Most of the food products were naturally contaminated, but, due to the narrow range of counts with pasteurised milk, chocolate and milk powders, some samples of these foods were artificially inoculated with different levels of either *E.coli*, *Enterobacter aerogenes* or *Klebsiella aerogenes*.

Table 5.2 Correlation between a conductance method and the VRBG method for enumeration of Enterobacteriaceae in food products

Type of product[a]	Number of positive samples	Correlation coefficient
Pasteurised Milk (n=50)	49	0.95
Cheese (n=50)	31	0.86
Milk powder (n=50)	46	0.94
Ice cream (n=50)	45	0.90
Chocolate (n=50)	41	0.90
Pâté (n=50)	31	0.96
Cooked meats (n=50)	28	0.96
Seafoods (n=50)	41	0.90

[a] All products obtained from retail outlets.

The correlation coefficients for the food products are also shown in Table 5.2. Detection times for the same numbers of organisms varied depending on the food product. They ranged from 7.3–10.3 h with 10^2 cfu/g and 1.8–3.8 h with 10^6 cfu/g. There was evidence that some foods behaved similarly in the assay and could be grouped without significantly affecting the correlation coefficient, i.e. liquid milk and ice cream.

In another study (unpublished results), naturally contaminated animal feed products were evaluated by the above methods. The correlation coefficient produced in Figure 5.7 was 0.88, confirming that conversion of the detection times to numbers of Enterobacteriaceae per is a reliable estimate of the true count in this instance. The results of the analyses of 1015 samples of animal feeds are summarised in Table 5.3. Of the 55 positive for Enterobacteriaceae, all were detected by the conductance procedure and confirmed by the ISO procedures; 49 were positive by the VRBGA method. Numbers ranged from 10 to $>10^4$ cfu/g. For samples containing low numbers of organisms, the conductance method detected more positive samples, indicating that it is more sensitive. Since the samples were taken during the whole production cycle of animal feeds, it is likely that some samples, contained sublethally damaged organisms, but there was no evidence from the data to suggest that this affected the overall trend

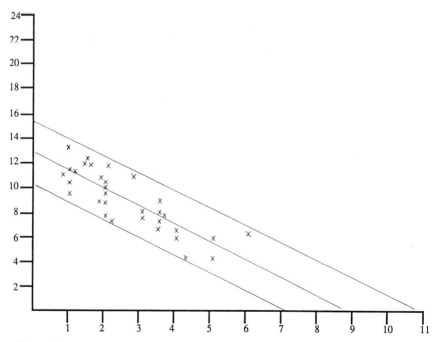

Figure 5.7 Calibration of conductance detection times with numbers of Enterobacteriaceae in animal feeds.
x axis, log no. Enterobacteriaceae count (VRBG);
y axis, detection time (h). This calibration curve was produced using 39 data points and gave a corre lation coefficient (*r*) of 0.879.

Table 5.3 Enumeration of Enterobacteriaceae in 1015 samples of animal feeds and proteins

Count determined by the VRBG agar method (cfu/g)	Number of samples in each group	
	VRBG	Conductance[a]
10	1	7
11–100	12	12
101–1000	13	13
1001–10 000	7	7
>10 000	16	16
Total	49	55

[a] Counts estimated from regression analysis.

of results. The conductance procedure is therefore a suitable screening method for monitoring plant and process hygiene during production of animal feeds.

5.8.2 Coliforms

Traditional methods for enumerating coliforms are either a most probable number (MPN) method or a pour plate technique using VRBA. These methods require dilutions of samples to be prepared, multiple tubes or plates to be inoculated, and incubation to be carried out at 30 or 37°C for 24 or 48 h. There is also a greater subjectivity in the interpretation of results and, since impedance methods resolve most of these problems, they have been readily adopted.

The first medium designed specifically for the detection and enumeration of coliforms by impedance was developed by Firstenberg-Eden and Klein (1983). The results of analysis of meat samples by a VRBA method and a Bactometer impedance method showed a correlation coefficient of 0.90. High numbers of coliforms i.e. >1000/g were detected in less that 6.5 h. Thus unsatisfactory samples were detected more rapidly than by conventional procedures. In a later study, Firstenberg-Eden and Eden (1984) demonstrated that this medium also gave reliable results with milk and mil-based products. Fryer and Forde (1989) used the same system to test powdered dairy products, e.g. skimmed-milk powder, cornflour and a starch product. They did not attempt to correlate detection time with numbers because the numbers of coliforms were always low. Hence, the authors used the procedure as a rapid screen to determine their presence or absence in 1 g or 0.1 g of product. In most cases, positive samples were detected by impedance within 12 h, in comparison with 48 h taken by the traditional enrichment method involving MacConkey broth.

Other analysers have also been used to detect coliforms. Irving *et al.* (1989) used both Bactometer and Malthus analysers for testing potable water. They used two media, one containing 0.1% sodium dodecul sulphate and the other that of Firstenberg-Edn and Klein (1983). Using the Bactometer, they reported that the capacitance signal showed the greatest changes, but also gave noisy signals. The impedance signal was found to be the most reliable. The Malthus analyser used in this study was a large volume cell (LVC) system. Each LVC was capable of testing 100 ml of sample, which is ideal for testing water samples. A special formulation called M20 medium was developed for these investigations. With this procedure, using conductance measurement, one coliform per 100 ml of water was detected in 12 h. However, to ensure complete absence, the test time was extended to 18 h. The RABIT analyser has also been used to determine coliform counts by both direct and indirect conductimetric methods. The direct method (using the manufacturer's coliform broth) was used to determine counts in pasteurised milk and proved to be a good alternative to the MPN method (Madden et al., 1990). The indirect conductimetric method used the same medium and the ability of coliform bacteria to ferment lactose, detected by a colour change in the medium, and to produce CO_2 which was detected by the negative conductance change in the indirect RABIT cell (Druggan et al., 1993). This method was evaluated with a variety of frozen meat products and bean sprouts. The highest counts were in frozen beefburger, which contained 87 cfu coliform/g (estimated by the VRBA method) and this was reproducibly detected by the analyser in about 9.6 h. Counts of 10 cfu/g of cottage pie were detected in 11.2 h, which confirms that this method is sensitive, rapid and a useful approach for the screening of food products.

A new medium has recently been developed by Malthus Instruments Limited, which detects coliform organisms by conductance. It contains lactose and a pH indicator which changes from purple to yellow to confirm the presence of coliforms. This medium has been tested with a range of naturally contaminated food products (unpublished results). Some samples of pasteurised milk, chocolate and milk powders were artificially inoculated with coliforms, to produce a range of counts covering 4-5 log cycles. In this study, the conductance tests were performed at 30°C and compared with the ISO (1978) method using VRBA incubated at 30°C. The results are summarised in Table 5.4. The correlation coefficients were all high and hence the detection time produces good estimates of the coliform counts. It is clear that separate calibration curves for each product are not needed. In this study, 10^2 cfu/g were detected in 8.8-11.0 h

Table 5.4 Correlation between a conductance method and the VRBA method for enumeration of coliforms in food products

Type of product[a]	Number of positive samples	Correlation coefficient
Pasteurised milk (n=50)	49	0.95
Cheese (n=50)	38	0.89
Milk powder (n=50)	43	0.98
Ice cream (n=50)	48	0.92
Chocolate (n=50)	41	0.97
Pâté (n=50)	39	0.91
Cooked meats (n=50)	49	0.96
Seafoods (n=50)	41	0.86

[a] All products obtained from retail outlets.

and 10^6 cfu/g in 2.4-4.7 h, depending on the product. Differences in results at low levels may reflect uneven distribution of coliforms in the products. Thus, correlation coefficients are only as good as the sampling allows, and distribution of organisms within the sample.

Regardless of the type of impedance analyser, these results confirm that these systems can be used to screen food products rapidly for the presence and enumeration of coliforms.

5.8.3 Escherichia coli

Methods for differentiation of *E.coli* from other coliforms are based on the ability of *E.coli* strains to produce acid and gas from lactose at 44°C, to produce indole at 44°C or ability to produce β-glucuronidase which can be detected by cleavage of chromogenic or fluorogenic substrates. These tests are usually used in conventional methods as confirmatory tests following a coliform screening procedure.

Impedance methods based on direct conductimetry using the Bactometer and Malthus analysers have been described (Easter and Kyriakides, 1991), but have not been widely used. Recently, an indirect conductimetric method has been described (Druggan *et al.*, 1993), which has been evaluated with meat-based products. Their method detects the presence or absence of *E.coli* by including methylumbelliferyl-β-D-glucuronide (MUG) in the medium. After the impedance assay, the tubes of media are exposed to ultraviolet light at 366 nm and examined for fluorescence. As part of this study, different numbers of *E.coli* were mixed with about 10^5 cfu/ml of *Klebsiella aerogenes*. They demonstrated that the lower limit of sensitivity of the fluorescence method was about 30 cfu *E.coli*/ml when in the presence of 10^4 cfu *Klebsiella aerogenes*/ml. The method successfully detected low numbers of *E.coli* in frozen meat products containing coliforms at 68-87 cfu/g.

A similar approach has been developed using direct conductimetry with a Malthus analyser. In this assay, 1 ml of 1:10 dilutions of food suspensions or supernatants from environmental swabs was inoculated into 2 ml of coliform medium containing MUG at 50 mg/l. It is thus possible to enumerate coliforms, confirm them by the colour change and then to test for the presence of *E.coli* by examining tubes for fluorescence. The results of these assays are shown in Table 5.5. The conductance protocol detected more samples containing coliforms than the VRBA method and the use of MUG indicated that 14 of 84 positive samples also contained *E.coli*. In this study, presence of coliforms and *E.coli* in the conductance tube was confirmed by ISO (1978) procedures. Thus, it is evident that MUG used in conjunction with impedance procedures is a reliable method for screening for the presence of *E.coli* in foods and environmental samples. The use of MUG in a traditional MPN system has been given approval by the Association of Official Analytical Chemists International (AOACI) (Feng and Hartman, 1982) and, therefore, its addition to media used in impedance systems should be acceptable.

Table 5.5 Detection of coliforms and *E.coli* by a conductance/MUG[a] assay on foods and swabs

Type of sample	Number of coliform positive		Number of *E.coli* positive	
	ISO method	Conductance	ISO method	Conductance with MUG[b]
Ready prepared meals (*n=36*)	6	9	0	2
Meats (*n=48*)	12	19	2	2
Sandwiches (*n=72*)	12	13	0	0
Fish (*n=15*)	6	8	0	2
Dairy (*n=45*)	1	1	1	0
Spices (*n=20*)	1	1	0	0
Environmental samples (*n=64*)	27	33	7	8
Total (*n=300*)	65	84	10	14

[a] MUG, methylumbelliferyl-β-D-glucuronide.
[b] Positive fluorescence.

A specific medium for *E.coli* has been developed (Ogden, 1993), which is based on the ability of *E.coli* strains to reduce trimethylamine-oxide (TMAO) and to utilise D-glucuronic acid. The test was done in a Malthus analyser at 37°C for up to 16 h. The medium has been evaluated with a range of food products, including raw and processed meats, cooked meats, shellfish and cheese. The correlation coefficient for these samples was 0.862 which, considering the range of food products tested, is acceptable. Ogden (1993) stated that the main advantages of the TMAO/glucuronic acid medium are that it is non-selective and it does not require incubation at an elevated temperature. Therefore, it allows the recovery of sub-lethally injured organisms. He also demonstrated that strains of *E.coli* O157 gave excellent response, which indicates that they can utilise the glucuronic acid even though they are known to give negative results with β-glucuronidase tests. In this study, some false-positive results were obtained by *Salmonella* spp. This medium requires further evaluation and comparison with approved traditional methods, but it promises to be a useful addition to the range of conductance media.

The above medium has been tested with some samples of shellfish and will be of interest to determine if the procedure can be used as an alternative to the MPN method described by West and Coleman (1986) for classification of bivalve shellfish. The EC Directive 491/91 states that shellfish can be classified according to the numbers of either *E.coli* or faecal coliforms, the distinction being that the latter produce gas in brilliant-green bile broth at 44°C, but are indole-negative at 44°C. A system utilising large volume cells in a Malthus incubator has been developed and validated in France (unpublished results). The procedure is being used routinely to estimate the numbers of faecal coliforms per 100 g of shellfish for classification of harvesting areas.

5.8.4 Enterococci

The presence of these organisms in heat-processed food and dairy products may indicate either inadequate pasteurisation or post-treatment contamination. Furthermore, under poor production conditions, enterococci may grow during the manufacture of dried milk powders. Neaves *et al.* (1989) evaluated several traditional media formulations for the detection of enterococci, using both Bactometer and Malthus analysers. From these studies they developed a medium containing urea, sodium azide and thallous acetate. A 1:10 dilution of the skimmed-milk powder was prepared in the above medium at 45°C (to help the sample to dissolve) and then inoculated into Bactometer modules (2 ml/well) and Malthus tubes (5 ml/cell). The assays were then incubated at 37°C for 24 h and the conductance responses monitored.

The results with artificially inoculated and naturally contaminated powders from conductance and conventional assays produced a correlation coefficient of 0.93, which indicates that the medium can be used to determine either presence or absence, or to enumerate enterococci in milk powders.

5.9 Detection of pathogens

5.9.1 Salmonella spp.

There have been numerous reports of media formulation for screening for the presence of salmonellas in food and feeds, probably because the potential payback from rapid automated assays more than outweighs the capital investment.

5.9.1.1 Easter and Gibson medium (Easter and Gibson, 1985) This medium uses sodium selenite as the selective agent, and metabolism of trimethylamine oxide (TMAO) to trimethylamine (TMA) to produce the conductance signal. Dulcitol is the primary carbon source. The medium arose from the authors' work on fish spoilage in which TMAO, an osmoregulant found in relatively large quantities in marine fish, is reduced to TMA. TMAO is a neutral uncharged molecule, while TMA is charged, being more basic than ammonia. There is a quantitative relationship between change in conductance and TMA production (Easter *et al.*, 1982). The electron donors for the reaction are NADH and formate and the former will arise from the fermentation of dulcitol, a property characteristic of most salmonellas. The concentration of selenite is 0.4%, but Leifson (1936), who originally suggested it, also used 0.2 and 0.8% for assays on sewage swab samples. All the salmonellas and foods tested by easter and Gibson (1985) gave large conductance changes in the instruments available to them (Malthus and Bactometer) and it is reported that the assay works well in the RABIT (Donaghy and Madden, 1992). Foods are pre-enriched in modified buffered peptone water containing appropriate inducers for the key enzyme substrates, dulcitol and TMAO, a strategy to reduce assay time. Short (6 h) pre-enrichments were found to be effective, but most users incubate overnight (16–20 h). Some positive responses were reported with some strains of *Citrobacter freundii*, which are closely related phylogenetically to salmonellas. Some dulcitol-negative salmonellas gave late or no responses in the medium. There has been an occasional instance of such results relayed to the originators of the medium formulation. Gibson (1987a) modified the pre-enrichment medium, to eliminate such results, by mannitol instead of dulcitol, as it proved to be a better inducer of dulcitol enzymes than dulcitol itself, even in apparently dulcitol-negative strains, and dimethylsulphoxide instead of TMAO, again a better and cheaper inducer. However, its reduced product, dimethylsulphide, has such an obnoxious smell that it is not used in many laboratories. All salmonellas tested have been TMA reducers. As this is a compound which in all probability salmonellas have never encountered, they may well induce the enzymes as part of a strategy for survival in anoxic conditions, as occur on broths etc.; some of the reductase enzymes are common to nitrite and nitrate reduction.

5.9.1.2 Lysine decarboxylation media. Ogden (1988) adopted a different approach to overcome problems with the above medium. From taxonomic data, he noted that most salmonellas could decarboxylate lysine and devised some media to exploit this. Of the selective agents he tested, selenite was the most successful. He also found that the best pre-enrichment medium was buffered peptone water containing glucose and lysine. It was then shown by Smith *et al.* (1989) that this recipe was also effective for inducing the enzymes for the TMAO-dulcitol medium, probably because the microaerobic conditions derepressed TMAO reductase, and glucose induced all of the enzymes for carbohydrate metabolism. Most salmonellas gave good conductance changes in this medium and there were no responses from strains of *Citrobacter* spp. However, some strains of *Hafnia alvei* gave responses in this medium, but not in the TMAO medium.

5.9.1.3 Easter and Gibson and Ogden media in combination. It has become a standard procedure to screen foodstuffs for the presence of salmonellas using the twin assays simultaneously. The conductance assays for *Salmonella* spp. using the Malthus instrument are the first automated machine-based system to have been given official status by the Association of Official Analytical Chemists International (AOACI), the methods authorisation body in the USA. This involved carrying out interlaboratory trials in accordance with their protocols, data analysis by their statisticians, and review and approval by various committees. Before a methods trial be undertaken, there are certain preliminaries to be done. First, of course, there has to be a method worth testing. The method was to be based on the assays published by Easter and Gibson (1985) and Ogden (1988), as described earlier. They have to be compared with the

AOACI reference method (Andrews *et al.*, 1992) which is laid down in their manual. The methods were tried by various laboratories who reported favourable experiences. Then a trial involving the analysis of a wide range of foodstuffs including naturally inoculated samples was done, the samples being analysed by both conductance and reference methods. In addition, the AOACI required that all positive results by the conductance assays be confirmed by conventional methods. The results for the 24 food types analysed are shown in Table 5.6.

Table 5.6 Comparison of AOAC and conductance methods for the detection of *Salmonella* in both inoculated and naturally contaminated food products

Product	Total tested	Positive AOAC	Malthus
Animal feed[a]	45	13	13
Duck eggs	50	43	40
Prawns	50	36	40
Minced beef	50	33	33
Skimmed-milk powder[a]	46	21	20
Entrée sauce	50	35	35
Broccoli	50	31	36
White sauce	100	51	59
Vegetable soup	50	29	36
Topping mix	50	34	30
Trout	50	33	28
Sausage meat	50	10	41[b]
Beefburgers	50	24	46[b]
Cream	50	40	38
Chicken	50	17	43[b]
Cocoa	50	12	15
Cake mix	100	11	5
Ground rice	100	12	9
Wheat flour	50	10	18[c]
Casein	100	13	0[b]
Dried yeast	100	13	4[c]
Mayonnaise	50	28	23
Pasteurised milk	50	38	40
Coconut	50	5	8
Totals	1441	592	660[b]

[a] Naturally contaminated product.
[b] Significant ($P < 0.01$).
[c] Significant ($P < 0.05$).

Of the 1441 samples tested, 592 were positive by the AOACI method and 660 by the Malthus assays. Most foods, except animal feeds and skimmed-milk powder (known as non-fat dried milk in the USA), were spiked with mixtures of various serovars of salmonellas (*S. enteritidis, S. typhimurium, S. aberdeen, S. agona, S. kentucky, S. chester, S. montevideo, S. thompson, S. anatum, S. eastbourne* and *S. infantis*) at either 1-5 cells per 25 g or 10-50 cells per 25 g. The Malthus method was more efficient at recovering salmonellas from the foods of high water activity, whereas the AOACI method was slightly better with some dried foods. These data were assessed by the AOACI who gave their approval for a collaborative trial to be undertaken. They required that 15 laboratories be included: 17 were recruited, but due to difficulties in distributing samples from the UK to the USA, some samples were tested by only 13 laboratories. Six foods were selected from those types used in the initial precollaborative trial. These were coconut, fish meal (animal feed), prawns, non-fat dried milk, liquid egg and minced (ground) beef. Preliminary trials were carried out to ensure that the laboratories could carry out both AOACI and Malthus assays. Some required training, especially for parts of the AOAC procedure, as it involved media and, for example, tube rather than slide agglutination techniques. The protocol was that summarised in Table 5.7.

Table 5.7 AOAC/BAM and Malthus conductance protocols

AOAC/BAM	Malthus
Pre-enrichment (16 h)	Pre-enrichment (16 h)
Culture in selenite and tetrathionate broths (24–48 h)	Culture in Malthus media (24–30 h)
Sub-culture onto selective agars	Sub-culture presumptive
(bismuth sulphite, Hektoen, XLD)	positives onto selective agars
Examine plates	
Further routine tests and serology	

Each laboratory received 15 samples of each food, five being controls, five being spiked at the low target level of 1–5 cells per 25 g, and five spiked at the higher target level of 10–50 cells per 25 g. The spiking serotypes were *S. agona* and *S. montevideo* for coconut, fish meal and non-fat dried milk, *S. enteritidis* and *S. typhimurium* for prawns and egg, and *S. typhimurium* for beef. The results are shown in Table 5.8. In all, 1245 samples were tested and 496 were positive by the AOAC procedure and 503 by the Malthus procedure confirmed by the AOAC tests. The methods were equivalent in that there were no significant differences in results obtained by either method. Laboratory performance was satisfactory. The full statistical analysis is shown in the report by GIbson *et al.* (1992), but these statistical methods were optimal for quantitative rather than qualitative data and for results derived from the same sample or pre-enrichment, that is, where there is a 'true' positive or negative test sample. In the collaborative trial, because the pre-enrichment step differed between the AOAC and the Malthus assays, two samples of 25 g were taken, one for each method of analysis. With the low spiking levels demanded by the AOAC protocols, it is quite conceivable for the portion tested by one method to contain *Salmonella* cells and that tested for the other not to.

Several other media formulations have been developed for use with the different analysers and some of these have been reviewed by Easter and Kyriakides (1991). Other lysine-based media have been developed for use with the Bactometer (Arnott *et al.*, 1988; Bullock and Frodsham, 1989; Blackburn,

Table 5.8 Detection of *Salmonella* in foods by AOAC/BAM and Malthus conductance methods

Food type	Level[a]	Samples analysed	Samples positive	
			AOAC/BAM	Malthus
Coconut	0	85	1	1
	L	75	54	54
	H	75	60	63
Fish meal	0	85	0	1
	L	80	22	29
	H	80	31	23
Prawns	0	65	0	0
	L	65	1	2
	H	65	17	15
Non-fat dried milk	0	85	0	0
	L	85	23	25
	H	85	79	81
Liquid egg	0	65	0	0
	L	65	57	60
	H	65	62	61
Minced beef	0	65	2	3
	L	45	45	45
	H	45	45	45

[a]L, low target level; H, high target level; 0, negative control.

1993) and have been used successfully with confectionery products and ingredients. Modifications to Ogden's (1988) original formulation have subsequently been reported by Ogden (1990) and Smith *et al.* (1990). These latter authors used the new formulation in a Malthus analyser and reported that it was highly satisfactory for testing animal feeds and proteins. Donaghy and Madden (1993) have evaluated an in house preparation of the Easter and Gibson formulation with a commercially available product using the RABIT analyser. These authors found that the impedance technique showed a detection rate (95%) equal to that of conventional enrichment for raw meats, but that, used singly, the medium was less sensitive than the traditional method for testing animal proteins. They reported that the commercially prepared medium was superior to the laboratory prepared medium, but that the false-positive level of 13.5% with the animal proteins was too high. Subsequently, these authors have described an indirect conductimetric assay using Rappaport-Vassiliadis (RV; Donaghy and Madden, 1993), an approach suggested previously by Bolton (1990). This technique could distinguish between *Salmonella* spp. and the closely related *Citrobacter* spp. and *Proteus* spp. which are known to give false-positive results in other conductance media. This technique was used to test processed animal protein and raw meats. Of the 73 samples tested, 37 were positive by the indirect RV method and 32 by the traditional RV method. This approach appears to be very promising, but additional studies are necessary to determine the sensitivity and specificity of the method for other food products.

An alternative approach has been described using a BacTrac analyser (Pless, 1993). A technique known as the impedance-splitting method (ISM) has been developed and monitors changes in the E values (section 5.5.2.). Using this method food samples were pre-enriched at 37°C for 14-16 h in peptone water supplemented with mannitol. A portion (0.1 ml) of the pre-enrichment culture was inoculated into 9.9 ml of a new medium (containing peptone, yeast extract, phosphate buffer, magnesium chloride (hydrated), malachite green oxalate and novobiocin) and incubated in a BacTrac analyser at 40°C for 22 h. A positive sample was indicated by a change in the E value of >15%. Of the 250 samples of poultry, eggs and minced meat tested, 122 samples were found to contain *Salmonella*. The ISM assay detected 119 of these positive samples, in comparison with 106 and 92 detected by conventional culture in selenite cystine broth and RV broth, respectively. Six of the 250 samples produced false-positive results using the ISM assay and on subculture these were all found to be due to *Enterobacter cloacae*.

The problems associated with high levels of false-positives with raw food products have recently been addressed by Malthus Instruments Ltd (Bolton and Powell, 1993). This company has produced a new twin tube assay with media dispensed in disposable cells. The procedure is outlined in Figure 5.8. Food samples are added to buffered peptone water containing glucose and lysine, as described in the AOACI approved method. These pre-enrichment cultures are then inoculated into Easter and Gibson's medium (1985) and Ogden's lysine medium (Ogden, 1988), both containing an additional antibiotic supplement. These media are incubated at 42°C and monitored for 30 h. This twin assay was evaluated using pure cultures of 52 different *Salmonella* serovars and gave a sensitivity of 100%. Additional testing with pure cultures of 56 different strains of Gram-negative and Gram-positive organisms showed that 1/5 strains of *Klebsiella* spp. tested gave a false-positive result. This resulted in a specificity of 96%.

The conductance assay was also used to test 338 samples of meat products comprising raw poultry meat, sausage meat, red meats, offal, burger meats, cooked meats and delicatessen products (Bolton and Powell, 1993). The method was evaluated by comparison with the ISO (1990) procedure and a modified USDA procedure, which used selenite cystine broth and tetrathionate broth (Difco 0104-17-6). All cells producing positive conductance responses and all selective enrichment media used in the two traditional procedures were subcultured onto modified brilliant-green agar (Oxoid CM329), Bacto bismuth sulphite agar (Difco 0073-01) and MLCB agar (Oxoid CM783) modified by the addition of 15 mg novobiocin/l. All suspect colonies were confirmed by biochemical and serological methods. The results are presented in Table 5.9. The new conductance procedure detected all 34 positive samples and gave an overall false-positivity rate of 4.1%. This improved selectivity is due to the combination of the antibiotic supplement and incubation at 42°C.

Following the AOACI approval for the Malthus *Salmonella* method and collaborative studies on milk-based products (Prentice *et al.*, 1989), the impedance/conductance methods have become widely adopted by industry and

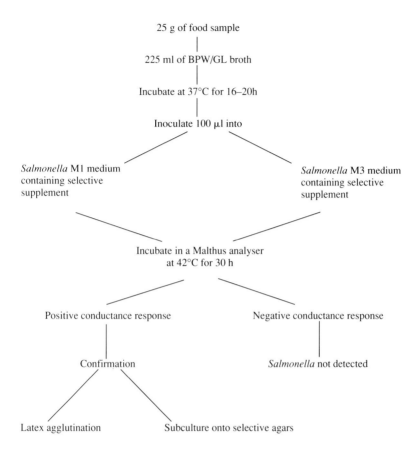

25 g of food sample

225 ml of BPW/GL broth

Incubate at 37°C for 16–20h

Inoculate 100 µl into

Salmonella M1 medium
containing selective
supplement

Salmonella M3 medium
containing selective
supplement

Incubate in a Malthus analyser
at 42°C for 30 h

Positive conductance response

Negative conductance response

Confirmation

Salmonella not detected

Latex agglutination

Subculture onto selective agars

Figure 5.8 Conductance protocol for detection of *Salmonella* spp. in raw foods.

Table 5.9 Detection of *Salmonella* in 338 samples of meat products by three methods

Result	Conductance assay	ISO procedure	USDA procedure
Positives	34	30	32
False–positives	14 (4.1%)	NA[a]	NA
False–negatives	0	4	2
Negatives	290	290	290

[a] NA, not applicable.

have now been given official approval by several international and national bodies. The Malthus procedure also appears in the Third Supplement 1992 to the 15th edition of Official Methods of Analysis (1990), AOACI and is listed as an acceptable method in the Bacteriological Analytical Manual (Andrews *et al.*, 1992). A list of approvals for impedance-based methods is presented in Table 5.10. Such approvals and recognition are necessary so that electrical methods can become established procedures and be used for regulatory purposes both nationally and internationally, and for intercompany trade.

Table 5.10 Approvals for impedance *Salmonella* methods

Approval body	Types of food product	Manufacturers[a]
AOAC	All products	Malthus Instruments Ltd
BSI	Dairy products	All
MAFF[b]	Animal feeds	All
USDA	Egg products	Malthus Instruments Ltd

[a] Malthus Instruments Ltd, Bactometer (bioMérieux), RABIT (Don Whitley Scientific Ltd), BacTrac (Sylabs, Austria).
[b] Processed Animal Protein Order 1989.

5.9.2 *Listeria spp.*

Listeria monocytogenes has emerged as an important foodborne pathogen and outbreaks of listeriosis have been associated with coleslaw, pasteurised milk, Mexican-style cheese, Swiss soft cheese (Anon., 1991). Other foods known to be associated with the transmission of *Listeria* include ice cream, cooked chicken, pâté,shellfish and raw vegetables (Anon., 1991). It may be unrealistic to have zero tolerance specifications for raw foods, but it has already been established by the World Health Organization (WHO) that packaged, heat-processed foods should be free of *L. monocytogenes* (Anon., 1991). Since *Listeria* spp. are frequently found in soil and the gastrointestinal tract of animals, it is not surprising that they contaminate food-production sites. Whilst it is important to test high risk foods for the presence of *L. monocytogenes* and sometimes to enumerate them, it is now accepted that the only way to control the problem is to reduce environmental contamination in food processing units. Ideally, therefore, methods should be able to detect *L. monocytogenes* in a variety of food products and *Listeria* spp. (particularly *L. monocytogenes* and *L. innocua*) in environmental samples.

Several investigators have used impedance systems for the detection of *Listeria* spp. Cordier (1989) used a Bactometer analyser to investigate the efficacy of disinfectants and detergents against *L. monocytogenes* and *L. innocua*. They also determined the activity of these antibacterial agents in the presence of milk and whey. Using a similar impedance analyser, Phillips and Griffiths (1989) evaluated 11 different selective formulations and demonstrated that a selective medium containing acriflavine, ceftazidine, nalidixic acid and aesculin gave good selectivity when testing pure cultures of *Listeria* spp. and other organisms commonly encountered in foods. They found that the capacitance signal was more reliable than the conductance signal.

Many of the traditional selective *listeria* media contain relatively high concentrations of lithium chloride and this can greatly affect the ability to measure impedance or conductance changes. Such high salt concentrations yield highly conductive media which may be outside of the measuring range of the analysers. This was overcome by using indirect conductimetry (section 5.10) and strains of *L. monocytogenes* were shown to give excellent responses using the RABIT (Bolton, 1990).

Methods for the detection of *Listeria* spp. in food samples and environmental samples have since been reported using Malthus analysers (Bolton, 1991; Bolton and Powell, 1992). In these studies, the primary enrichment broth for dairy and non-meat foods was *Listeria* enrichment broth (FDA formulation; Anon., 1988). Primary enrichment cultures were then inoculated into a selective detection medium which was incubated at 35°C. In this medium *Listeria* spp. including *L. monocytogenes* gave characteristic conductance responses. Of 189 foods, including cheese, milk, seafood and vegetables, tested by the conductance method and by the FDA procedure (Anon., 1988), 32 samples were found to be positive and all were detected by the conductance procedure.

More recently, Malthus Instruments Ltd have produced a *Listeria* testing kit consisting of a primary enrichment broth (PALCAM formulation) and a *Listeria* detection medium in disposable cells and selective supplement. The method is shown in Figure 5.9 and includes a 24 h primary enrichment step in PALCAM broth at 30°C, followed by inoculation into the *Listeria* detection medium. This is incubated in a Malthus analyser at 35–37°C for 30 h and only those cells which show a conductance response are subcultured to confirm the presence of *listeria* spp. Typical conductance curves are shown in Figure 5.10. The conductance procedure has been compared with the USDA method (Anon., 1989) for meat samples and environmental samples and the results are presented in Table 5.11. The meat samples comprised 45 raw red meats, 16 raw poultry, 34 cooked meats, 31 delicatessen and 26 pâté. Of these, 63 samples were positive and produced 89 isolates of *Listeria* spp. Interestingly, 46 out of 63 samples were naturally contaminated with *L. monocytogenes*. In this study, the conductance procedure gave results equivalent to those found by the USDA procedure.

In another study (unpublished results), 182 environmental samples were collected from processing units producing sandwiches, cheese, cooked meats and burger meat. Samples from utensils, containers, surfaces and drains were taken using sterile sponges (4 cm × 4 cm × 1 cm) moistened in a neutralising broth (peptone water 1 g/l, lecithin 7 g/l, Tween 80 5 g/l and sodium thiosulphate 6 g/l). These were placed into 30 ml of neutralising broth and transported to the laboratory within 2 h of collection. Each sample and neutralising broth was placed in a stomacher bag and stomached for 2 min. The expressed supernatant from the sponges was used for the comparison. For each sample, 10 ml of fluid were added to 90 ml of Malthus primary enrichment broth and 10 ml to the

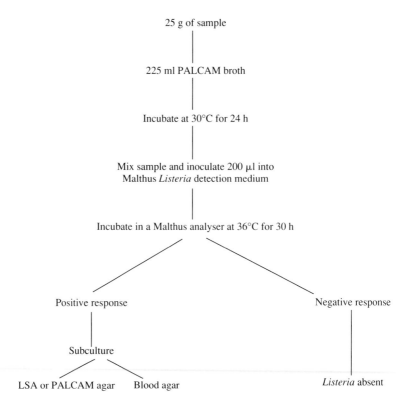

Figure 5.9 Conductance method for detection of *Listeria* spp. in meat products.

University of Vermont medium (UVM 1) broth for the USDA method (Anon., 1989). The remainder of the conductance procedure was identical to that shown in Figure 5.9. The results are shown in Table 5.11. Seventy eight samples were positive for *Listeria* spp. Both methods detected 70 positive samples and each missed eight positive samples. This was probably due to a combination of the low numbers of *Listeria* in the samples and the effect of taking two discrete volumes for testing in each of the enrichment broths, only one of which contained *Listeria* spp.

Recently, an impedance method has been described which enumerated *Listeria* spp. in artificially contaminated cheese (Hancock *et al.*, 1993). This method utilised a glucose-enriched nutrient broth supplemented with proflavine hydrochloride 21 mg/l and moxalactam 20 mg/l. The limit of detection in this medium was about 10^3 cfu/g of cheese. Currently, there are no results from testing of naturally contaminated samples. Some workers, but not regulatory

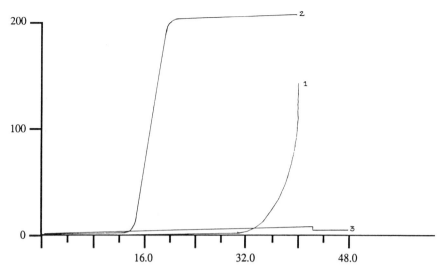

Figure 5.10 Conductance responses produced by *Listeria* spp. in Malthus *Listeria* detection medium. *x* axis, time (h); *y* axis, conductance (μS); (1) *L. monocytogenes*, (2) *L. innocua*, (3) negative sample.

Table 5.11 Detection of *Listeria* species in 152 samples of meat products and 182 environmental samples

Type of sample	Result	Conductance test	Traditional method[a]
Meat products	Positive	63	60
	False–positive	23 (15%)	22 (14%)
	Negative	66	70
Environmental samples	Positive	70	70
	False–positive	4 (2%)	7 (4%)
	Negative	100	97
	False–negative	8	8

[a] USDA.

bodies, have stated that it is not relevant to require zero tolerance of *L. monocytogenes* in all foods and that a quantitative estimation used in association with the risks of consuming the food is more pertinent. From the above mentioned studies it is clear that impedance methods can be used successfully to screen samples for *Listeria* spp. to meet regulatory requirements.

5.9.3 Campylobacter *spp.*

Campylobacter jejuni and *C. coli* are the major cause of bacterial enteritis in developed countries. Since the early 1980s, the number of human infections in the UK has exceeded those of salmonellosis. In the UK, there were about 35 000 cases of laboratory diagnosed *Campylobacter* infections during 1991 (Pearson and Healing, 1992). Infection with these organisms is primarily associated with consumption and handling of poultry products, although red meats, milk and water have all been implicated in outbreaks. However, the source of most of the sporadic cases is unknown.

Campylobacters are microaerophilic organisms and several selective enrichment broths and agars have been developed to recover them from faeces, food and water (Hunt, 1992). Isolation procedures can be labour-intensive, complicated and can take up to 4 days to obtain positive and negative results. There is, therefore, a need for simpler, more user friendly techniques which are automated and rapid.

A conductance protocol has been described using a malthus analyser (Bolton, 1991). The assay was a two-stage procedure. Initially, samples were inoculated into a selective primary enrichment broth incubated at 42°C for 24 h and then 100 μl were inoculated into 10 ml of a specially formulated detection medium which was incubated at 42°C for 30 h in the analyser. The conductance protocol was tested against a UK Public health laboratory Service method using Preston enrichment broth incubated at 42°C for 48 h and sub-cultured onto Preston agar (Bolton, 1991). The samples tested comprised chicken meat, chicken giblets, pork and beef. Of the 213 samples tested, 56 were positive for campylobacters, 48 being detected by the traditional method and 50 by the conductance method. The latter method gave a false-positivity level of <1% and had the advantage that presumptive positive and negative results were available 24 h sooner than with the traditional technique. The conductance method was also more sensitive than the traditional method for testing raw milk and naturally contaminated water samples (unpublished data).

A test procedure has become commercially available from Malthus Instruments Ltd using a similar two-stage approach (Bolton and Powell, 1993). The method is summarised in Figure 5.11. Meat samples are added to primary enrichment broth incubated at 37°C for 4 h and then at 42°C for 20 h. The medium is used in screw-top containers and has been designed so that it does not need to be incubated in a microaerobic environment. Following primary enrichment, samples were inoculated into the detection medium dispensed in disposable cells and incubated in a Malthus analyser at 42°C for 40 h. A special algorithm has been developed to determine positive and negative responses and has, therefore, removed the subjectivity associated with the interpretation of conductance responses. This procedure was evaluated against a FDA procedure using Humphrey's medium (Hunt, 1992). The results are summarised in Table 5.12. Out of 330 samples of meat products tested, comprising raw poultry meat,

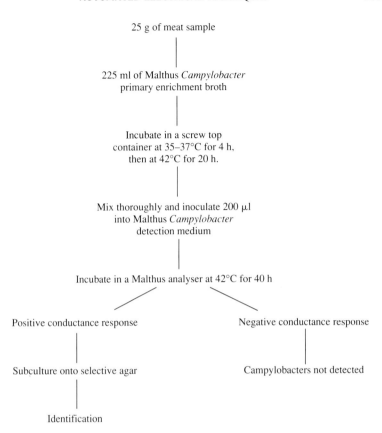

Figure 5.11 Conductance method for detecting *Campylobacter* spp. in meat products.

Table 5.12 Detection of 78/330 *Campylobacter*-positive samples by two methods

Result	Malthus	FDA
Number of positives	76	64
Number of false–positives	12	0

sausage meat, red meats, offal, burger meats and cooked meats, 78 were positive for campylobacters, 76 by the conductance method and 64 by the FDA procedure. This difference is statistically significant (McNemars test $P<0.01$). In additional studies using pure cultures of 163 strains of thermophilic campylobacters and 47 strains of non-*Campylobacter* organisms, the procedure was shown to have a sensitivity of 100% and a specificity of 97% (unpublished results). These qualities make the conductance protocol an ideal screening test for the presence of campylobacters in meat products.

5.9.4 Other pathogens

Electrical methods have been developed and used to investigate several other pathogens, e.g. *Staphylococcus aureus, Yersinia enterocolitica, Vibrio parahaemolyticus* and *Clostridium botulinum*.

Methods to detect and enumerate *Staphylococcus aureus* have been reviewed by Easter and Kyriakides (1991). Several traditional media formulations (Baird-Parker, mannitol-salt, tryptic soya, Vogel Johnson, Giolitti-Cantoni and Chapman Stone) have been evaluated, but proved to be unsatisfactory for detection by any of the measuring criteria. Some success has been described for meat products (Kahn and Firstenberg-Eden, 1985) with skimmed-milk powder (Prentice and Neaves, 1987) and cheese (Khayat *et al.*, 1988). These methods have not become widely adopted for screening general food products, primarily because of the failure to suppress competing flora.

Yersinia enterocolitica is a well-established pathogen of the gastrointestinal tract, but the number of cases in the UK is less than those reported in parts of Scandinavia and North America. Traditional methods for the isolation of *Y. enterocolitica* usually take longer than 7 days (Walker and Gilmore, 1986) and therefore more rapid procedures are necessary. A novel medium has been developed using a Malthus analyser, for the detection of *Y. enterocolitica* in raw milk (Walker, 1989). This medium is based on a peptone, mannitol, salt broth with urea and supplemented with cefsulodin, irgasan and novobiocin as selective agents. *Yersinia enterocolitica* is urease-positive and it is this property which results in the conductance change in the medium. The main advantage of the procedure is the ability to screen large numbers of samples within 48 h. There are no reports of this medium being used to test other food products.

A selective botulinum medium (SBM) has been evaluated for the detection of *Clostridium botulinum* (Gibson, 1987b). It was used in a Malthus analyser to investigate the growth of *Cl. botulinum* in pork slurries. Although there was a good correlation between plate counts and detection times, it was concluded that the conductance method did not give accurate estimates of the counts. It was, however, a useful method of determining whether a particular slurry sample contained growing *Cl. botulinum* cells. Furthermore, if used as a screening procedure, presumptive positives were recognised 24 h earlier than with traditional methods.

Vibrio parahaemolyticus is the major cause of food poisoning in Japan and is particularly associated with coastal fish and shellfish. Gibson *et al.* (1984) demonstrated the potential of the Malthus system to detect *V. parahaemolyticus* in artificially contaminated raw fish. The medium investigated was TCBS broth and in this medium *V. parahaemolyticus* present at 10^0–10^5 cfu/g assay were detected in 2–7 h. Competing organisms gave detection times longer than 8 h.

5.10 Indirect conductimetry

5.10.1 Principle

Detection of CO_2 by indirect conductimetry was first described by Owens *et al.* (1989). CO_2 is produced as the microbes metabolise carbohydrates and some amino acids. The gas dissolves in the aqueous phase until the solubility product is exceeded and the volatile gas diffuses into the head space. The CO_2 is measured by electrodes placed in an alkaline solution of KOH or NaOH. As the CO_2 is evolved, it reacts with the alkaline solution, forming carbonates whose conductivity is less than that of the original solution. This can be explained by the following reaction:

$$CO_2 + 2OH^- \rightarrow CO_3^{2-} + H_2O$$

The pH of the assay cultures is important with regard to detection times, as the amount of CO_2 in solution is governed by the Henderson-Hasselbach equation which relates to the ionisation constant of the carbonate formed in the culture, the pH and the relative amount of volatile CO_2 available. The original development work was performed using a Malthus large volume cell system, but, subsequently, the RABIT system has been adapted for indirect conductimetric assays (Bolton, 1990). Malthus Instruments Ltd have produced a 10 ml, disposable, indirect cell system.

A diagrammatic representation of the two systems is shown in Figure 5.12. Model A represents the Malthus system and Model B the RABIT system. A key factor in the design of the systems is that the measuring electrodes (placed in the

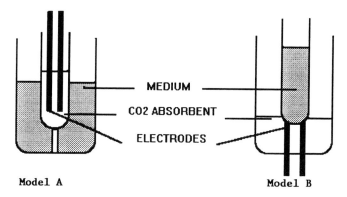

Figure 5.12 Indirect conductimetric assay systems. Model A, Malthus system; model B, RABIT system.

alkaline solution) are completely separate from the chamber holding the growth medium. The indirect conductimetric method has the following advantages over the direct conductimetric method.

1. Conventional media formulations can be used and do not need optimising to produce the conductance response.
2. Media containing high salt concentrations which are outside of the measuring capability of the direct method can be used, because the measuring electrodes are not in contact with the growth medium.
3. Some microorganisms produce very small or no detectable conductance changes by the direct method. These organisms, if they produce CO_2 (i.e. yeasts), can be detected by the indirect method.
4. It can be used with sample types which may physically interfere with the electrodes in the direct method.

Aqueous solutions of 0.03 and 0.04M KOH were used by Owens *et al.* (1989), a 0.04M KOH semi-solid medium was used by Bolton (1990) and a 0.06M KOH semi-solid medium by Druggan *et al.* (1993). Malthus Instruments recommend a 0.1M KOH aqueous solution for their assay system. It is essential that these alkaline solutions are heated to de-gas them and are stored in well-sealed glass bottles prior to use.

5.10.2 Detection of yeasts and moulds

The potential to detect yeast was demonstrated by Owens *et al.* (1989). These authors successfully detected strains of *Saccharomyces cerevisiae, Zygosaccharomyces bailii, Rhodotorula* sp. and *Torulopsis* sp. using growth in modified MCS medium (Owens *et al.*,1989). Subsequently, detection of yeast cells in fruit juice has been reported by Druggan *et al.* (1993). These authors tested one strain of *Zygosaccharomyces bailii* and three strains of *Rhodotorula* sp. in either wort broth or carbon base ammonium tartrate (CBAT) broth. All tests were performed at 30°C. With the *Rhodotorula* strains, 10^4 organisms were detected in 16.7–24.6 h, vs. 10^5 organisms detected in 13.3 – 21.3 h in the CBAT medium. Detection times for *Z. bailii* were inversely proportional to the number of organisms in the inoculum. Interestingly, with the *Z. bailii* the detection time was shorter in CBAT, from the lowest inoculum (*c.* 10^2 cfu/ml). The detection time was 46.5 h in comparison with 63.3 h in wort broth. This effect was not evident at higher inocula.

Similar studies with yeast have been performed using the Malthus system (unpublished results). In these investigations, yeasts such as *candida albicans, Saccharomyces bayanus, S. cerevisiae, Rhodotorula* sp. and *Z. bailii* were all detected rapidly in either CBAT medium or malt extract broth. Inocula containing $10^2 – 10^3$ cfu/ml gave detection times of 13.6 – 16.6 h.

Studies have also been conducted using the following moulds: *Aspergillus candidus, Aspergillus niger, Penicillium chrysogenum, Rhizopus stolonifer* and

Mucor circinelloides. The detection times produced following growth in potato dextrose (PD) medium and Sabouraud dextrose (SD) medium used as broths and as agar slopes at 30°C are shown in Table 5.13. The detection times produced by

Table 5.13 Detection times(h) of moulds tested by indirect conductimetry

Test strain	Potato agar	Dextrose broth	Sabouraud agar	Dextrose broth	Inoculum (cfu/ml)
Aspergillus candidus	18.3	ND[a]	18.2	ND	2000
Aspergillus niger	17.0	ND	18.0	ND	50
Penicillium chrysogenum	18.2	ND	18.6	44.6	3000
Rhizopus stolonifer	15.3	ND	15.1	ND	6000
Mucor circinelloides	10.6	ND	15.0	ND	900

[a]ND, not detected.

growth on agar slopes are 10.6 – 18.6 h, depending on the strain, vs. 43 h in broth media. Since fungi are fundamentally aerobic organisms, which proliferate on surfaces, perhaps these results with the agar slopes are expected, but an additional benefit is that the CO_2 produced by the growing mycelium is evolved directly into the headspace with little absorption into the solid medium. Therefore, surface growth methods used in conjunction with indirect conductimetry are optimal for the detection of fungi. A typical indirect response produced by an *Aspergillus niger* strain is shown in Figure 5.13.

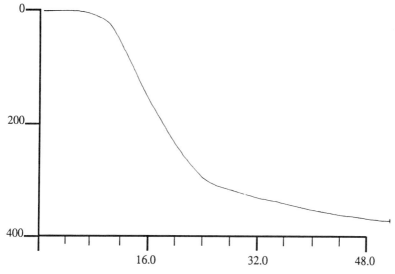

Figure 5.13 Indirect conductance response produced by *Aspergillus niger* grown on Sabouraud dextrose agar. *x* axis, time (h); *y* axis, conductance (μS).

5.10.3 Other applications

Bolton (1990) demonstrated the potential use of this technique for detection of a wide range of bacteria and the potential to develop detection methods using conventional media. Strains of *Staphylococcus aureus, Listeria monocytogenes, Enterococcus faecalis, Bacillus subtilis, E. coli, Pseudomonas aeruginosa, Aeromonas hydrophila* and *Salmonella typhimurium* were all detected in Whitley Impedance broth. More recently, indirect conductimetric assays have been described for the detection of other organisms in foods, i.e. coliforms, *E. coli, Clostridium tyrobutyricum* (Druggan *et al.*, 1993) and *Salmonella* (Donaghy and Madden, 1993). A more exciting application of this technique is for the assay of microbial biofilms (Holah *et al.*, 1990; Druggan *et al.*, 1993). These studies have shown that it is possible to determine the effect of biocide treatment on organisms in biofilms. This is unique because the assay monitors the response of cells in an intact biofilm and does not require their destruction to produce viable counts.

5.11 Future trends

This review of impedance assays has confirmed the wide range of applications now associated with this rapid and automated technology. Currently, there is no other technique which can perform quantitative assays for spoilage organisms and indicator organisms and can detect such a wide range of pathogens. This, coupled with the official approvals and recognition of the technology, has confirmed the acceptability of impedance monitoring systems. These systems are widely used in industry and the role of impedance microbiology in food-quality control has recently been reviewed (Arnott, 1993).

Perhaps the best is still to come. These systems are now being used to produce data for predictive models of spoilage organisms and pathogens in food. In particular, the indirect method is suitable for this task, because it can be applied to a wide range of organisms, at different temperatures, in media of differing a_w levels. This application is one originally suggested over 15 years ago (Richards *et al.*, 1978) and is illustrated by the work of Gibson (1985).

Impedance methods can also be used in conjunction with other rapid techniques to improve the speed, sensitivity and specificity of the assays. An immunocapture system and latex agglutination test have been used in combination with an impedance *Salmonella* assay (Bird *et al.*, 1990; also, refer to chapter 4). More recently, one of the authors (F.J.B.) has demonstrated that a nucleic acid probe method (Gen Probe) can be successfully used for rapid confirmation of *L. monocytogenes* from presumptive positive conductance assays. The use of such assays will undoubtedly become more common and as we progress into the next generation of impedance systems it is likely that we will see other novel and unique applications being described.

Acknowledgements

The authors would like to thank all of the manufacturers for their co-operation and for providing access to recent information.

References

Andrews, W.H., Bruce, V.R., June, G., Satchell, F. and Sherrod, P. (1992) Salmonella, in *FDA Bacteriological Analytical Manual* (7th edn), AOAC, Arlington, Virginia. pp. 51–69.

Anon. (1988) Revised method of analysis, in *Bacteriological Analytical Manual* (Food and Drug Administration). Federal register 53, Number 211, FDA, USA.

Anon. (1989) *FSIS Method for Isolation and Identification of* L. monocytogenes *from Processed Meat and Poultry Products* (Laboratory Communication No 57). USDA-FSIS. Microbiology Division, Beltsville, Maryland.

Anon. (1991) *Listeria monocytogenes*–Recommendations by the National Advisory Committee on Microbiological Criteria for Foods. *Int. J. Food Microbiol.*, **14**, 185–246.

Arnott, M.L. (1993) Impedance microbiology in food quality control, in *Instrumentation and Sensors for the Food Industry* (ed. Kress-Rogers, E.). Butterworth-Heinmann, Oxford, pp. 499–519.

Arnott, M.L., Gutteridge, C.S., Pugh, S.J. *et al.* (1988) Detection of salmonellas in confectionery products by conductance. *J.Appl. Bacteriol.*, **64**, 409–420.

Bird, J.A., Easter, M.C., Hadfield, S.G. *et al.* (1990) Rapid *salmonella* detection by a combination of conductance and immunological techniques, in *Rapid Microbiological Methods for Foods, Beverages and Pharmaceuticals* (eds Stannard, C.J., Petitt, S.B. and Skinner, F.A.). Blackwell Scientific, Oxford, pp 165–183.

Blackburn, D. ed W. (1993) Rapid and alternative methods for the detection of salmonellas in foods. *J. Appl. Bacteriol.*, **75**, 199–214.

Bolton, F.J. (1990) An investigation of indirect conductimetry for detection of some food-borne bacteria. *J. Appl. Bacteriol.*, **69**, 655–661.

Bolton, F.J. (1991) Conductance and impedance methods for detecting pathogens, in *Rapid Methods and Automation in Microbiology and Immunology* (eds Vaheri, A., Tilton, R.A. and Balows, A.). Springer-Verlag, Berlin, pp. 176–181.

Bolton, F.J. and Powell, S.J. (1992) Detection of *Listeria* in food and food processing environments. *Eur. Food drink Rev.*, **spring**, 43–48.

Bolton, F.J. and Powell, S.J. (1993) Rapid methods for the detection of *salmonella* and *Campylobacter* in meat products. *Eur. Food Drink Rev.*, **Autumn**, 73–81.

Bullock, R.D. and Frodsham, D. (1989) Rapid impedance detection of salmonellas in confectionery using modified LICNR broth. *J. Appl. Bacteriol.*, **66**, 385–391.

Cady, P. (1975) Rapid automated bacterial identification by impedance measurements, in *New Approaches to Identification of Microorganisms* (eds Heden, C.G. and Illeni, T.). Wiley, London, pp. 74–99.

Connolly, P.J., Lewis, S.J. and Corry, J.E.L. (1988) A medium for the detection of yeasts using a conductimetric method. *Int. J. Food Microbiol.*, **7**, 31–40.

Cordier, J.L. (1989) Impedimetric determination of activity of disinfectants and detergents on *Listeria*: preliminary study. *Int. J. Food Microbiol.*, **8**, 293–297.

Donaghy, J.A. and Madden, R.H. (1992) Impedance detection of *Salmonella* in processed animal protein and meat. *Int. J. Food Microbiol.*, **16**, 265–269.

Donaghy, J.A. and Madden, R.H. (1993) Detection of *Salmonella* in animal protein by Rappaport-Vassiliadis broth using indirect impedimetry. *Int. J. Food Microbiol.*, **17**, 281–288.

Druggan, P., Forsythe, S.J. and Silley, P. (1993) Indirect impedance for microbial screening in the food and beverage industries, in *New Techniques in Food and Beverage Microbiology*, (eds Kroll, R.G. and Gilmour, A.). Technical Series 31, Blackwell, London, pp. 115–130.

Easter, M.C. and Gibson, D.M. (1985) Rapid and automated detection of *Salmonella* by electrical measurements. *J. Hyg. Camb.*, **94**, 245–262.

Easter, M.C. and Kyriakides, A.L. (1991) Impedimetric methods for the detection of foodborne pathogens, in *Rapid Methods and Automation in Microbiology and Immunology* (eds Vaheri, A., Tilton, R.C. and Balows, A.). Springer-Verlag, berlin, pp. 490–502.

Easter, M.C., Gibson, D.M. and Ward, F.B. (1982) A conductance method for the assay and study of bacterial trimethylamine oxide reduction. *J. Appl. Bacteriol.*, **52**, 357–365.

Feng, P.C.S. and Hartman, P.A. (1982) Fluorogenic assays for immediate confirmation of *Escherichia coli. Appl. environ. Microbiol.*, **43**, 1320–1329.

Firstenberg-Eden, R. and Eden, G. (1984) *Impedance Microbiology*. Research Studies Press, Letchworth.

Firstenberg-Eden, R., Klein, C.S. (1983) Evaluation of a rapid impedimetric procedure for the quantitative estimation of coliforms. *J. Food Sci.*, **48**, 1307–1311.

Fryer, S.M. and Forde, K. (1989) Electrical screening of powdered dairy products, in *Rapid Microbiological Methods for Foods, Beverages and Pharmaceuticals* (eds Stannard, C.J., Petitt, S.B. and Skinner, F.A.). Blackwell Scientific, Oxford, pp. 143–153.

Gibson, D.M. (1985) Predicting the shelf-life of packaged fish from conductance measurements. *J. Appl. Bacteriol.*, **58**, 465–470.

Gibson, D.M. (1987a) Some modifications to the media for rapid automated detection of salmonellas by conductance measurement. *J. Appl. Bacteriol.*, **63**, 299–304.

Gibson, A.M. (1987b) Use of conductance measurements to detect growth of *Clostridium botulinum* in a selective medium. *Lett. Appl. Microbiol.*, **5**, 19–21.

Gibson, D.M. (1989) Optimization of automated electrometric methods, in *Rapid Microbiological Methods for Foods, Beverages and Pharmaceuticals* (eds Stannard, C.J., Petitt, S.B. and Skinner, F.A.). Blackwell Scientific, Oxford, pp. 87–99.

Gibson, D.M. and Jason, A.C. (1993) Impedance techniques for microbial assay, in *Instrumentation and Sensors for the Food Industry* (ed. Kress-Rogers, E.). Butterworth-Heinmann, Oxford, pp. 457–498.

Gibson, D.M., Ogden, I.D. and Hobbs, G. (1984) Estimation of the bacteriological quality of fish by automated conductance measurements. *Int. J. Food Microbiol.*, **1**, 127–134.

Gibson, D.M., Coombs, P. and Pimbley, D. (1992) Automated conductance methods for the detection of *Salmonella* in foods: collaborative study. *J. A. O. A. C.*, **75**, 293–302.

Hancock, I., Bointon, B.M. and McAthry, P. (1993) rapid detection of *Listeria* species by selective impedimetric assay. *Lett. Appl. Microbiol.*, **16**, 311–314.

Holah, J.T., Higgs, C. Robinson, S. *et al.*. (1990) A conductance based surface disinfection test for food hygiene. *Lett. Appl. Microbiol.*, **11**, 255–259.

Hunt, J.M. (1992) *Campylobacter*, in *FDA Bacteriological Analytical Manual* (7th edn). AOAC, USA., pp. 77–94.

Irving, T.E., Stansfield, G. and Hepburn, B.W.T. (1989) Electrical methods for water quality testing, in *Rapid Microbiological Methods for Foods, Beverages and Pharmaceuticals* (eds Stannard, C.J., Petitt, S.B. and Skinner, F.A.). Blackwell Scientific Oxford, pp. 119–130.

ISO (1978) *Enumeration of Coliforms-Colony Count Technique at 30°C* (ISO4832). International Standards Organisation.

ISO (1985). *Enumeration of Enterobacteriaceae* (ISO7402). International Standards Organisation.

ISO (1990) *General Guidance on Methods for the Detection of Salmonellae* (ISO.DIs6579). International Standards Organisation.

Jason, A.C. (1983) A deterministic model for monophasic growth of batch cultures of bacteria. *Antonia van Leuwenhoek*, **49**, 513–536.

Kahn, P. and Firstenberg-Eden, R. (1985) An impedimetric method for the estimation of *Staphylococcus aureus* concentrations in raw ground beef, in *1985 Annual Meat*. Institute of Food Technology, Atlanta.

Khayat, F.A., Bruhn, J.L. and Richardson, C.H. (1988) A survey of coliforms and *Staphylococcus aureus* in cheese using impedimetric and plate count methods. *J. Food Prot.*, **51**, 53–55.

Leifson, E. (1936) New selenite enrichment media for the isolation of typhoid and paratyphoid (*Salmonella*) bacillus. *Am. J. Hyg.*, **24**, 423–432.

Madden, R.H., Watts, M. and Gilmour, A. (1990) Impedance as an alternative to MPN for total coliform enumeration in pasteurised milk. *J. appl. bacteriol.*, **69**, xvii.

Mossel, D.A.A. (1985) Media for Enterobacteriaceae. *int. J. Food Microbiol.*, **2**, 27–32.

Neaves, P., Waddell, M.J. and Prentice, G.A. (1989) A medium for detection of Lancefield group D cocci in skimmed milk powder by electrometric methods, in *Rapid Microbiological Methods for Foods, Beverages and Pharmaceuticals* (eds Stannard, C.J., Petitt, S.B. and Skinner, F.A.). Blackwell Scientific Oxford, pp. 203–211.

Ogden, I.D. 91988) A conductance medium to distinguish between *Salmonella* and *Citrobacter* spp. *Int. J. Food Microbiol.*, **7**, 287–297.

Ogden, I.D. (1990) *Salmonella* detection by a modified lysine conductance medium. *Lett. Appl. Microbiol.*, **11**, 69–72.

Ogden, I.D. (1993) A conductance assay for the detection and enumeration of *Escherichia coli*. *Food Microbiol.*, **10**, 321–327.

Owens, J.D., Miskin, D.R., Wacher-Viveros, M.C. (1985) Sources of conductance changes during bacterial reduction of trimethylamine oxide to trimethylamine in phosphate buffer. *J. Gen. Microbiol.*, **131**, 1357–1361.

Owens, J.D., Thomas, D.J., Thompson, R.S. *et al.* (1989) Indirect conductimetry : a novel approach to the conductimetric enumeration of microbial populations. *Lett. Appl Microbiol.*, **9**, 245–249.

Pearson, A.D. and Healing T.D. (1992) The surveillance and control of *Campylobacter* infection. *Communicable Disease Report, CDR Review*. **12** R133–R139.

Petitt, S.B. (1989) A conductance screen for Enterobacteriaceae in foods, in *Rapid Microbiological Methods for Foods, Beverages and Pharmaceuticals* (eds Stannard, C.J., Petitt, S.B. and Skinner, F.A.). Blackwell Scientific, Oxford, pp. 131–141.

Phillips, J.D. and Griffiths, M.W. (1989) An electrical method for detecting *Listeria* spp. *Lett. Appl. Microbiol.*, **9**, 129–132.

Pless, P. (1993) Rapid detection of *Salmonella* by means of a new impedance-splitting method. *J. Food Prot.* (in press).

Prentice, G.A. and Neaves, P. (1987) Detection of *Staphylococcus aureus* in skimmed milk powder using the Malthus 128H microbiological analyser. *5th Int Symp Rapid Methods in Microbiology and Immunology*, Florence, Brixie Academic Press, Brescie.

Prentice, G.A., Neaves, P., Jervis, D.I. *et al.* (1989) An inter-laboratory evaluation of an electrometric method for detection of salmonellas in milk powders, in *Rapid Microbiological Methods for Foods, Beverages and Pharmaceuticals* (eds stannard, C.J., Petitt, S.B. and Skinner, F.A.). Blackwell Scientific, Oxford, pp. 155–164.

Richards, J.C.S., Jason, A.C., Hobbs, G. *et al.* (1978) Electronic measurement of bacterial growth. *J. Phys. E: Sci. Inst.*, **11**, 560–568.

Smith, P.J., Boardman, A. and Shutt, P.C. (1989) Detection of salmonellas in animal feeds by electrical conductance. *J. Appl. Bacteriol.*, **76**, 575–588.

Smith, P.J., Botton, F.J., Gaynor, V.E. *et al.* (1990) Improvements to a lysine medium for detection of *Salmonella* by electrical conductance. *Lett. Appl. Microbiol.*, **11**, 84–86.

Stewart, G.N. (1899) The changes produced by the growth of bacteria in the molecular concentrations and electrical conductivity of culture media. *J. Exp. Med.*, **4**, 235–247.

Walker, S.J. (1989) Development of an impedimetric medium for the detection of *Yersinia enterocolitica* in pasteurised milk. *IDF Seminar on Modern Microbiological Methods.* Santander, Spain.

Walker, S.J. and Gilmour, A. (1986) A comparison of media and methods for the recovery of *Y.enterocolitica* and *Y.enterocolitica*-like bacteria from milk containing raw milk microfloras. *J. appl. Bacteriol.*, **60**, 175–184.

West, P.A. and Coleman, M.R. (1986) A tentative national reference procedure for isolation and enumeration of *Escherichia coli* from bivalve molluscan shellfish by most probable number method. *J. Appl. Bacteriol.*, **61**, 505–516.

6 Modern methods for the detection of viruses in foods

S. H. MYINT

6.1 Introduction

Despite viruses being well-documented causes of foodborne outbreaks of illness, there has been a lack of interest in developing methods for detecting these important pathogens in the foodstuffs concerned. Documentation of actiology has, usually, been on epidemiological grounds, coupled with the specific viral diagnosis in those who succumb to illness. From health and economic viewpoints this must be short-sighted, as prevention of illness by food-testing must be a better approach. This chapter aims to review methods that have been used to detect viruses in food.

6.2 Background

During the years 1973-1987, there were 2841 outbreaks of foodborne disease with confirmed actiology reported to CDC, Atlanta, USA (Dean and Griffin, 1990). Of these, 25 (0.88%) were due to viruses, 15 attributed to Norwalk-like agents (see below). Similar prevalence figures are reported in the UK (Galbraith et al., 1987; Cooke, 1990). This almost certainly represents under-reporting due to the lack of utilisation of appropriate diagnostic facilities for viruses. During the same period there were 4167 outbreaks of unknown aetiology, many of which were likely to be viral in origin on clinical and epidemiological criteria. As diagnostic facilities for viruses improve, it is likely to become apparent that viruses will become recognised as, at least, of equal significance in morbidity associated with foodborne illness to the better studied bacterial pathogens.

Many foods have been incriminated as carriers of pathogenic viruses, but shellfish have been the single most commonly incriminated vehicle. Molluscs, such as mussels and oysters, take up virus when exposed to contaminated waters. With few exceptions, whatever the foodstuff, inadequate cooking or food-handling has been responsible for contamination. In the UK, the Ministry for Agriculture, Fisheries and Food (MAFF) introduced recommendations in 1988 that molluscs should be heated to, at least, 90°C for 90 s before consumption.

In theory, any virus that can enter the human body via the gastrointestinal tract can be transmitted by food. It is, however, those viruses that are found in faeces and, predominantly, cause local gastrointestinal illness that are most commonly associated with foodborne illness (Table 6.1). A common property of

Table 6.1 Viruses most commonly associated with foodborne illness

Norwalk and related viruses
Non-Norwalk Caliciviruses
Hepatitis A
Poliovirus
Echovirus
Coxsackie A and B
Other enteroviruses
Rotaviruses
Astroviruses (probably)
Enteric adenoviruses (probably)
Hepatitis E (probably)

these agents that allows them to be transmitted by this route is their ability to withstand the extremes of pH found in the gastrointestinal tract.

Norwalk and related agents are associated with 28–67% of major outbreaks of viral gastroenteritis (Hedberg and Osterholm, 1993), although only a fraction of these appear to be foodborne (Pether and Caul, 1983); nevertheless these agents are the main recognised cause of foodborne illness in the UK (Riordan, 1989). Norwalk virus was first identified from an outbreak of acute gastroenteritis in Norwalk, OH in 1968 (Kapikian *et al.*, 1972). They are small-round viruses with a diameter of about 28 nm. Under electron microscopy their surface structure is more amorphous than those of other small-round viruses in stools and this is used as a distinguishing feature. They possess a single major structural polypeptide with a molecular weight of, approximately, 60 000 and a single-stranded RNA genome. A number of closely related agents (collectively, they have been termed small-round structured viruses or SRSV) have been identified since 1968, most of which have also been named for the location of the associated outbreak: these include Snow Mountain, Hawaii and Otofuke agents. Classification of these agents is somewhat confused, as different monoclonals are used in the UK and USA for antigenic characterisation. At least two of these agents, Norwalk and Snow Mountain, have now had their nucleotide sequence determined. These data show regions of dissimilarity, which explains some of the confusion in their antigenic classification. It is also apparent from this nucleotide sequence data that they are very closely related to caliciviruses and may thus be reclassified in the near future (Lambden *et al.*, 1993).

Caliciviruses have an identical size and genome length to Norwalk-like viruses and result in cross-reacting antibodies to these agents after infection. They are, however, distinguished by characteristic cup-shaped depressions on their outer surface. Another group of recently discovered viruses which are similar in size, and more likely to cause foodborne illness than is currently recognised, are astroviruses. These are distinguished from the Norwalk and caliciviruses, using electron microscopy, by their 'star' shape. There are at least seven different serotypes of astrovirus.

Infection caused by Norwalk virus generally has an incubation period of 24–48 h and manifests as nausea, vomiting, diarrhoea and abdominal cramps. Spontaneous resolution occurs after 1–2 days. Caliciviruses cause an identical illness, but the age-distribution of those affected seems to be different: typically, caliciviruses infect infants and the elderly, but Norwalk has a predilection for older children and adults. Astroviruses have a similar clinical and epidemiological profile to caliciviruses, but the illness tends to be milder. All three viruses have a worldwide distribution and occur throughout the year.

Before the discovery of these small-round viruses, the most frequently reported agent of virally-induced foodborne illness was hepatitis A. This is a small non-enveloped single-stranded RNA-containing virus belonging to the Picornaviridae. It was formerly classified as enterovirus 72, but recently has been redesigned as the sole member of the hepatovirus genus. There is only one serotype. It is the commonest identifiable cause of faecal-oral transmitted infectious hepatitis. Infection tends to lead to an incubation period of 14–45 days, which is then followed by a prodromal illness of malaise, myalgia and headache that heralds jaundice; it should be noted, however, that those who are most susceptible, children under the age of 5 years, tend not to become jaundiced. Another cause of faecal-oral transmitted infectious hepatitis has recently been characterised: hepatitis E virus. This has caused large waterborne epidemics (Ray et al., 1991) and is likely to be a foodborne hazard. It causes an illness clinically indistinguishable from hepatitis A, but has a mortality rate of 20% in pregnant women (Bradley, 1992).

Other members of the Picornaviridae (enteroviruses, coxsackie A and B and polioviruses) have been associated with foods, but, with the exception of polioviruses, not necessarily with disease causation. A dozen or more outbreaks of foodborne poliomyelitis have been recorded, though not in the last decade in developed countries. A combination of widespread vaccination and improved hygiene are the likely explanations for the rarity of this potential source of transmission.

Rotaviruses are the commonest cause of virally-induced diarrhoeal illness in children. Transmission appears to be more by person-to-person and fomite transmission, as a result of poor hygiene, than by foods, though the latter occurs. Transmission by the waterborne route is better documented. These viruses are so-called because they look like a wheel ('rota' being Latin for wheel) and are larger than most other enteric viruses at 70 nm in diameter. They have a segmented genome of 11 double-stranded RNAs encased within a double-layered capsid. In common with all enteric viruses, there is no envelope. There are at least seven serotypes, but most human isolates are serotypes 1-4. Disease tends to occur in infancy, and may be severe; mortality is well documented. Reinfections also occur later in life, though the disease tends to be milder. Much work is currently being directed towards the development of a vaccine to these viruses.

6.3 Preparation of samples

6.3.1 General principles

Samples should be collected under 'clean' conditions and either refrigerated for processing within 48–72 h or stored at –70°C.

There are three main steps in sample processing.

1. *Liquefaction.* Virus needs to be separated from the foodstuff. This is most commonly achieved by release into solution prior to being grown in tissue culture or further manipulated for immunoassay or gene detection. Samples are liquefied by homogenisation, vigorous shaking, stirring or by stomacher into solution. A buffered solution is used to maintain integrity of the virus and this is particularly important if infectious virus is wanted for cell culture. The simplest, but most laborious, methods are grinding with a pestle and mortar (most effectively in the presence of sterile sand), and cutting with a scalpel knife. Automated homogenisers and ultrasonic baths are the most efficient and can allow the processing of large numbers of samples most easily. These devices are not inexpensive and produce aerosols which should be contained; they also require cleansing between sample batches.

2. *Clarification.* Virus released into solution is recovered by centrifugation, adsorption onto columns or filtration, to remove particulate food matter. The removal of food matter is particularly important for virus detection by cell culture, as most foods are toxic to standard cell lines used. More recently, chemical methods (ethyl ether or freon) and immunomagnetic separation have been used; the latter also acts to concentrate virus.

3. *Concentration.* Virus is usually present in low numbers in food samples and concentration may have to be effected. Most often required with large samples, this can be achieved by ultrafiltration, ultracentrifugation, or membrane dialysis (using an osmotic attractant such as polyethylene glycol and/or dextran).

6.3.2 Methods

The precise method used for processing samples depends on a number of parameters which include the type of foodstuff (volume of the sample, the consistency of the sample), the detection method used and the organism that is sought. Few researchers have attempted to evaluate directly one sample processing method against another and have used a single method that has worked (Bouchriti and Goyal, 1992). It is also clear from the published literature that the same method applied in different laboratories to the same food-type has had different success rates; in the author's experience it is best to try a number of methods and adapt one that gives the best recovery for a particular foodstuff and a specific virus. It is beyond the scope of this review to detail all the published methods, but some representative methods are given below for different foodstuffs.

6.3.2.1 Shellfish.   Shellfish contain components which are directly toxic to cell culture and a number of methods have been developed to reduce this toxicity.

6.3.2.1.1 Ethyl ether extraction (Metcalf and Stiles, 1965)   Shellfish are washed in cold running water to remove contaminating material. The meat is homogenised and particulate matter removed by centrifugation. The supernatant is then mixed with ethyl ether, kept overnight at 4°C and then recentrifuged. The aqueous layer can then be inoculated directly into cell culture or concentrated by ultracentrifugation, the pellet resuspended in a small volume of buffer, prior to inoculation into cell culture:

1. homogenise shellfish meat;
2. centrifuge at 2–8000 × *g* for 30 min;
3. add an equal volume of ethyl ether;
4. incubate at 4°C overnight;
5. recentrifuge; and
6. inoculate aqueous layer directly into cell culture or concentrate by ultra-centrifugation.

6.3.2.1.2 Freon extraction (Hermann and Cliver, 1968a)   Freon is a lipid solvent and is thus best not used for enveloped viruses; it has the inherent advantage, however, of disassociating antigen–antibody complexes and of removing most of the food components toxic to cell culture. Shellfish meat is homogenised in a buffer solution containing freon and either serum or bentonite. After centrifugation, clarified supernatant can be used directly in cell culture or, after concentration, by ultracentrifugation. This method has also been used for detecting enterovirus experimentally inoculated into cottage cheese, frozen chocolate eclairs, ground beef, potato salad, frozen chicken pie, raw carrots and frozen strawberries. The sensitivity of the method is such that, for a 50% probability of detecting virus, the food samples need to be contaminated with at least 3–4 pfu per sample:

1. place 25 g food in Sorvall 530 cup;
2. add 100 ml, glycine–sodium hydroxide pH 8.8 buffer:
   20 g magnesium chloride hexahydrate,
   10 g bentonite,
   100 ml freon;
3. homogenise with Omni-mixer;
4. centrifuge at 8000 × *g* for 30 min;

5. collect supernatant, repeat steps 2–4 if the supernatant is still turbid; and

6. use supernatant directly in cell culture or concentrate with polyethylene glycol (Carbowax 20 000, Union Carbide) inside dialysis tubing, and centrifugation at 275 000 × *g* for 2 h; discard supernatant and resuspend pellet in 0.5 ml phosphate-buffered saline with 2% agamma chicken serum.

6.3.2.1.3 Flocculation-filtration (Kostenbader and Cliner, 1972, 1973, 1981) Flocculation of food solids with polyelectrolyte solutions, followed by removal by filtration or by centrifugation, has been used successfully to recover 80–100% of 10^4 pfu enteroviruses/ml inoculated into shellfish and ground beef. The principle of the method is that food substances that can be toxic to cell culture are better removed than by centrifugation alone. A commercially available flocculant, Cat-floc (polydimethyldiallyl ammonium chloride, Calgon, Pittsburgh) appears to be superior to others available.

1. Add 20 g oyster tissue to 100 ml 0.09M glycine-sodium hydroxide buffer pH 8.8;

2. homogenise;

3. add 2 ml 1% Cat-floc;

4. stir for 15 min;

5. allow to settle;

6. filter through a series of cheesecloth, glass-fibre pre-filter and 0.45 µm membrane filter;

7. concentrate to 5 ml in Amicon 202 filter holder with PM30 membrane; and

8. the sample is ready for inoculation into cell culture.

6.3.2.1.4 Acid precipitation (Konowalchuk and Spiers, 1972) This has been shown to aid recovery of 50–60% of coxsackie B5 inoculated into oysters, clams and mussels.

1. Homogenise shellfish meat and liquor for 1 min;

2. centrifuge for 30 min at 15 000 rpm;

3. to the supernatant add hydrochloric acid to pH 3.0–3.5;

4. recentrifuge and adjust pH of supernatant to 6.6;

5. either concentrate by ultrafiltration or dilute 1:4 in foetal calf serum; and

6. inoculate into cell culture.

6.3.2.1.5 Adsorption with alkaline elution and ultrafiltration (Sobsey *et al.*, 1975). Virus is first adsorbed to homogenised oyster meat, eluted from this by adjustment of pH and then concentrated by ultrafiltration. Recovery rates of between 16 and 76% of inoculated virus have been achieved with this method. The precise pH used for elution and adsorption is dependent on the virus sought. The original basic method of Sobsey *et al.* (1975) is given below. In this experimental protocol, it is the oyster meat itself which is contaminated.

1. Homogenise 100–400 g oyster meat in distilled water (1:7 w/v);
2. adjust pH and conductivity to 5.5 and <1500 mg NaCl/l;
3. centrifuge at 1900 × g for 10 min;
4. discard supernatant;
5. resuspend precipitate in 0.05M glycine (pH 3.5, 5000 g NaCl/l).
6. recentrifuge;
7. adjust pH of supernatant to pH 7.5;
8. filter through 0.2 µm filter;
9. concentrate by ultrafiltration; and
10. sample ready for inoculation into cell culture.

6.3.2.1.6 Adsorption with alkaline elution and precipitation (Sobsey *et al.*, 1978, 1980; Ellender *et al.*, 19890; Sullivan *et al.*, 1984; Cole *et al.*, 1986). This is a modification of the above procedure which affords similar virus recovery rates, but is less time consuming. It has been, itself, modified by addition of a flocculant at the elution step and by removing the filtration step. A basic procedure without these modifications is given here and the reader is encouraged to refer back to the original reports for details of these modifications.

1. Homogenise 100–500 g oyster meat in distilled water (1:7 w/v) with added virus/contaminated material;
2. adjust pH and conductivity to 5.0 and <1500 mg NaCl/l,
3. centrifuge at 1900 × g for 10 min;
4. discard supernatant;
5. resuspend precipitate in 0.05M glycine (pH 7.5, 5000 g NaCl/l);
6. recentrifuge;
7. clarify by filtration through Millipore AP25 fibre-glass filter;
8. adjust pH of eluate to 4.5;
9. centrifuge and resuspend precipitate in 0.1M Na_2HPO_4, pH 7.4; and
10. use directly for cell culture.

6.3.2.1.7 Elution-precipitation (Goyal *et al.*, 1979; Landry *et al.*, 1980; Vaughn *et al.*, 1980; Richards *et al.*, 1982).   Virus is directly eluted from food, rather than adsorbed, and then concentrated. One recent comparative study has shown methods based on this technique to yield greater percentage virus recoveries than those that incorporate an initial adsorption step. A number of different methods based on this principle have been described. Two of these which have also incorporated flocculation with a polyelectrolyte are given here. The first of these can detect down to 1.5 pfu of virus/ml in shellfish, with a recovery of over 90%, the second yielded a 79% recovery of poliovirus from oysters.

Method 1:
1. homogenise 100–400 g oyster meat in glycine-sodium hydroxide buffer pH 9.5;

2. add 10 ml 1% Cat-Floc;
3. stir for 5 min then let stand for 15 min;
4. centrifuge at 144 000 $\times$ g for 15 min;
5. acid precipitate (see above) at pH 3.5 with 3% beef extract;
6. centrifuge at 10 440 $\times$ g for 15 min;
7. resuspend pellet in 15–20 ml 0.1M Na_2HPO_4, pH 9.5; and
8. add antibiotics and inoculate directly into cell culture.

Method 2:
1. homogenise oysters with an equal volume of water;
2. add 0.5% Cat-floc;
3. adjust pH to 9.0 with sodium hydroxide;
4. centrifuge at 1800 $\times$ g for 20 min;
5. dilute supernatant with water;
6. acid precipitate at pH 4.5 (see above) in the presence of 0.2% milk;
7. resuspend pellet in 0.15M Na_2HPO_4, pH 9.0; and
8. add antibiotics and inoculate directly into cell culture.

6.3.2.2 Animal products.

6.3.2.2.1 Comminuted (ground) meat. Many of the methods used for shellfish have been applied to animal meat but low recovery of virus has been common. Good (over 90%) recovery of poliovirus from comminuted meat has been achieved with a method based on freeze-thawing followed by filtration (adapted from Idziak and Srivasta (1975)):

1. take five 1 g samples of the meat and add to a plastic pouch;
2. store at −18°C overnight;
3. thaw at room temperature for 30 min;
4. freeze at −18°C for 4 h;
5. thaw at room temperature for 30 min;
6. divide into 5 1-g aliquots;
7. place each aliquot into a 17 $\times$ 100 mm tube with 4 ml MEM medium (containing non-essential amino acids, Hank's buffered salt solution without phenol red, 25 000 units penicillin G/ml, 25 mg streptomycin/ml and 250 units bacitracin/ml adjusted to pH 8.5 with NaOH);
8. shake manually, adjust pH to 8.0;
9. agitate in reciprocating shaker for 1 h;
10. store at −2°C for 16 h;
11. return to room temperature; and
12. filter; wash twice in MEM and use in cell culture.

An alternative and shorter method results in about 50% recovery of virus (Tierney *et al.*, 1973):

1. mix 100 g ground beef and 100 ml of HAMEM (MEM with non-essential amino acids in Hank's salts with 2% foetal calf serum and 2.5 mg/(ml) MgCl$_2$.6H$_2$O, 100 ug diethylaminoethyl-dextran sulphate, 4740 units penicillin G, 5000 ug streptomycin sulphate, 250 ug tetracycline hydrochloride, 5 ug amphotericin B, adjusted to pH 8.5 with NaOH) in a sterile plastic bag;
2. mechanically shake for 15 min;
3. pass contents through either 4 g glass wool or 4 g woven fibre-glass pretreated with HAMEM and placed in a glass funnel; and
4. inoculate clarified meat slurry into cell culture.

6.3.2.2.2 Pork samples. A simple method based on flocculation, clarification and then concentration with polyethylene glycol or ultrafiltration has yielded 53–68% recovery of poliovirus from pork samples (Deng, 1991). The best recovery utilised ultrafiltration and this method will be given here:

1. mince 20 g pork sample with scissors;
2. suspend in 100 ml ice-cold 0.09M glycine-NaOH, pH 8.8;
3. homogenise four times, 30 s each time, at 16 000 rpm in an ice bath;
4. add 2 ml 1% (w/v) Cat-floc;
5. stir for 5 min;
6. stand for 15 min at 4°C;
7. centrifuge at 8000 × g for 20 min at 4°C;
8. pass supernatant under vacuum through sterile G6 filter (<1.5 µm); and
9. reduce sample size to 5 ml using ultrafilter (Diaflo PM 30, Amicon) of 62 mm diameter and molecular weight cut-off of 30 kDa.

6.3.2.2.3 Milk. As most milk in developed countries is now pasteurised, this is not a foodstuff that would merit processing. A standard method based on simple filtration (Kalitina, 1978) has been used for the detection of inoculated poliovirus and an organic flocculation method (Hassen *et al.*, 1991) has been used to recover both poliovirus and echovirus. Recovery rates of between 26 and 46% are recorded.

Filtration
1. hold 100 ml raw milk in refrigerator for 24 h;
2. discard cream;
3. add 1–2 ml concentrated HCl to pH 4.4;
4. filter curd through three layers of cheesecloth manually;
5. add 3M NaOH to pH 7.0;
6. filter through Whatman GF/F filter, 110 mm diameter;
7. filter through 0.2 µm cellulose filter; and
8. inoculate into cell culture.

Organic acid flocculation:
1. add concentrated HCl to 500 ml milk to pH 4.5;
2. centrifuge at 3000 rpm for 10 min;
3. add 15 g extract of beef and ferric chloride to a final concentration of 0.5mM;
4. dissolve completely the beef extract;
5. add HCl to pH 3.5;
6. stir for 30 min;
7. centrifuge at 3000 rpm for 10 min;
8. resuspend pellet in 10 ml Na_2HPO_4, pH 9.5;
9. add antibiotics and adjust pH to 7.2; and
10. inoculate into cell cultures.

6.3.2.3 Vegetables and fruit. There is a paucity of published methods for processing these food types. Methods have, however, been described for recovering virus from lettuce (cliver *et al.*, 1983) and frozen strawberries (Hermann and cliver, 1968b). Neither of these methods has been used successfully in a 'field' setting, only with experimentally contaminated food.

Lettuce:
1. cut lettuce leaves into 2 cm squares;
2. add to 100 ml glycine–sodium hydroxide buffer pH 8.8;
3. add 0.02 g Cat-floc;
4. stir for 5 min, leave standing for 15 min;
5. filter through 11 cm Whatman GF/F filter under vacuum;
6. filter through 0.2 µm cellulose filter under vacuum;
7. concentrate by ultrafiltration (62 mm Amicon ultrafilter mol.wt. 30 000); and
8. inoculate into cell culture.

Frozen strawberries:
1. thaw 25 g strawberries;
2. add to 100 ml glycine–sodium hydroxide buffer pH 10.8;
3. add 20 g $MgCl_2.6H_2O$ and 10 g bentonite (final pH should be 8.5);
4. homogenise for 5 min in cold;
5. centrifuge at $8000 \times g$ for 30 min;
6. discard sediment;
7. extract for at least 10 h at room temperature against 30 g polyethylene glycol, mol.wt. 20 000, till volume is about 35 ml;
8. centrifuge at $275\,000 \times g$ for 2 h;
9. resuspend pellet in 0.5 ml phosphate-buffered saline with 2% foetal calf serum; and
10. inoculate into cell culture.

6.3.3 Immunomagnetic separation

Magnetic beads are coated with either antibody or gene probe. These are then incubated with liquefied sample so that binding of specific target takes place to the probe. Bound target can then be removed by magnetic separation of the beads. The virus or viral genome can then be detected by other means. This has been applied to the separation of bacteria from food and will, almost certainly, become a standard method for the processing of food samples for virus and viral nucleic acid. The technique encompasses separation, concentration and specificity of diagnosis (in that only a specific target is available for subsequent analysis). This method has been used with oligonucleotide-coated magnetic beads to recover, successfully, astrovirus nucleic acid from experimentally contaminated milk (unpublished data from the author's laboratory); the method is based on the manufacturer's instructions (Dynal, 1992) and so will not be detailed here.

6.4 Cell culture

As in diagnostic clinical virology, the 'gold' standard of virus isolation in food virology is isolation of virus in cell culture. Many of the viruses that cause 'food poisoning' will not grow in standard cell lines and strains and those that do tend to require quite fastidious cell lines. Table 6.2 lists those cell lines useful for the isolation of the common enteric viruses. The principles of cell culture for viruses from food are the same as for viruses from human material: best yields are obtained at a temperature of 37°C, in roller tubes and high multiplicity of infection. Early work on isolating viruses from food has also shown that an inoculum of ≤1 ml and an adsorption time to cells of 2 h also help to optimise virus isolation (Metcalf *et al.*, 1980). For many of these viruses, prior treatment with trypsin may be required before infection occurs. As already mentioned, it is important to have removed from inocula those substances in food which are toxic to cells.

Precise identification of the virus isolated may be inferred from specific cytopathic effect in a unique cell line, or may require further neutralisation tests. As most enteric viruses are fastidious and are grown in unusual cell lines, the latter confirmatory test is not routinely necessary.

As food microbiology evolved, it became evident that the mere presence of an organism did not necessarily imply that the food was hazardous to human health. The concept of 'infectious load' is clearly important and for many bacterial pathogens this is reasonably well defined. This is not so clearly the case with foodborne viruses, but it might be expected that quantitation of viral load in food may be more meaningful than mere detection. There are two basic quantitative assay methods used: quantal, where the titre is given as a 50% tissue culture infective dose ($TCID_{50}$) per unit volume; and plaque assays where a titre is

Table 6.2 Cell lines used for enteric viruses

Virus	Cell lines	Comments
Norwalk and caliciviruses	None	Cell culture is not yet available though there have been reports of successful adaptation of caliciviruses to growth in human embryonic kidney (HEK) cells.
Hepatitis A	Primary monkey kidney (PMK)	Not yet reliable.
Poliovirus	PMK MRC-5	Will grow in a number of primate and human embryonic lung fibroblast lines used commonly in diagnostic laboratories.
Echoviruses	PMK	Not all serotypes will grow, notably type 21.
Coxsackie A and B	PMK (for A7, 9 and 16) Rhabdomyosarcoma (RD) (for Coxsackie A but not all types) HEp-2 (for Coxsackie B) HeLa	As not all types can be grown in a single cell line, inoculation into infant mouse brain is often used for primary isolation.
Other enteroviruses	BGM cells	Probably the best single all round cell line for cultivating the widest range of enteroviruses.
Rotaviruses	LLC-MK2 IB-RS-2 MA104	Not routinely available and primary isolation not always successful with any of these cell lines.
Astroviruses	CaCo-2 cells	Colon cancer line. Poor for primary isolation. Used only in specialist laboratories.
	HT-29 cells	Same comments as for CaCo-2.
	HEK cells	Used by some researchers as cell line of choice for primary isolation.
Enteric adenoviruses	PMK Graham 293	Primate and human kidney cell lines. High multiplicity of infection and prior treatment with trypsin required.
	Chang conjunctival cells	Not routinely available.
Hepatitis E	None	Not yet determined to have foodborne transmission.

expressed as plaque forming units (pfu) per unit volume. These methods are described in standard textbooks of diagnostic virology (Lennette *et al.*, 1989; Lennette, 1994).

There is little published data on the minimal infectious doses of viruses, though the minimal infectious dose of polioviruses has been examined in humans. For poliovirus 1, it appears to be about 2 pfu/ml saline suspension, for poliovirus 2, 100 $TCID_{50}$, and for poliovirus 3, 1 $TCID_{50}$ (Larkin, 1980).

6.5 Immunoassays

Many immunoassays are now available for detecting foodborne viruses. These are dependent on the inherent ability of living systems to produce antibodies against foreign substances (antigens) which are specific for that antigen. It is the specificity of this interaction that makes it a useful diagnostic tool. These antigens can be the proteins, lipid constituents or nucleic acids of a virus; as the lipid constituent is host-derived, it is not useful as a means of detecting the virus.

Immunoassays can be in a number of method formats: enzyme-linked immunosorbent assay (ELISA), radioimmunoassay (RIA), immunodiffusion, immunoblotting, latex agglutination (LA) and countercurrent immunoelectrophoresis (CIE). Many of these methods have been applied to detect human enteric viruses in clinical specimens and for the detection of plant viruses in food crops (Tables 6.3 and 6.4). The methods used are not specific to foodborne agents and are described in standard manuals (Rose, 1992). Immunoassays offer ease of use, relative low cost and the feasibility of testing large numbers of specimens rapidly. They are, however, insensitive when compared to culture and gene amplification methods. Perhaps their biggest drawback is that they do not differentiate between 'dead' antigen and viable infectious organisms.

Table 6.3 Immunoassays used to diagnose common foodborne enteric viruses

Virus	Method
Norwalk	RIA
	Biotin-avidin immunoassay
	FIA
Caliciviruses	RIA
	EIA
Rotavirus	EIA
	Latex agglutination
	(CIE has also been used in the past)
Enteric adenoviruses	EIA
	RIA
Hepatitis A	RIA
	EIA
Hepatitis E	EIA

Table 6.4 Immunoassays used to detect viruses in food crops[a]

Virus	Method type	Food(s)
Beet necrotic yellow vein virus	ELISA	Beet
Cauliflower mosaic virus	RIA	Cauliflower
Citrus tresteza virus	Immunodiffusion	Citrus fruit
Cucumber mosaic virus	Immunodiffusion	Tobacco
Elongated potato virus	Immunodiffusion	Potato
Isometric plant viruses	Immunodiffusion	Cowpea, cucumber, bean, squash
Pea seed borne mosaic virus	Immunodiffusion	Pea
Potyviruses	Immunoblot	Maize, sorghum, sugar cane
Soybean mosaic virus	RIA	Soya bean
Zucchini yellow mosaic virus	Immunodiffusion	Squash

[a] Adapted from Samarajeewa et al. (1991).

6.6 Nucleic acid hybridisation

Non-amplification nucleic acid hybridisation methods have been developed for a number of viral and non-viral foodborne pathogens. They were developed in the hope that detection was more rapid than those based on culture. They are, however, limited by lack of sensitivity and ar unlikely to become routine in the era of gene amplification technology.

6.6.1 Principles

All replicative organisms, whether prokaryotes or eukaryotes, contain DNA, RNA, or both. The basic building bricks of these nucleic acids are the four nucleosides: adenosine (A), cytidine (C), guanosine (G) and either thymidine (T) in DNA or uridine (U) in RNA. RNA is, with few exceptions, made up of a single chain of phosphorylated nucleosides (nucleotides). DNA usually exists as a duplex with one chain being 'complementary' to the other. This complementarity arises from the specific binding that exists between A and T (or U) and C and G. Gene probes are nucleotide (nt) sequences, labelled to enable their detection, which are complementary to a sought nucleotide sequence. Hybrids can be formed that consist of two strands of DNA, two of RNA, or one of each.

There are five principal variants in any hybridisation reaction: the probe, the conditions of hybridisation, the hybridisation format, the labelling or detection system, and the target sequence(s).

6.6.1.1 Probes and probe selection. In principle, probes are designed from known nucleic acid sequences. They may be many hundreds of nucleotides in length or short sequences of only tens of nucleotides. The former tend to be products of cloned nucleic acid, the latter synthesised chemically. 'Long' probes are generally used in direct probe techniques and short oligonucleotide sequences used for gene amplification methods. In either case it is a prerequisite that the target sequence is known and unique to the agent sought. Long or 'full-length' probes have greater specificity because of their length and complexity. They can also be labelled more extensively, thus increasing the level of sensitivity. Synthetic oligonucleotides, however, may be easier and less costly to generate. They can also be synthesised in an already modified form, for example with detector molecule(s), and have a far faster rate of hybridisation. Most useful oligonucleotide probes are 18–30 bases in length, have 40–60% G+C, are without intrastrand complementary regions and have less than 70% homology with non-target regions. Computer software is now available to aid the design of these oligonucleotide probes. Probes may also consist of DNA or RNA. If RNA targets are sought, the latter type of probe is more sensitive.

6.6.1.2 Hybridisation conditions. These determine both the kinetics and stringency of hybridisation. The variables used are temperature, salt concentration, formamide concentration and the use of hybridisation accelerators. Other major factors influencing the rate of hybridisation are the temperature at which annealing is allowed to occur, the monovalent cation concentration and the viscosity of the milieu.

6.6.1.3 Hybridisation format. Essentially, there are two formats in common use. Single-phase hybridisation implies a liquid environment, and two-phase (or mixed-phase) hybridisation has either the probe or target bound to a solid matrix. Screening methods which use nucleic acid target bound to a solid matrix (a nitrocellulose or nylon membrane) are commonly used for screening large numbers of specimens: these are the so-called slot-blot and dot-blot hybridisation methods. Liquid phase hybridisation is used in the technologies employing gene or probe amplification (see below). Most non-amplification-based probe methods use a two-phase system.

6.6.1.4 Probe labels. The probe must be labelled so that when it is bound to the target it can be detected. Traditionally, radioactive labels (^{32}P or ^{35}S) have been used, but these are being replaced on safety grounds by labels such as alkaline phosphatase, biotin, digoxigenin and luminescent molecules which can be detected by enzymatic reaction, colorimetrically or by auto-radiography.

6.6.1.5 Target processing This is, often, the most critical step for a method to work and is particularly pertinent to foods. Proteins and, to a lesser extent, lipids are removed so that they do not interfere with the hybridisation reaction. The method employed must enable a good recovery of nucleic acid. Different methods are, usually, employed if the target nucleic acid is RNA or DNA, as RNA is more easily susceptible to degradation; most enteric viruses are, of course, RNA-containing. Initial steps in food sampling may involve any of the procedures outlined in the section on sample processing, but then further 'purification' of nucleic acid is required. There are two basic methods: the first involves digestion of proteins and then extraction with phenol; the second involves lysing the cells and recovery of nucleic acid by differential precipitation in salts.

6.6.2 Methods applied to foods

There are few nucleic acid hybridisation methods applied to the detection of viruses in foods described in the literature. A method applied to the detection of hepatitis A and rotavirus in shellfish has, however, been described and is representative of the methods that can be applied (Zhou *et al.*, 1991). In this method there was greater than 90% recovery of virus. The procedure given is that used for detecting hepatitis A using ssRNA probes produced as SP6 transcripts from a cloned vector:

1. add 10–100 g shellfish meat to 10 volumes 10% tryptose phosphate broth in 0.05M glycine pH 9.0–9.5;
2. homogenise in a Waring blender for 40 s at 20 000 rpm;
3. shake homogenate at 100 rpm in mechanical shaker;
4. sonicate for 2 min;
5. centrifuge at 15 000 $\times$ g for 30 min at 4°C;
6. discard pellet, adjust pH of supernatant to 7.2–7.4 with HCl;
7. add polyethylene glycol (PEG 6000) to a final concentration of 8% w/v;
8. stir for 2 h at 4°C;
9. centrifuge at 10 000 $\times$ g for 30 min;
10. discard supernatant, resuspend pellet in 15-20 ml 0.15M Na_2HPO_4, pH 9.0–9.5;
11. sonicate for 30 s;
12. shake for 20 min at 100 rpm;
13. centrifuge at 14 000 $\times$ g for 30 min;
14. adjust pH of supernatant to pH 7.4;
15. add penicillin (150 µg/ml), gentamicin (5 µg/ml) and kanamycin (100 µg/ml);
16. treat with 0.1% sodium dodecyl sulphate (SDS) and proteinase K (400 µg/ml) for 1 h at 37°C;
17. extract proteins with phenol–chloroform–isoamylalcohol (25:24:1) three times;
18. mix five parts of aqueous phase with two parts of 5% cetyl-trimethylammonium bromide (CTAB) in 0.4M sodium chloride;
19. incubate at room temperature for 15 min;
20. centrifuge at 23 300 $\times$ g for 30 min;
21. resuspend pellet in 1M NaCl;
22. extract with one volume of chloroform;
23. centrifuge at 7000 $\times$ g for 15 min;
24. repeat chloroform extraction if interphase present;
25. ethanol precipitate with three volumes of ethanol for at least 1 h at –20°C;
26. centrifuge at 12 000 $\times$ g for 30 min at 4°C;
27. resuspend pellet in 1 ml sterile distilled water;
28. treat with DNAase I for 15 min at 37°C;
29. denature nucleic acid (heat and quench on ice) and apply to dry nylon membrane pre-treated with 20 $\times$ SSC (1 $\times$ SSC+0.15M NaCl, 0.015M sodium citrate);
30. air-dry and bake at 80°C under vacuum;
31. prehybridise membranes for 2 h at 50°C in (4$\times$ SSC, 2.67 $\times$ Denhardt's solution, 0.1% SDS, 1.83mM EDTA, 83ug tRNA, 50% formamide);
32. hybridise in same buffer containing ^{32}P-labelled ssRNA probes at 50°C for 24–36 h;
33. wash filters three times in 3 $\times$ SSC/0.2% SDS, followed by three times in 2 $\times$ SSC at 50°C; and
34. dry and autoradiograph.

The major drawback of these direct probe methods when applied to foods is the lack of sensitivity. This method was only able to detect as few as 10^6 hepatitis A virions which is several logs higher than the required sensitivity. The author has found, however, that steps 1–27 of this method are suitable for subsequent detection of viral RNA by reverse transcription-polymerase chain reaction (RT-PCR) (unpublished data).

6.7 Gene amplification

Gene amplification techniques are of two principal types, based on whether target or probe sequences are amplified.

6.7.1 Target amplification

The polymerase chain reaction (PCR) has found widespread application in the few years since its introduction (Ehrlich *et al.*, 1991). The basis of the method is shown in Figure 6.1. Two oligonucleotide primers are allowed to anneal to regions flanking the target region of DNA, which has been dissociated into its two complementary strands. These primers are then extended by DNA polymerase. Using a thermostable DNA polymerase, it is possible to apply many temperature cycles of dissociation, annealing and primer extension without the addition of additional enzyme. Typically, 30–50 cycles are used and the entire process is automated in a heating block. After 30 cycles there is a million-fold amplification of target sequences and this product DNA can be visualised by gel electrophoresis. The length of the product is that of the two primers plus the intervening region. It is usual to confirm the specificity of the primer product by internal probing with an oligonucleotide.

'Nested' PCR involves two rounds of amplification, the second round utilising primers complementary to sequences produced in the first round. This adaptation increases the degree of amplification and specificity.

Amplification of RNA sequences, as in most enteric viruses, requires the step of reverse transcription to make cDNA copies which can then be amplified. Reverse transcription is usually achieved by an enzyme such as AMV or MMLV reverse transcriptase, but an alternative approach is to use a thermostable reverse transcriptase, such as rTth polymerase, which has both reverse transcriptase and DNA polymerase properties.

The specificity of priming has also been a problem, with priming of non-target sequences resulting in primer artefacts. Recent developments such as the addition of the thermostable polymerase at high temperature only, the so-called 'hot start', and the addition of DNA binding proteins may be the solution to this (Saiki *et al.*, 1988). The hot start also seems to reduce the incidence of 'primer–dimers', a product which consists essentially of primer and complementary sequences only.

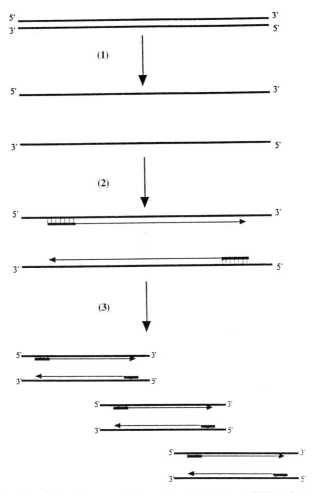

Figure 6.1 Principles of the polymerase chain reaction. (1) Heat target DNA to denature strands. (2) Add primers, nucleotides and thermostable Taq polymerase. (3) Repeat steps 1 and 2 n times.

However, the major problem with such an exquisitely sensitive technique is that of contamination (Ehrlich, 1989). This can result from previous PCR reactions (product carry-over) or stray DNA, particularly plasmids. The most important measure to prevent this is attention to strict handling procedures. These include separation of preparation areas from those for analysis of reaction products and clinical material or plasmid DNA, the use of dedicated and positive-displacement pipettes and pre aliquoting of reagents. In addition to these precautions, short-wave ultraviolet irradiation of the PCR reaction mix prior to addition of the template can be used to reduce or eliminate contaminating nucleic acids. Another strategy, which is commercially available, uses substituted nucleotides. Deoxyuridine triphosphate (dUTP) is used in place of deoxythymidine triphosphate (dTTP). In subsequent PCR mixes, the enzyme uracil-N-glycosylase (UNG) enables the excision of uracil from any contaminating DNA containing uracil. The resultant abasic polynucleotides cannot function as templates and are susceptible to alkaline hydrolysis at the high temperatures used in a PCR reaction. The enzyme does not excise uracil from RNA so this can still be used as template.

There have been a number of modifications of the basic PCR protocol, including quantitation by blotting and probing products, adaptation of the procedure to microtitre plates, and modification of primers so that colorimetric detection is possible. Commercial kits for PCR diagnosis are now becoming available. Perhaps the biggest obstacles that still need to be overcome are the cost and labour-intensiveness of these methods; some of these modifications may prove to be of help.

An alternative gene amplification technique is transcription-based, TAS (transcription-based amplification system), which is applicable to RNA targets (Kwoh *et al.*, 1989). A cDNA copy is made of the RNA target using an appropriate primer and reverse transcriptase. This primer is sequence specific and contains a polymerase binding site. This partial duplex is then denatured and a second primer and additional reverse transcriptase used for second cDNA strand synthesis. The resultant double-stranded DNA is then used as a template for T7 RNA polymerase, with the result that 10–100 RNA transcripts are produced. These RNA transcripts can then be used for a second cycle, and so on. The sensitivity of the method is of the same order as that of PCR, but, currently, is proving difficult to automate.

A modification of the TAS is the self-sustained sequence replication or 3SR (Guatelli *et al.*, 1990). It differs from TAS in that it is isothermal (37°C) and the RNA target is degraded. Advantages are the lack of requirement for temperature cycling and its greater speed than PCR.

6.7.2 Probe amplification

There are two systems that show promise. The first of these is based on the ability of an RNA-dependent RNA polymerase, Qβ replicase, to synthesise 10^6–10^9 copies of a template sequence (Lizardi *et al.*, 1988). The probe sequences are inserted into the multiple cloning site (MCS) of a Qβ phage-derived plasmid containing a T7 promoter. T7 RNA polymerase is then added to produce RNA transcripts containing the probe sequence. These are hybridised to their target, and after elution, amplified by the addition of Qβ replicase. This is several times more rapid than the thermal cycling of a PCR reaction. The other major advantage is that, because the amount of amplified RNA depends on the amount hybridised, the method is quantifiable. The major difficulty is non-specific hybridisation. Once this is overcome, the Qβ replicase system may offer a worthy alternative to PCR.

Perhaps the most exciting development since automated PCR is the ligase chain reaction (LCR; Barany, 1991). This technique has been developed consequent to the cloning of the thermostable enzyme, Taq ligase. The method relies on complete complementarity at the junction of two contiguous oligonucleotides hybridised to a target sequence. A single base mismatch means that ligation will not occur. The use of four oligonucleotides, two complementary to each target strand, and repeated cycles of denaturation, hybridisation and ligation results in multiple copies of ligated oligonucleotides. It has been shown that sub-picomolar quantities of lambda phage DNA can be detected by this method. Developments have paralleled those of PCR, but have been even more rapid. Labelling of one oligonucleotide with biotin, and the other with a reporter molecule, allows non-isotopic detection without electrophoresis. 'Nested LCR' utilises, as in its PCR equivalent, two rounds of amplification. This may improve the signal to background ratio. Perhaps the most significant development for the detection of infectious disease is a combination of PCR and LCR, 'PLCR'. Primary amplification of RNA or DNA is with PCR/RT-PCR with subsequent LCR. The exquisite sensitivity of PCR is combined with the specificity of LCR. Automation of this procedure has much potential for the outline detection of infectious agents.

6.7.3 Applications of PCR to foods

Only PCR has been applied to the detection of viruses in foods. Athough there are obvious, theoretical advantages, there are a number of practical difficulties that need to be overcome. The first is that there are substances in food which inhibit the PCR when above critical concentrations (Table 6.5). The second is that the quality, and quantity, of template is important in a successful PCR method; the number of steps commonly used in virus extraction of foods may lead to loss of viral nucleic acid and contamination of the final product with a number of inhibitory chemicals. Three methods for extracting nucleic acid from

Table 6.5 Components in foods inhibitory to PCR[a]

Substance	Inhibitory concentration (in 100 μl PCR mix)
Casein hydrolysate	1 mg
DNA	0.4 mg
L-Lysine	0.5 mg
Ovalbumin	0.25 mg
Sodium chloride	1 mg
Sucrose	10 μg

[a] Adapted from Rossen et al. (1992).

foods are given here. Two of these methods are based on polyethylene glycol concentration (Atmar, *et al.*, 1993; Yand and Xu, 1993) and have been applied to the detection of poliovirus and hepatitis A, the third has been used to extract DNA viruses (Andersen and Omiecinski, 1992).

6.7.3.1 Method to detect poliovirus in oysters.
1. Add 15–20 g of oyster meat to 160 ml 0.2M glycine–sodium chloride buffer pH 9.2;
2. add 4 ml antifoam B and blend at low speed four times for 30 s intervals;
3. add 60 ml ice-cold trifluoroethane and homogenise;
4. remove homogenate and wash container with 50 ml glycine–sodium chloride buffer, add this wash to the homogenate;
5. adjust pH to 9.5, add 1.5 ml Cat-floc T;
6. stir for 5 min and leave standing at room temperature for 15 min;
7. centrifuge at 11 000 × *g* at 4°C for 20 min;
8. harvest aqueous phase and adjust pH to 7.4;
9. add 75% (w/v) PEG-6000 in PBS to a final concentration of 8%;
10. stir for 2 h at 4°C;
11. centrifuge for 20 min at 11 000 × *g* at 4°C;
12. resuspend pellet in 0.15M disodium hydrogen phosphate pH 9.3;
13. adjust pH to 9.0 –9.5;
14. sonicate for 30 s;
15. shake for 10 min at 250 rpm;
16. sonicate again;
17. centrifuge for 10 min at 15 000 × *g* at 4°C;
18. adjust pH to 7.5;
19. digest with proteinase K (50ug/ml) in 10mM Tris–HCl pH 7.5, 5mM EDTA and 0.5% SDS at 56°C for 30 min;
20. extract twice with phenol–chloroform–isoamylalcohol (68:18:14);
21. ethanol precipitate aqueous phase and resuspend in sterile distilled water;
22. add CTAB and NaCl to 1.4% and 0.11M, respectively;
23. incubate at room temperature for 20 min;
24. centrifuge at 12 100 × *g* for 30 min;
25. resuspend pellet in 1M NaCl;
26. dilute in water and ethanol to 0.32M and 35%, respectively;
27. centrifuge at 12 000 × *g* for 6 min;
28. wash pellet three times in 65% STE buffer (0.01M NaCl, 0.05M Tris-HCl pH 7, 1mM EDTA) and 35% ethanol; and
29. resuspend in RT buffer for RT–PCR.

Using these conditions and a PCR cycle of 94°C for 60 s, 49°C for 90 s and 72°C for 60 s, the limit of sensitivity was 10 pfu (30–300 virions). The authors also describe PCR methods for hepatitis A and Norwalk virus with the use of the same nucleic acid extraction procedure.

6.7.3.2 Method to detect hepatitis A in contaminated clams.
1. Homogenise 20–50 g clam samples in 10 volumes of PBS;
2. centrifuge at 5000 rpm for 10 min at 4°C;
3. add PEG 6000 and NaCl to aqueous phase to 8% and 0.4M, respectively;
4. incubate overnight at 4°C;
5. centrifuge at 10 000 rpm for 20 min at 4°C;
6. resuspend pellet in 1/20th volume of 50mM Tris–HCl pH 7.6/140mM NaCl buffer;
7. vortex and extract twice with chloroform;
8. add 1/10th volume of 50% (w/v) trichloroacetic acid (TCA) to give final volume of 5% TCA;
9. mix thoroughly and boil for 10 min;
10. remove insoluble debris by centrifugation at 18 000 rpm for 20 min at 4°C;
11. add 1/10th volume sodium acetate pH 5.2 and 2.5 volumes ethanol;
12. precipitate and recover nucleic acid by centrifugation; and
13. use in RT-PCR.

After reverse transcription using AMV reverse transcriptase, PCR was carried out with 35 cycles of 93°C for 30 s, 55°C for 45 s and 72°C for 60 s. The authors claim that the method was more sensitive than a 'standard' culture method combined with immunofluorescence.

6.7.3.3 Method to detect bacterial plasmids in foods.
1. Add 25 g samples to 225 ml phosphate-buffered dilution water (62.5mM KH_2PO_4, pH 7.2);
2. homogenise for 2 min;
3. to 300ul homogenate, add 300ul lysis buffer (0.2M NaOH, 1% SDS);
4. vortex slowly for 10 min at 50°C;
5. neutralise with 300μl 2.55M potassium acetate pH 4.8;
6. centrifuge at 10 000 × g for 30 s;
7. pass 600 μl supernatant through 1 ml Magic Minipreps purification column;
8. wash column with 2 ml wash solution (100mM NaCl, 10mM Tris-HCl pH 7.5, 2.5mM EDTA, 50% ethanol);
9. elute DNA in 100μl 50mM NaOH;
10. neutralise with 100mM HCl and 10mM Tris-HCl pH 7.0; and
11. use in PCR reaction.

This method was applied to 14 different experimentally contaminated foods: water, bean sprout, lettuce, oyster meat, terrific broth, cooked shrimp, cheese, crab meat, coconut milk, tofu, dried mil and egg powder, mushroom and macadamia nuts. It was able to detect as few as 10^3 cfu/ml.

6.8 Other methods

Other methods used for the detection of viruses, such as electron microscopy and immune-electron microscopy, have been used by food researchers in the past, but primarily to compare them with better tests. They are unlikely ever to play a significant role in food virology as they are insensitive, a minimum of 10^6 virions/ml being required. The use of animal culture work is sensitive, but has, largely, been replaced by cell culture methods and is, in any case, considered by many to be too drastic for this type of work.

6.9 Why look for viruses in foods?

The principle that properly cooked food should not contain infectious micro-organisms applies to viruses as much as it does to other infectious agents (with the possible exception of prions). Routine testing of foods is also laborious and is not cost-effective, even more so when compared to bacteriological testing. As with diagnostic clinical virology, the results may also have taken a number of weeks, which does not compare favourably with the few days for bacteriology results. In the past, epidemics of food poisoning have led to attempts to discover the culprit in food samples, but, in practice, it has been more fruitful to seek the pathogen in those affected or food-handlers. The advent of gene amplification techniques may, however, increase the feasibility of testing foods for viruses. With this feasibility it is now important to identify situations in which food testing may be worthwhile.

1. Shellfish and other seafood; many bivalve molluscs are cultivated in sewage contaminated waters and are eaten uncooked. Uncooked, or lightly cooked, fish are a delicacy in parts of the world.
2. Other uncooked food; uncooked vegetables have been occasionally incriminated as vectors of foodborne disease. Uncooked meats, such as beef tartar, are a specialist taste and, as such, cases of related food poisoning are not well documented, but are a theoretical risk.
3. The sale of pre-prepared foods which may be handled after cooking is on the increase and there is little doubt that this will become a major source of viral food poisoning. The advent of cooked-chilled foods in hospitals and schools is likely to be a factor in this increase unless hygiene standards can be maintained.
4. The import of tinned and dried foods from countries with lower standards of hygienic practices is becoming more common. Sampling of these foods may prevent transmission of infection.
5. Although they are not viruses, prions can be handled in the laboratory in the same manner. They are resistant to standard cooking methods and with the increased awareness of causation of disease in man and their presence in foodstuffs, consideration will, no doubt, be given to testing for these agents in the future.

6.10 Future prospects

Food virology has been practised by only a handful of experts in the last few decades, notably in the USA and Australia. With the use of more reliable, rapid and sensitive techniques it will probably become part of the armamentarium of many public health laboratories. It can only be for the good that the methodology will advance rapidly in line with other fields.

References

Andersen, M.R. and Omiecinski, C.J. (1992) Direct extraction of bacterial plasmids from food for polymerase chain reaction amplification. *Appl. Environ. Microbiol.*, **58**, 4080–4082.

Atmar, R.I., Metcalf, T.G., Neill, F.H., *et al.* (1993) Detection of enteric viruses in oysters by using the polymerase chain reaction. *Appl. Environ. Microbiol.*, **59**, 631–635.

Barany, F. (1991) Genetic disease detection and DNA amplification using cloned thermostable ligase. *Proc. Nat. Acad. Sci. USA*, **88**, 189–193.

Bean, N.H. and Griffin, P.M. (1990) Foodborne disease outbreaks in the United States, 1973–1987: pathogens, vehicles and trends. *J.Food Prot.*, **53**, 804–17.

Bouchriti, N. and Goyal, S.M. (1992) Evaluation of three methods for the concentration of poliovirus from oysters. *Microbiologica*, **15**, 403–408.

Bradley, D.W. (1992) Hepatitis E: epidemiology, aetiology and molecular biology. *Rev. Med. Virol.*, **2**, 19–28.

Cliver, D.O., Ellender, R.D. and Sobsey, M.D. (1983) Methods to detect viruses in foods: testing and interpretation of results. *J. Food Prot.*, **46**, 345–357.

Cole, M.T., Kilgen, M.B. and Hackney, C.R. (1986) Evaluation of methods for extraction of enteric virus from Louisiana oysters. *J. Food Prot.*, **49**, 592–595.

Cooke, E.M. (1990) Epidemiology of foodborne illness: UK *Lancet*, **336**, 790–793.

Deng, M.Y. (1991) Processing pork samples for virus detection: a preliminary study of methods. *J. Food Prot.*, **54**, 28–31.

Dynal A/S (1992) Oslo, Norway. Manufacturer's instructions.

Ehrlich, H. (1989) *PCR Technology: Principles and Applications for DNA Amplification*. Stockton Press, New York.

Ehrlich, H.A., Gelfand, D. Sninsky, J. (1991) Recent advances in the polymerase chain reaction. *Science*, **252**, 1643–1651.

Ellender, R.D., Mapp, J.B., Middlebrooks, B.L. *et al.* (1980) Natural enterovirus and fecal coliform contamination of Gulf Coast oysters *J. Food Prot.*, **43**, 105–110.

Galbraith, N.S., Barrett, N.J. and Sockett, P.N. (1987) The changing pattern of foodborne disease in England and Wales. *Public Health*, **101**, 319–328.

Goyal, S.M., Gerba, C.P. and Melnick, J.L. (1979) Human enteroviruses in oysters and their overlying waters.

Guatelli, J.C., Whitfield, K.M., Kwoh, D.Y., *et al.* (1990) Isothermal, *in vitro* amplification of nucleic acids by a multienzyme reaction modelled after retroviral replication. *Proc. Nat. Acad. Sci. USA*, **87**, 1874–1878.

Hassen, A., Hachicha, R., Jedidi, N. *et al.* (1991) Une methode de recherche des enterovirus dans le lait. *Arch. Inst. Pasteur Tunis*, **68**, 261–268.

Hedberg, C.W. and Osterholm, M.T. (1993) Outbreaks of food–borne and waterborne viral gastroenteritis. *Clin. Microbiol. Rev.*, **6**, 199–210.

Hermann, J.E. and Cliver, D.O. (1968a) Methods for detecting food–borne enteroviruses. *Appl. Microbiol.*, **16**, 1564–1569.

Hermann, J.E. and Cliver, D.O. (1968b) Methods for detecting foodborne enteroviruses. *Appl. Microbiol.*, **16**, 1564–1569.

Idziak, E.S. and Srivasta, A.N. (1975) The isolation of virus from comminuted meat. *Can. Inst. Food Sci. Technol J.* **8**, 219–220.

Kalitina, T.A. (1978) A processing method for the detection of poliomyelitis virus in milk. *Vopr.*

Pitan., **6**, 70–73.

Kapikian, A.Z., Wyatt, R.G., Dolin, R. *et al.* (1972) Visualisation by immune electron microscopy of a 27 nm particle associated with acute infectious nonbacterial gastroenteritis. *J. Virol.*, **10**, 1075–1082.

Konowalchuk, J. and Spiers, J.I. (1972) Enterovirus recovery from laboratory–contaminated samples of shellfish. *Can. J. Microbiol.*, **18**, 1023–1029.

Kostenbader, K.D. and Cliver, D.O. (1972) Polyelectrolyte flocculation as an aid to recovery of enteroviruses from oysters. *Appl. Microbiol.*, **24**, 540–543.

Kostenbader, K.D. and Cliver, D.O. (1973) Filtration methods for recovering enteroviruses from foods. *Appl. Microbiol.*, **26**, 149–154.

Kostenbader, K.D. and Cliver, D.O. (1981) Flocculants for recovery of food-borne viruses. *Appl. Environ. Microbiol.*, **41**, 318–320.

Kwoh, D.Y., Davis, G.R., Whitfield, K.M., *et al.* (1989) Transcription-based amplification system and detection of amplified human immunodeficiency virus type 1 with a bead-based sandwich hybridisation format. *Proc. Nat. Acad. Sci. USA*, **86**, 1173–1177.

Lambden, P.R., Caul, E.O., Ashley, C.R. and Clarke, I.N. (1993) Sequence and genomic organisation of a human small-round structured (Norwalk-like) virus. *Science*, **259**, 516–519.

Landry, E.F., Vaughn, J.M. and Vicale, T.J. (1980) Modified procedure for extraction of poliovirus from naturally infected oysters using Cat-floc and beef extract. *J. Food Prot.*, **43**, 91–94.

Larkin, E.P. (1980) Foods as vehicles for the transmission of viral diseases, in *Indicators of Viruses in Water and Food* (ed. Berg, G.). Ann Arbor Science, Ann Arbor, Michigan, pp. 299–328.

Lennette, E.H. (1994) *Procedures for Viral, Rickettsial and Chlamydial Infections*, 8th edn. APHA, New York.

Lennette, E.H., Halonen, P. and Murphy, P.A. (1989) *Laboratory Diagnosis of Infectious Diseases: Principles and Practice, Vol. II. Viral, Rickettsial and Chlamydial Diseases*. Springer-Verlag, New York.

Lizardi, P.M., Guerra, C.E., Lomeli, H. *et al.* (1988) Exponential amplification of recombinant RNA-RNA hybridisation probes. *Biotechnology*, **6**, 1197–1202.

Metcalf, T.G. and Stiles, W.C. (1965) The accumulation of enteric viruses by the oyster, *Crassostrea virginica. J. Infect. Diseases*, **115**, 68–76.

Metcalf, T.G., Moulton, E. and Eckerson, D. (1980) Improved method and test strategy for recovery of enteric viruses from shellfish. *Appl. Environ. Microbiol.*, **39**, 141–152.

Pether, J.V.S. and Caul, E.O. (1983) An outbreak of food-borne gastroenteritis in two hospitals associated with a Norwalk–like virus. *J. Hug.*, **91**, 343–350.

Ray, R., Aggarwal, R., Salunke, P.N., *et al.* (1991) Hepatitis E virus genome in stools of hepatitis patients during large epidemic in north India. *Lancet*, **338**, 783–784.

Richards, G.P., Coldmintz, D. and Babinchak, L. (1982) Rapid method for extraction and concentration of poliovirus from oyster tissues. *J. Virol. Meth.*, **51**, 285–291.

Riordan, T. (1989) Virus gastroenteritis. *PHLS Microbiol. Digest*, **6**, 130.

Rose, N.R. (1992) *Manual of Clinical Laboratory Immunology.* ASM, Washington.

Rossen, L. Norskov, P. Holmstrom, K. *et al.* (1992) Inhibition of PCR by components of food samples, microbial diagnostic assays and DNA-extraction solutions. *Int. J. Food Microbiol.*, **17**, 37–45.

Saiki, R.K., Gelfand, D. Stoffel, S. *et al.* (1988) Primer-directed enzymatic amplification of DNA with a thermostable polymerase. *Science.*, **239**, 486–494.

Samarajeewa, U., Wei, C.I., Huang, T.S., *et al.* (1991) Application of immunoassay in the food industry. *Crit. Rev. Food Sci. Nutr.*, **29**, 403–434.

Sobsey, M.D., Wallis, C. and Melnick, J.L. (1975) Development of a simple method for concentrating enteroviruses from oysters *Appl. Environ. Microbiol.* **29**, 21–26.

Sobsey, M.D., Carrick, R.J. and Jensen, H.R. (1978) Improved method for detecting enteric viruses in oysters. *Appl. Environ. Microbiol.*, **36**, 121–128.

Sobsey, M.D., Hackney, C.R., Carrick, R.J. *et al.* (1980) Occurrence of enteric bacteria and viruses in oysters. *J. Food Prot.*, **43**, 111–113.

Speirs, J.I., Pontefract, R.D. and Harwig, J. (1987) Methods for recovering poliovirus and rotavirus from oysters. *Appl. Environ. Microbiol.*, **53**, 2666–2670.

Sullivan, R., Peeler, J.T., Tierney, J.T., *et al.* (1984) Evaluation of a method for recovering poliovirus 1 from 100–gram oyster samples. *J. Food Prot.*, **47**, 108–110.

Tierney, J.T., Sullivan, R., Larkin, E.P. *et al.* (1973) Comparison of methods for the recovery of virus inoculated into ground beef. *Appl. Microbiol.*, **26**, 497–501.

Vaughn, J.M., Landry, E.F., Thomas, M.Z., *et al.* (1980) Isolation of naturally occurring enteroviruses from a variety of shellfish species residing in Long Island and New Jersey marine embayments. *J. Food Prot.*, **43**, 95–98.

Yang, F. and Xu, X. (1993) A new method of RNA preparation for detection of hepatitis A in environmental samples by the polymerase chain reaction. *J. Virol. Meth.*, **43**, 77–84.

Zhou, Y–J., Estes, M.K., Jiang, X., *et al.* (1991) Concentration and detection of hepatitis A virus and rotavirus from shellfish by hybridisation tests. *Appl. Environ. Microbiol.*, **57**, 2963–2968.

7 Luminescence techniques for microbiological analysis of foods

A.L.KYRIAKIDES and P.D.PATEL

7.1 Introduction

Luminescence techniques based on the detection of light generated by enzyme-mediated reactions represent some of the fastest assays currently available for the detection of microbial contaminants and are some of the few rapid technologies to be currently utilised extensively in the food industry. Bioluminescence is a phenomenon of widespread occurrence, observed in a number of ocean-dwelling bacteria, some insects and a variety of other organisms. Techniques based on this phenomenon can be divided into two broad categories: (i) ATP bioluminescence and (ii) bacterial bioluminescence. Chemiluminescence, on the other hand, has been widely exploited in the clinical field, particularly in the study of cellular luminescence and diagnostic immunoassays utilising chemiluminescent labels such as lucigenin and acridinium esters (Weeks and Woodhead, 1988). Both chemiluminescent labels and chemiluminescent substrates, e.g. AMPPD (Patel, 1991) have also recently been used for the detection of the foodborne pathogens *Listeria monocytogenes* (nucleic acid probe) and *Salmonella* (enzyme immunoassay), respectively. This chapter focuses on the use of bioluminescence techniques for food microbiological analysis.

ATP bioluminescence assays have been available for many years, but until recently have not found widespread use, due to a number of factors, including poor specificity, poor reagent stability, high reagent cost and high capital cost of instrumentation. The recent resurgence of ATP bioluminescence has been principally due to the exploitation of its most potent application, the rapid assessment of hygiene or cleaning efficiency of food contact surfaces. Development of reagents generating comparatively stable light signals, introduction of low-cost luminometers and provision of training and technical support by commercial reagent and instrument manufacturers have all assisted in its success.

Bacterial bioluminescence is several years behind its ATP counterpart and few assays are commercially available. Potential applications for this specific, sensitive and rapid technique are wide ranging and the development of commercial protocols is widely anticipated.

To achieve sensitive detection, ATP and bacterial bioluminescence have a common need for instrumentation capable of detecting low amounts of light. A range of photometers or luminometers is commercially available for this purpose and a number of them will be referred to in this chapter, but, for a more detailed assessment of these systems, the reader is referred to an excellent review published by Stanley (1992).

This chapter aims to review current and future applications of both ATP and bacterial bioluminescence, placing particular emphasis on commercial systems that have exploited the strengths of these technologies.

7.2 ATP bioluminescence

ATP bioluminescence is based on a reaction that occurs naturally in the North Amercian firefly, *Photinus pyralis*. The biochemistry of the bioluminescence reaction was first described in the late 1940s (McElroy, 1947). The reaction, catalysed by the *Photinus* enzyme, luciferase, uses the chemical energy contained in the energy-rich molecule ATP to drive the oxidative decarboxylation of luciferin with the resultant production of light. The detailed mechanisms of this rapid reaction have been comprehensively reviewed (DeLuca and McElroy, 1978) (Figure 7.1). Peak light output occurs within 0.3 s and, because almost one photon of light is emitted for every ATP molecule consumed, a linear relationship exists between the concentration of ATP and light output, over a wide range of ATP concentrations (Stanley, 1982). Living cells, including microorganisms, use ATP as a major carrier of chemical energy and since the luciferase enzyme provides a mechanism for the assay of this ATP, a rapid system exists for detecting microorganisms. Although the level of ATP within living cells can vary to a certain extent depending on environmental factors, the ATP pool within a cell is normally constant (Stanley, 1989). Thus, the measurement of cellular ATP is a good indicator of biomass.

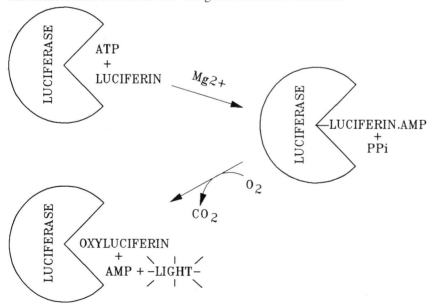

Figure 7.1 Schematic representation of the ATP bioluminescence reaction. ATP, adenosine triphos phate; Luciferin.AMP, luciferyl adenylate; PPi, pyrophosphate; AMP, adenosine monophosphate.

ATP bioluminescence, like any enzyme-based reaction, has certain conditions under which it performs optimally, which include a pH of 7.75 and a temperature of 20–22°C (McElroy and Strehler, 1949). Samples containing compounds that interfere with the enzyme reaction are collectively known as quenching agents, since they reduce the amount of light emitted or detected. Common causes of quenching include shifts in pH, changes in temperature, turbidity of samples and presence of metal ions (Zn^{2+}, Cd^{2+}, Hg^{2+}) (Lundin, 1982). Practical application of ATP bioluminescence in the food industry has necessitated the development of protocols to overcome these factors. One such approach commonly used in ATP assays is internal standardisation. Quenching effects are compensated for by taking light readings (ATP assays) before and after the addition of a known quantity of ATP. The degree of quenching can be assessed from the relative amount of light generated by the ATP standard and this used to correct the original sample light output.

7.2.1 ATP bioluminescence applications in modern food hygiene

The hygiene status of food-production surfaces is an essential element in the manufacture of high-quality, safe foods. Cleaning of such surfaces is conducted either automatically using clean in place (CIP) systems or manually, depending on the nature of the surface to be cleaned. Cleaning generally includes both a deterging stage to remove product residues and a sanitising stage to destroy microorganisms. Like any critical control point, a system of verification should be operated to ensure that cleaning is effective. Inadequate cleaning may result in the failure to remove microorganisms or product residues. Microorganisms constitute a direct risk to the quality and safety of the batch of food that next comes into contact with the improperly cleaned surface. Product residues, on the other hand, provide sufficient nutrients to allow initially low levels of microorganisms to rapidly grow into high populations in the period between cleaning and reuse of the surface, which again jeopardises the microbiological integrity of the following batch of food. Traditional approaches of monitoring cleaning efficiency have always concentrated on agar plating techniques using cotton-tipped swabs to sample a production surface, followed by enumeration of microbial contaminants using non-selective or selective agars. Like all plating techniques, a period of time is required before the microscopical cells divide sufficiently to produce macroscopical colonies and this incubation period may vary from 1 to 7 days, depending on the microorganism of interest to a particular industry. These techniques serve a very useful purpose for retrospective trend analysis of the microbiological cleanliness of surfaces. Trends can be monitored and poor results can be used to focus attention on particular areas for improved cleaning. However, the techniques only monitor the microbiological aspects of poor cleaning, do not detect the presence of product residues and cannot generate results sufficiently quickly to prevent contaminated surfaces being reused.

ATP bioluminescence, in contrast, can turn this retrospective monitoring into a proactive assessment of all aspects of cleaning efficiency. ATP is present in all viable microorganisms, but, importantly, it is also present in large quantities in most food residues (Sharpe et al., 1970). Application of an ATP bioluminescence assay for the assessment of cleaning efficiency of food contact surfaces, therefore, allows both microorganisms and product residues to be detected. Another major advantage of this approach is in the rapidity of the ATP assay which generates results within a few minutes. Improperly cleaned surfaces can therefore be detected prior to reuse and remedial action, in the form of recleaning, can be effected quickly. The protocols for ATP bioluminescence hygiene monitoring are very simple and involve relatively few stages. Production surfaces are sampled using cotton-tipped swabs and any microorganisms or product residues captured by the swab are transferred to a cuvette containing either a diluent or an extractant. The extractant ruptures microbial cells and the ATP inside these cells, together with ATP from product residues, is released into solution. Detection of ATP is achieved by the addition of the luciferin/luciferase enzyme, which results in light production. The amount of light is subsequently quantified using a luminometer. The level of light generated by the reaction in these assays is a relative measure of the amount of ATP present on the surface, which may have been derived from microorganisms, product residues or both. The greater the amount of light, the more microorganisms and/or product residues present on the surface and therefore the less efficiently the surface was cleaned.

A variety of commercial kits is available for hygiene monitoring using ATP bioluminescence (Table 7.1). It is important to note that the amount of light displayed as relative light units (RLU) on the display of the luminometer is not constant between the manufacturers of reagents and instruments, and levels of light are a factor of instrument sensitivity, display configuration and reagent preparation. A comparison of light output from some commercially available luminometers has been reported (Jago et al., 1989). Levels of ATP can vary markedly between different surfaces and the levels that represent inefficient cleaning on a synthetic conveyor belt may differ from that on a stainless steel fermentation vat. Prior to the introduction of ATP bioluminescence for hygiene monitoring, it is essential to establish the levels of ATP associated with improperly cleaned surfaces and with surfaces cleaned efficiently (Table 7.2). Targets can then be established which relate to acceptable and unacceptable levels of ATP.

ATP bioluminescence hygiene monitoring has greatly benefited from the development and introduction of a range of low cost, portable luminometers that are as sensitive as the more expensive parent instruments. Most reagent/luminometer combinations are capable of detecting quantities of ATP in the range of 0.1–1 pg (1 pg=10^{-12} g) in the assay cuvette, although recently we have seen the introduction of more sensitive, portable luminometers (Multi-Lite, Biotrace Ltd), capable of detecting 30 fg ATP (1 fg=10^{-15} g), which is roughly

Table 7.1 Some commercially available reagents and luminometers for hygiene assessment using ATP bioluminescence

Manufacturer	Extractant	Luminometer	Light detection	Display	Ease of use	Portable	Cost(£)	Assay[a]
Biotrace Ltd	Cold detergent	Multi-Lite	Solid-state[b]	RLU[c]	Technician	Yes	6500	S+P+C
Biotrace Ltd	Cold detergent	Uni-Lite	Solid-state[b]	RLU	Operative	Yes	4900	S+P
Lumac BV	Cold detergent	M1500Light	Photomultiplier	RLU	Technician	Yes	5209	S+P+C
Bioorbit[f]	Cold detergent	1253	Photomultiplier	SCU[d]	Technician	Yes	3500	S+P+C
Lab M	Cold detergent	HY-LiTE	Photomultiplier	RLU	Operative	Yes	4500	S[e]

[a] S, swabs; P, pipettes; C, cuvettes.
[b] Avalanche photodiode photon counter.
[c] RLU, relative light units.
[d] SCU, standardised count units.
[e] Assay occurs in 'pen' which contains extractant, buffer and luciferin/luciferase reagent.
[f] Distributed by Rhône-Poulenc.

Table 7.2 Levels of ATP on clean and unclean surfaces [a]

Surface	ATP bioluminescence result (RLU)[b]	
	Unclean	Clean
Cheese fermenter[c]	323	8
Cheese conveyor[d]	2359	29
Meat slicer[c]	9577	14
Milk road tanker[c]	1052	4

[a] All results are after subtraction of instrument and reagent background; swabbing area
  10 cm × 10 cm; source: Kyriakides (unpublished data).
[b] Biotrace HMK and M3 luminometer.
[c] Stainless-steel.
[d] Synthetic.

equivalent to 30 bacterial cells. The principal difference between the array of commercially available systems for ATP bioluminescence hygiene monitoring is in the simplicity with which they can be used. All systems (reagents and luminometers) are adequate in their ability to assess cleaning efficiency, but the two simplest systems suitable for use by non-skilled personnel are the HY-LiTE from Lab M and the Uni-Lite from Biotrace Ltd (Figure 7.2). HY-LiTE is the only system that has no pipetting stages, since the entire assay is performed in a unique pen. Individually packaged swabs and diluent are provided for sampling the product contact surfaces and the captured microorganisms and/or product residues are resuspended in a diluent vial. The shaft of the pen (Figure 7.2) is dipped into the diluent and a precise volume is captured in spiral rings, cut into the plastic of the shaft. The rings are impregnated with a microbial extractant which, upon rehydration, ruptures and releases ATP from microbial cells and product residues. The shaft is then forced into the internal chamber of the pen by striking the base of the shaft on a solid surface. This internal chamber contains a buffer which neutralises the extractant, ready for the addition of a dehydrated luciferase enzyme contained behind a plastic seal in the upper part of the pen. The cap of the pen is used to pierce the seal and the pen is shaken to rehydrate the enzyme and expose it to any ATP. Light output is quantified using a dedicated luminometer, which displays results in RLU after compensation for temperature effects. This system is undoubtedly one of the most simple and complete rapid methods ever to have been developed and has additional design features that ensure its compatibility with food industry use. Plastic moulded casing provides for disinfection, glass components are non-existent, vials are colour coded to make them highly visible and the pen even has an internal metallised seal to enable detection by metal detectors. No work has yet been published using the instrument, but early indications are that it will see widespread use in the food and water industries. The only system to rival the HY-LiTE for ease of use is the Uni-Lite from Biotrace Ltd, which is also a dedicated system for hygiene monitoring. Swabs premoistened with a microbial extractant are used to sample production surfaces. Microbial ATP is extracted

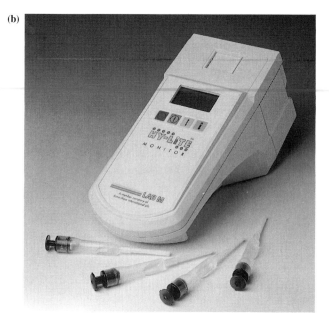

Figure 7.2 Luminometers for ATP bioluminescence hygiene monitoring. (a) Uni-Lite (Biotrace Ltd); (b) HY-LiTE (Lab M).

and recovered from the surface, together with any ATP from product residues. Luciferin/luciferase, previously dispensed into eppendorf tubes, is then drawn directly into the cotton tip of the swab, which is then replaced into its transparent sleeve. The entire unit is then inserted into the sample chamber of the dedicated luminometer and the amount of light quantified. Whilst other commercially available systems for hygiene assessment undoubtedly provide similarly accurate and reliable results as the two mentioned above, it is the simplicity of use that makes these two stand apart.

ATP bioluminescence has been used extensively in a number of industries for hygiene assessment (Simpson *et al.*; Kyriakides *et al.*, 1991; Bautista *et al.*, 1992). Simpson *et al.* (1989) described a bioluminescence protocol for the rapid assessment of brewery plant hygiene (road tankers, filling heads and fermentation vessels). The assay, subsequently modified and marketed by Lumac BV, demonstrated the ability of the ATP assay to highlight unclean surfaces not detectable by traditional methods on sensitive product contact surfaces such as filling heads (Table 7.3). Comparisons between ATP biolumi-

Table 7.3 ATP bioluminescence hygiene assay of brewery cask racking filling heads demonstrating the ability to detect product residues, not detectable by traditional techniquesa

Filling head	ATP result (fmol ATP/swab)[a]		Plate count (cfu/swab)	
	(Precleaning)	(Postcleaning)	(Precleaning)	(Postcleaning)
4	440	9	<10	<10
8	106	6	<10	<10
12	2685	11	<10	<10
16	2398	5	<10	<10
18	2220	6	<10	<10

[a] Assays using Lumac 2010a; source: Simpson *et al.* (1989).
[b] ATP results calculated from relative light units using internal standardisation (1fmol=1 $\times$ 10^{-15} mol).

nescence hygiene monitoring assays and plate count assessments consistently demonstrate the powerful ability of the former technique to detect a greater number of improperly cleaned surfaces than the latter assay (Kyriakides *et al.*, 1991; Bautista *et al.*, 1992). This is due to the ability of the ATP assay, to detect both microorganisms and product residues, either of which may be a manifestation of improper cleaning. It is important to note that there are occasions when the plate count assay detects improperly cleaned surfaces not detected by the ATP assay, due to the superior sensitivity of the plate count technique. Surfaces where low levels of microorganisms are present in the absence of product residues fall below the detection threshold of the ATP assay (*c.* 10^3 bacteria per swab) and are, therefore, not detected (Kyriakides *et al.*, 1991). This does not, however, constitute a major problem, since it is the authors experience that improperly cleaned surfaces, tested soon after cleaning, are usually contaminated with high levels of product residues rather than microorganisms, a fact that reinforces the strength of ATP assays for such purposes since the plate count assay would fail to detect any such surfaces. In addition, it is clearly of greater technical validity to accept a limited number of false-negative results and use an ATP assay which highlights major cleaning deficiencies before they become major product deficiencies, rather than use plate count assays which generate

204 RAPID ANALYSIS TECHNIQUES IN FOOD MICROBIOLOGY

greater numbers of false-negative reactions (in comparison to the ATP assay), provide limited information (microbial contamination and not product residues) and are always retrospective.

In a study to evaluate the potential use of ATP bioluminescence for assessing the cleaning efficiency of milk road tankers and milk reception tanks (silos), over 750 assays were conducted on cleaned, rinsed and uncleaned vessels, to establish criteria for interpretation of ATP assays (Table 7.4). Comparing the results of swabs taken from adjacent surfaces and assayed by the plate count and ATP bioluminescence techniques, respectively, there were 62 occasions where the ATP assay detected improperly cleaned surfaces not detected by the plate count assay and only 27 occasions where the converse was true (Kyriakides, unpublished data).

Table 7.4 ATP bioluminescence criteria for assessing cleaning efficiency of milk road tankers and reception tanks (silos)[a]

Criteria[b]	ATP assay (RLU per assay)[c]
Efficient clean	<5
Moderate clean	5–15
Poor clean	16–69
Unclean	>70

[a] All results represent values after subtraction of instrument and reagent background; swabbing area 10 cm × 10 cm; source: Kyriakides (unpublished data).
[b] Criteria established using >750 assays of clean, rinsed and unclean surfaces.
[c] Biotrace HMK and M3 luminometer.

The principal use of ATP bioluminescence hygiene monitoring systems is for routine verification of cleaning of production equipment. Using this approach, trend analysis provides an additional mechanism, if applied correctly, for continuous improvement, since target levels can be reduced as consistency is achieved in results. For such precise application it is essential to standardise assay conditions as much as possible and this should include defining areas for swabbing, defining methods of swabbing and defining surfaces to be swabbed. The rapid feedback achieved with these assays also provides for excellent training aids to demonstrate the effect of poor and efficient cleaning (Slater, 1993). Additional benefits of such systems are derived from their use in trouble-shooting. Systems are portable and can, therefore, be taken to factories requiring rapid assistance in identifying potential sources of contamination where traditional methods may have revealed poor end-product results. A large number of potential sources of contamination can be sampled and results generated immediately. Potential sources of contamination can be identified, remedial action can be taken and the results of this action can be assessed using further ATP analyses. An example of this is shown in Table 7.5, associated with the investigation of a dairy filling line where high microbial counts were detected in product using standard plating techniques (Kyriakides, unpublished data). To investigate the potential source of contamination using traditional techniques would have taken days, but, using ATP assays, a number of areas where microorganisms or product residues remained after cleaning were rapidly identified on the basis of comparatively high levels of ATP. The speed of the assay also enabled corrective action to be monitored, to ensure improvements were effective (Table 7.5). ATP assays cannot replace the need for good equipment design and efficient cleaning, but they can assist in identifying areas requiring greater attention and in rapidly highlighting major deficiencies which would not be apparent visually.

Table 7.5 Rapid identification of sources of contamination using ATP bioluminescence[a]

Surface	ATP bioluminescence result (RLU)[b]	
	Cleaned before remedial action	Cleaned after remedial action[c]
Valve (prefilling heads)	463	3
Filling head 1	13	0
Filling head 2	18	0
Filling head 3	24	0
Filling head 4	14	0

[a] All results after subtraction of instrument and reagent background; source: Kyriakides (unpublished data).
[b] Biotrace HMK and M3 luminometer.
[c] Remedial action involved disassembly of valve, full manual clean and reassembly.

The adverse effect of quenching agents on the ATP bioluminescence reaction has already been discussed and assays for hygiene monitoring can be subject to the influence of these factors, which principally relate to the effect of terminal disinfectants on the enzyme. However, most procedures involve processing a swab in comparatively large volumes of extractant or diluents which dilute the effect of the quenching agents prior to exposure to the luciferase enzyme. The Uni-Lite system from Biotrace Ltd is the only notable exception to this, as the swab is exposed directly to the luciferase and enzyme protection is presumably achieved by the use of disinfectant neutralisers such as lecithin or Tween 80 in the enzyme preparation (Schram, 1991). Quenching of the light signal has been reported, but the interpretation of the results (i.e. clean or dirty) was not affected by the phenomenon (Kyriakides *et al.*, 1991).

Some food-production environments are not suitable for the application of ATP bioluminescence hygiene monitoring, e.g. milk-powder production or flour mix environments where dry cleaning procedures do not attempt to remove all product residues (Coleman, 1992). In addition, some highly refined foods such as sugar may contain little residual ATP (Coleman, 1992), whilst some fish-processing environments may be prone to retention of light-emitting compounds presumably from marine bacteria (Rigarlsford, 1992). Like any other technology, it is an essential prerequisite to gain an understanding of how the technique performs in the environment where it is proposed to be applied before launching into full-scale use.

ATP bioluminescence hygiene monitoring is one of the strongest applications of the technology and its use by many food industry establishments may open the door for a wide variety of other ATP applications.

7.2.2 ATP bioluminescence for raw material testing

The production of high-quality, safe foods is intrinsically linked to the use of high-quality raw materials. Raw material control should form an integral part of every company's quality assurance considerations and, whilst the need to test each batch of raw material should be lessened by active supplier quality assurance programmes, there is, nevertheless, a need to frequent verify the quality of consignments. Like many aspects of food microbiological quality control, raw material will have been accepted, processed and the final product possibly consumed prior to the generation of microbiological data using plate counts. A good example of this is pasteurised milk, where the short shelf-life dictates the need for milk to be processed and retailed within 1 or 2 days. ATP bioluminescence assays have always faced a formidable challenge when attempting to use them for microbial enumeration in foods. This is principally due to the presence in foods of high levels of ATP that is not associated with viable microorganisms, collectively referred to as non-microbial ATP. This may exist in the form of free ATP, ATP within somatic cells or ATP associated with other compounds. In ATP bioluminescence hygiene assays, this ATP is used to the benefit of the application, but for assays requiring microbial enumeration or detection, this must be removed prior to assay. The amount of ATP present in foods varies significantly, but the ratio of non-microbial to microbial ATP in a variety of foods can be quite formidable (Table 7.6).

Table 7.6 Levels of non-microbial ATP in a variety of foods and beverages

Sample	ATP per gram of sample[a] (femtogram, fg)[b]
Raw milk[c]	10^6
Beer[c]	10^7
Raw meat[d]	10^8
Orange juice[d]	10^8
Ice cream[d]	10^7

[a] 1 femtogram is taken to be equivalent to one bacterial cell.
[b] 1 femtogram = 1×10^{-15} g.
[c] Data taken from Kyriakides (unpublished data).
[d] Data taken from Sharpe et al. (1970).

7.2.2. Raw milk quality.   The bacteriological quality of raw milk in the UK forms the basis of both an incentive payment scheme to farmers and a contractual agreement with the processor. Road tankers deliver milk to dairies for subsequent processing, but prior to acceptance of the raw milk a number of analytical checks are performed (Anon., 1985a). A dye reduction test (resazurin test) for the rapid assessment of the bacteriological quality of raw milk forms part of this analytical assessment and represents a 10 min tanker rejection criterion. The resazurin test (Anon., 1985b) relies on the reduction of a coloured substrate to a colourless end-product within a short period of time, as a result of the reducing activity associated with unacceptably high microbial populations in the raw milk. However, due to industry changes associated with the implementation of chill storage and distribution systems, the microflora typically associated with milk spoilage has been selectively modified from Gram-positive, fermentative bacteria to Gram-negative, psychrotrophic bacteria (Robinson, 1990). These bacteria perform notably poorly in dye reduction tests (Learoyd et al., 1992), due to their metabolic incompatibility with the assays and, therefore, the use of such techniques is essentially meaningless. Rapid alternative methods for the assessment of raw milk quality have been developed, with the most notable being the direct epifluorescent filter technique (DEFT) developed by Pettipher (1983). The technique originally found quite widespread use for the rapid (30 min) enumeration of bacteria in raw milk, but has become less popular recently and is infrequently used nowadays.

ATP bioluminescence assays for the detection and enumeration of bacteria in raw milk have always had to contend with the presence, in milk, of high quantities of non-microbial ATP. The ATP is present in a number of forms; as free ATP, as ATP associated with casein micelles and as ATP in somatic cells, derived from the cows' udder (Griffiths, 1989). In spite of these apparent difficulties, ATP bioluminescence assays have been developed for the assessment of raw milk quality. Early attempts at enumerating bacteria in raw milk exploited the use of weak detergents, incubated with milk, to selectively rupture and release ATP from somatic cells, whilst leaving the bacterial cells intact. The released ATP, together with any other free ATP, was then enzymically degraded using an ATP hydrolysing enzyme, apyrase (Bossuyt, 1981). Subsequent bacterial detection was performed by addition of stronger detergents to release microbial ATP, followed by the luciferin/luciferase enzyme. As a consequence of performing all assay stages in the same cuvette, the light reading was extremely time-dependent. This was principally due to the activity of the apyrase, which actively degraded the bacterial ATP following the addition of bacterial extractant. It has recently been reported that poor stability of light output seen in these early ATP assays was also due to an incompatibility of the detergent-based bacterial extractant with the luciferase enzyme, which resulted in luciferase degradation and loss in light output. Such problems were attributed to the need for extractants to be strong enough to break open bacterial cells and release ATP, but which could not then be sufficiently neutralised to prevent denaturation of the enzyme, luciferase (Simpson and Hammond, 1989). Nevertheless, Bossuyt (1981) demonstrated an ATP assay for the enumeration of bacteria in raw milk with a correlation coefficient of 0.93, when compared with the plate count assay, for both bulk milk tankers (48 samples) and ex-farm milk (209 samples). The 20 min assay had an effective sensitivity of 10^5 cfu/ml.

This was further refined into a 5 min test to assess the microbiological quality of tanker milk (ATP platform test) (Bossuyt, 1982), with a sensitivity in excess of 10^6 cfu/ml. The protocol was principally designed for the rapid detection of raw milk contaminated with unacceptably high bacterial counts.

A significant step forward to improve the sensitivity of the ATP assay was introduced by adaptation of the enzyme pretreatment stage used in the DEFT to allow filtration of milk (Pettipher, 1983). Pettipher had shown that addition of a cocktail consisting of a proteolytic enzyme (trypsin) and a weak detergent (Triton X-100) added to milk and incubated at 50°C broke down protein, lysed somatic cells and emulsified fats, allowing the milk to be filtered. Webster *et al.* (1988) adapted this procedure to develop at 20 min ATP bioluminescence filtration protocol for the enumeration of bacteria in raw milk, with a sensitivity approaching 10^5 cfu/ml. Samples of milk, pretreated with trypsin and Lubrol, were incubated at 50°C to lyse somatic cells and aid filtration. The procedure enabled volumes of milk up to 50 ml to be filtered through a 0.4 μm membrane filter prior to extraction and assay of microbial ATP. Filtration of milk confers a multitude of benefits to the ATP assay, which include concentration of microorganisms, removal of quenching factors and reduction of non-microbial ATP. A filtration approach was also used in the development of the commercially available ATP-F test from Lumac BV. Samples of milk preincubated with a somatic cell extractant (NRS) and apyrase (Somase) for 4 min are filtered through a 47 mm diameter, 0.65 μm positively charged nylon membrane filter. Bacteria are concentrated on the membrane filter, whilst residual ATP and apyrase are rinsed away by subsequent filtration of sterile diluent. Bacterial extractant (L-NRM) is then used to release the ATP from the microbial cells on the filter and a sub-sample of the ATP extract assayed for the ATP content. Waes *et al.* (1989) reported an excellent correlation between this ATP procedure and the standard plate count assay ($r=0.865$, $n=247$), with a detection limit as low as 2×10^4 cfu/ml. Similar results have been reported by other workers (Van Crombrugge, 1989). However, the use of large filters, together with the need for stainless-steel filtration manifolds, made the technique somewhat cumbersome. The most notable ATP bioluminescence system currently available for enumeration of bacteria in raw milk is that developed by Griffiths *et al.* (1991), which has subsequently been modified and improved to form the Milk Microbial ATP Kit marketed by Biotrace Ltd. Like the assays already described, samples of milk are treated with a somatic cell extractant (Somex A) and apyrase (Somex B), to release and degrade somatic and free ATP. After a 5 min incubation at 37°C, a large proportion of this non-microbial ATP is degraded and any residual ATP, together with apyrase, is removed by filtration through a novel miniaturised manual filtration device containing a small (13 mm) nylon mesh filter. Bacteria are retained on the filter and further traces of ATP are removed by rinsing with a diluent. Bacterial ATP is then released by adding the filter to a bijoux containing bacterial extractant (Bactex) and the ATP is detected by addition of the luciferin/luciferase enzyme. The light generated from this reaction is quantified in a luminometer and the number of bacteria originally present in the raw milk can be estimated using a standard curve relating bacterial counts (cfu/ml) to ATP results (RLU). The protocol takes approximately 10 min and benefits from the use of new generation reagents that generate a relatively constant light signal which ensures compatibility with low cost, manual luminometers. In addition, since filtration and rinsing stages are performed in the barrel of a 10 ml syringe, but for the need of a 5 min incubation at 37°C, the system has almost become entirely portable. The principle aim of the raw milk enumeration systems are not to generate accurate counts reflecting precise bacterial numbers but, instead to generate results to allow raw milk to be graded into broad categories such as poor, acceptable and good. This provides the dairy quality manager with information sufficiently quickly to enable decisions to be taken relating to the fate of particular consignments of raw milk.

Griffiths *et al.* (1991) showed this procedure to give accurate results when related to standard plate counts with a sensitivity approaching 10^4 cfu/ml (Figure 7.3). The assay was also capable of categorising milk into plate count ranges of log 4, log 5, log 6 and log 7 with an accuracy of greater than 90%. Further work using UK and Canadian milk confirmed these earlier findings (Bautista, 1992).

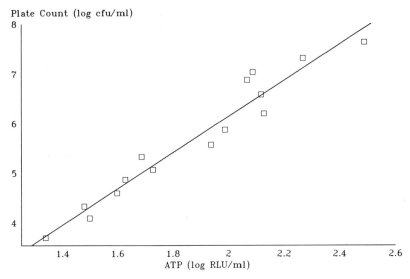

Figure 7.3 Relationship between ATP assay (mean log RLU) and plate count (mean log cfu) for raw milk samples. Results represent the mean values of 240 milk samples divided according to plate count range. ATP assay using Milk Microbial ATP Kit (Biotrace Ltd). (Adapted from Griffiths *et al.*, 1991).

Reybroeck and Waes (1992) compared the ATP test kits from Biotrace Ltd (Milk Microbial ATP Kit) and Lumac BV (ATP-F test) using 189 farm milk samples and reported effective sensitivities of 1.5×10^4 cfu/ml and 3.0×10^4 cfu/ml, respectively. They observed that the somatic cell count, fat, protein and lactose content of the milk did not influence the assay. Interestingly, recent studies investigating parameters associated with the assessment of milk by these two kits suggest marked differences in light output associated with minor temperature fluctuations (Reybroeck and Schram, 1992). For example, the Biotrace and Lumac luciferase enzyme reagents emitted 86 and 24.1% less light, respectively, at 18 than at 22°C.

A different approach has been taken in another commercially available system for the enumeration of bacteria in raw milk from Promega Corp. Instead of filtration, the Enliten test utilises a concentration procedure for bacteria in milk based on centrifugation at $12\,000 \times g$ in the presence of a special proprietary milk clarification reagent. Following centrifugation and washing stages, the microorganisms, concentrated in a pellet, are enumerated by extraction of their ATP using a detergent, followed by the addition of the enzyme and detection of light in a luminometer. The reported sensitivity is 10^4 cfu/ml (Martin, 1991), although little work has been published using this assay.

Automation is not a common sight for routine ATP bioluminescence applications in the food industry, but an instrument has been developed and is commercially available for the enumeration of bacteria in raw milk. The Bactofoss (Foss Electric Ltd) (Figure 7.4) automates the entire process of sample analysis, which involves bacterial capture on a filter strip, followed by

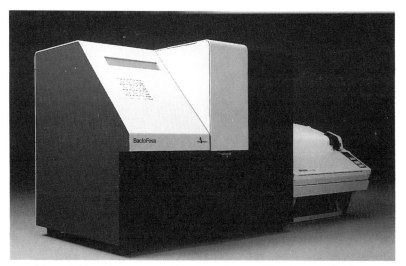

Figure 7.4 Bactofoss: automated instrument for raw milk and meat applications.

somatic cell lysis, washing, ATP extraction, enzyme addition, light detection and bacterial quantification. The assay takes approximately 90 s, with each sample being analysed in 3 min, and has a reported sensitivity of 10^4 cfu/ml and a correlation coefficient of 0.92 when compared with plate counts (Andersen, 1989). The instrument is used routinely in a number of countries including

Belgium, Italy, Japan, New Zealand and the USA, but the principal reason for the lack of market penetration in the UK has been the high price of the instrument, which, in excess of £25K, was outside the scope of most dairy budgets for raw milk assessment. However, the Bactofoss has recently been relaunched at a more competitive price (<£20K) with more cost-effective material costs and this may result in some uptake of the system.

7.2.2.2 Enzyme detection. Psychrotrophic bacteria present in raw milk are selectively favoured as a consequence of the strict adherence to the chill chain from milking to the dairy. The most commonly occurring bacterial psychrotrophs found in raw milk are Gram-negative rods and of these approximately 50% belong to the genus *Pseudomonas* (Bramley and McKinnon, 1990). These organisms are capable of producing a variety of enzymes, including proteases and lipases, which are known to be significantly more heat resistant than the vegetative cells. Pasteurisation and ultra-high temperature (UHT) treatment is sufficient to destroy the vegetative cells, but the thermostability of these enzymes ensures that they remain active even after these severe heat treatments. Following high-temperature short-time (HTST) pasteurisation, Griffiths *et al.* (1981) reported residual enzyme activity of 66 and 59% for protease and lipase, respectively, and even after UHT treatment respective residual activities of 29 and 40% were reported. The enzymes are capable of selectively degrading their substrates, milk protein and fats, to subsequently cause product defects. Longer shelf-life products such as UHT milk are evidently more susceptible to these problems, since storage at ambient temperatures for long periods of time provides for greater enzyme action. Defects reported in dairy products have included gelation and bitterness of UHT milk (Gilmour and Rowe, 1990) and poor cheese yield (Hicks *et al.*, 1986).

An ATP bioluminescence assay has been developed by Rowe *et al.* (1991) for the specific detection of bacterial protease in milk and milk products, based on the novel approach of luciferase degradation. Luciferase is a key component of the ATP bioluminescence reaction and, under conditions where the concentration of luciferase is limited, any reduction in its activity will result in lower levels of light emission. In the assay for bacterial protease, a sample of milk is incubated with the luciferase reagent, during which time any protease selectively degrades the luciferase. Upon addition of luciferin and ATP, light is produced and the amount of light is effectively related to the amount of protease originally present in the sample, since higher protease concentrations will result in greater luciferase degradation and, therefore, lower light output. An inverse relationship, therefore, exists between protease concentration and light output and the amount of protease can be quantified by reference to a standard curve of protease concentration versus light. Potentially, the higher the protease concentration the greater the deleterious effect on the product during the shelf-life, due to the degradative effect on milk proteins. The protease kit has been commercialised and is currently marketed by Biotrace Ltd.

7.2.2.3 Raw meat and fish. ATP bioluminescence protocols have been developed for the enumeration of microorganisms on fresh meat. Meat, like most other foods, has extremely high levels of non-microbial ATP derived from somatic cells. Direct analysis for bacteria in the presence of meat is not possible, due to the interference from this non-microbial ATP and protocols must first remove microorganisms from the meat before application of an ATP assay. Stannard and Wood (1983a) described a complete procedure for the separation and concentration of bacteria from fresh meat samples using a centrifugation, ion exchange and filtration protocol. A meat homogenate was centrifuged (2000 × *g* for 10 s) to remove large particulate matter and the supernatant was subjected to cation exchange resin treatment. This selectively removed, by adsorption, protein and other interfering compounds and enabled the supernatant to be filtered. The pH of the ion exchange separation system is an important determinant of the characteristic adsorption/desorption behaviour of microorganisms and the intrinsic food components. Figure 7.5 shows the interaction of

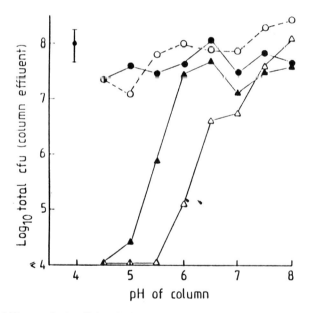

Figure 7.5 Differences in the affinity of microorganisms for large columns (1.6 × 6.5 cm) of cation exchange resin Bio-Rex 70. *E. coli* NCTC 9001 (○); Ps. fluorescens NCTC 10038 (●); *Staph. aureus* NCTC 8530 (△); *L. plantarum* NC1B 8960 (▲); bar = inoculum level. (Mean and SD are shown, *n* = 32.) (Adapted from Patel, 1984.)

Gram-negative and Gram-positive bacteria with the cation exchange resin, Bio-Rex 70 (Patel, 1984). Around pH 6.0, both types of bacteria exhibit lower binding affinity with the cation exchanger. The subsequent filtration step both separated and concentrated microorganisms from the supernatant and provided for excess non-microbial ATP to be rinsed through the filter. Bacterial ATP was extracted on the membrane filter and assayed using the luciferase enzyme. The relatively simple method gave results within 20–25 min and was capable of discriminating between orders of magnitude of bacteria on meat, showing a good correlation between traditional plating methods for bacterial counts in excess of 10^6 cfu/g.

Two alternative techniques to the solid ion exchanger have also been developed, one of which involved the use of a soluble cation exchange polymer, Cat-floc 243, to flocculate the meat particles while allowing the intrinsic microbial flora to remain in suspension (Patel, 1984). The microbial cells were then concentrated by centrifugation prior to measurement of microbial ATP (Figure 7.6). Figure 7.6 also shows that a double centrifugation procedure, incor-

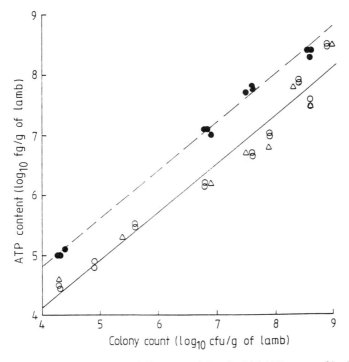

Figure 7.6 Relationship between total viable count and the microbial ATP content of lamb and the effect of nutrient medium on this relationship. Flocculation–centrifugation ($\bigcirc$); centrifugation procedures, solid regression line: $y = 0.99 + 0.79\ x$ ($I = 0.98$) ($\triangle$); after incubation in nutrient medium, broken regression line: $y = 1.62 + 0.80\ x$ ($I = 0.99$) ($\bullet$). (Adapted from Patel, 1984.)

porating a brief centrifugation step ($200 \times g$ for 30 s) to remove large partic-ulate matter, followed by further centrifugation ($2000 \times g$ for 10 min) of the supernatant to concentrate microorganisms, can give a good correlation ($r=0.98$) between the microbial ATP content and colony count. This technique has been successfully used as an alternative to colony count for the prediction of shelf-life of chilled foods (Stannard and Williams, 1987). Kennedy and Oblinger (1985) used a novel 'filter tip', consisting of a polyethylene mesh (100 μm pore size) attached to a standard 1 ml pipette, to sample meat homogenates. The filtrates were treated with a somatic cell extractant to release the ATP from the meat cells and an ATP hydrolysing enzyme (apyrase) to destroy the non-microbial ATP. Samples were then assayed for their microbial ATP and the resultant light output was related to the standard plate count. Correlation coefficients of 0.98 were achieved between ATP results and plate counts at 20°C for the 75 samples of beef over plate count ranges of log 4.8–log 9.9 cfu/g. However, the levels of non-microbial ATP present on the filter remained very high, even after treatment, and represented over 60% of the total ATP at plate counts below log 6 cfu/g. This necessitated the use of both a total ATP assay and a non-microbial ATP assay (i.e. without the addition of bacterial extractant) to determine the correct microbial ATP value (i.e. total ATP minus non-microbial ATP). Such approaches are notoriously inaccurate, due to subsample and reagent variation. Ward et al. (1986), following a similar principle, were able to demonstrate excellent correlation between the ATP and plate count assays for fresh fish using both surface swab and flesh samples. Correlation coefficients in excess of 0.96 were demonstrated in this 1 h procedure over a plate count range of log 4–log 8.6 cfu/g or square inch. Littel et al. (1986) used a double filtration procedure on samples of macerated meat products (beef, chicken, etc.). Samples were pretreated to remove coarse meat particles by filtration through glass wool and the solution was then further treated by filtration through a dual-filter system present in the barrel of a 10 ml syringe. The primary filter was designed to trap food particles (somatic cell filter), whilst allowing microbial cells to pass through to the secondary filter (bacterial filter), which separated and concentrated the bacteria. The bacterial filter was treated with a somatic cell extractant and apyrase and finally rinsed with a buffer prior to extraction of bacterial ATP and assay. This treatment was shown to remove in excess of 99.99% of somatic ATP, whilst recovering over 90% of the bacteria originally present in the sample. Relating actual plate count results to predicted bacterial colony forming units (using ATP results), the procedure showed a correlation coefficient in excess of 0.97 for actual counts in beef and chicken over the range 5×10^4–5×10^8 cfu/g.

A common problem with the ATP technique for microbial determination is the effect of stress (e.g. filtration and centrifugation) on the levels of intracellular ATP (Stanley, 1989). Figure 7.6 shows that the level of ATP within microbial cells can be increased by brief incubation of the cells in a glucose-containing medium. The overall analysis time for the flocculation method was approximately 35 min, while an additional 10 min was required for the resuscitation step using the glucose medium. The correlation values between the ATP levels and plate counts for beef and lamb were high ($r=0.98$ and 0.99, respectively) at microbial levels ranging from approximately log 4–log 9 cfu/g (Patel, 1984).

Clearly, the ATP bioluminescence assay can be used for rapid assessment of bacterial populations, even in complex foods such as meat and fish. The automated Bactofoss system, in addition to being used for raw milk bacterial enumeration, has also been reported to be applicable for rapid enumeration of bacteria on raw meat following homogenisation of the sample with a reported sensitivity of 10^4 cfu/g and a correlation coefficient of 0.95 (Eriksen and Olsen, 1989).

7.2.3 ATP bioluminescence for end-product testing

7.2.3 Brewing applications   The brewing industry was not slow in recognising the potential application of ATP bioluminescence for the rapid detection of spoilage microorganisms. The principal application for ATP bioluminescence in the brewing industry has been for the rapid assessment of microbiological contamination in filtered or pasteurised beer. Spoilage of these products is principally caused by a group of yeasts collectively referred to as non-brewing or 'wild yeasts' (Campbell, 1987) and a narrow group of Gram-positive cocci (*Pediococcus* spp.) and rods (*Lactobacillus* spp.) (Priest, 1987). ATP bioluminescence assays are incapable of distinguishing bacteria from yeasts and, since the latter contain levels of ATP which are approximately 100 times that of the former, the nature and extent of microbial contamination in ATP assays where either or both of these organisms may be present will be unclear. In the context of pasteurised or filtered beer, this may not actually be a major problem, since any contamination by spoilage bacteria in final product is considered to represent a potential spoilage concern, given the long shelf-life of these products. Pasteurised, filtered beer is a harsh environment for the application of ATP assays and direct analysis of beer is not advocated, since the low pH (<4), variable colour and presence of other quenching agents significantly affect the luciferase enzyme reaction. In addition, levels of non-microbial ATP in beers can be extremely high and variable (Table 7.7). To achieve the sensitivity required, traditional methods employ filtration of large volumes of beer (100–250 ml) prior to detection of microbial contamination by incubation of these membrane filters on agar plates, which may take in excess of 4 days for the growth of wild yeasts and up to 7 days for the detection of lactic acid bacteria.

Table 7.7 Non-microbial ATP in a variety of pasteurised beers[a]

Beer	Non-microbial ATP (fg/ml)	Quenching[b] (%)
Lager 1	6×10^4	51
Lager 2	1×10^7	70
Bitter 1	9×10^5	48
Bitter 2	2×10^7	55

[a] All assays using Lumit PM (Lumac BV) and 2010a luminometer; source: Kyriakides (unpublished data).
[b] Quenching determined by direct addition of ATP to sample of beer in presence of luciferin/luciferase. Percentage quenching in relation to distilled water control.

ATP bioluminescence must also employ membrane filtration to allow a high proportion of the non-microbial ATP and quenching agents to be rinsed away whilst retaining and concentrating the microorganisms for detection. However, the need to detect extremely low levels of contaminants (fewer than 10 yeasts or bacteria per membrane filter) necessitates the development of strategies to amplify the ATP signal from such low levels of microorganisms and further reduce the background signal from residual non-microbial ATP.

Hysert *et al.* (1976) clearly demonstrated the potential for highly sensitive assays using ATP bioluminescence with a theoretical sensitivity using a liquid scintillation counter of 10 yeast cells. However, interference from the high level of non-microbial ATP in the reagents and beer prevented the practical demonstration of this sensitivity and, in fact, statistically significant correlation between plate count and ATP could not be established. Avis (1988) reported a detection limit for yeast in pasteurised packaged beer of 1 cfu per 250 ml, using an ATP bioluminescence assay. Filtration was used to concentrate yeast from beer and the membrane filters were then incubated for 20 h to amplify the yeast numbers and, thereby, allow detection. Simpson *et al.* (1989) used positively charged nylon membrane filters to concentrate yeast cells from bright beer (filtered, unpasteurised beer), applied apyrase to reduce non-microbial ATP levels and then used novel reagents to inactivate the apyrase and extract microbial ATP. Detection of the low levels of released ATP was achieved using a sensitive luciferase enzyme preparation of poor stability but high sensitivity (Lumit HS, Lumac BV). The assay was able to detect low levels of yeast (fewer than 10 cfu per membrane or sample volume) within 60 min. This rapid approach did not allow the differentiation of yeast and bacteria and, therefore, the ATP results could only give an estimate of the microbial contamination which may be either yeast (sensitivity 1–10 per sample) or bacteria (sensitivity 100–1000 per sample). Nevertheless, the use of such rapid methods on continuous in-line filters taken throughout the pasteurisation or filtration process can give rapid feedback on process efficiency.

ATP bioluminescence has also been applied for selective detection of brewery spoilage microorganisms. Ono *et al.* (1987) used a filtration procedure to concentrate wild yeast and *Lactobacillus* spp. from beer, followed by incubation of membrane filters on selective agars for 16-24 h, and was able to demonstrate sensitive detection of 1 cell per 100 ml of original sample. Kilgour and Day (1983) reported a 15 min test for the detection of bacteria or yeasts in swab or equipment rinses where levels of 1 yeast or 100 bacteria per ml could be detected using ATP bioluminescence, providing at least 1 l of sample was filtered (i.e. 10^3 yeast cells or 10^5 bacteria per assay.)

7.2.3.2 Beverages. Spoilage of carbonated beverages principally occurs as a result of the growth of acid- and osmo-tolerant contaminant yeast. Low levels of contamination represent spoilage concerns, due to the long shelf-life of these products and, as a consequence, many suppliers resort to the use of preservatives to guard against such eventualities. Carbonated beverages are reported to have very low levels of non-microbial ATP (LaRocco *et al.*, 1986) and, since they can be easily filtered, procedures for the detection of yeast contamination using ATP bioluminescence have been developed. Using artificially inoculated cola samples, LaRocco *et al.* (1985) demonstrated excellent relationships ($r=0.96$) between a filtration ATP assay and the standard plate count for *Saccharomyces rouxii*, with a detection limit of 1–10 cells/ml (sample volume, 50–100 ml). Similar relationships were found using a mixture of other yeast species (*S. cerevisiae, Candida albicans* and *Torulopsis kefyr*) ($r=0.94$). Littel and LaRocco (1986) used the same protocol to establish a frequency distribution of non-microbial ATP levels using 240 culturally negative cola samples and set a statistically derived cut-off value for the detection of microbiologically contaminated samples. Non-microbial ATP values ranged from 0 to 867 fg ATP/ml and, using a screening value of 200 fg ATP/ml, the assay gave false-positive results on less than 3% of samples. Using these criteria on artificially contaminated cola samples, less than 11% gave false-negative results in the presence of < 5 cfu/ml and 2% at 10 cfu/ml.

7.2.3.3 Fruit juice. Patel and Williams (1985) examined the relationship between total viable counts and ATP measurements for six food-spoilage psychro-trophic yeasts (*Metschnikowia pulcherrima, Rhodotorula glutinis, Kluyveromyces marxianus, K. lactis, Candida pseudotropicalis* and *C. holmii*) grown at 4, 10 and 15°C. The results showed a good agreement ($r > 0.97$) between the ATP content and viable counts, the relationship being temperature-dependent.

Fruit juice contains high levels of non-microbial ATP derived from pulp cell debris. In addition, the very low pH (< 3) of these products is capable of significant luciferase inhibition. Removal of the non-microbial ATP prior to the extraction of microbial ATP has been attempted with limited success. Graumlich (1985) reported that neutralisation of orange juice prior to assay resulted in unacceptable dilution of the sample and concentration by filtration was not feasible, due to pulp cell debris. A protocol was developed which involved centrifugation (1000 $\times$ *g* for 10 minutes), resuspension of the sample in Tris buffer, treatment with an enzyme to destroy pulp cells (NRS, Lumac BV) and enzymic removal of ATP (apyrase) prior to microbial ATP analysis. Orange juice samples analysed immediately after reconstitution of the juice concentrate gave a very poor correlation between plate count and ATP results ($n=53$, $r=0.58$), attributable to the presence of stressed cells in stationary phase. Reconstituted juice tested after 24 h at 25°C gave a better correlation ($r=0.92$, $n=94$), with an effective sensitivity of 10^3 active yeast cells per ml of orange juice. Stannard and Wood (1983b) used a combination of centrifugation and enzyme treatment to demonstrate rapid enumeration of yeast levels above 10^4 cfu/ml in pasteurised orange juice.

7.2.3.4 Pasteurised milk and cream. The sensitivity of the ATP assay previously described for raw milk is principally limited by the levels of somatic cell and micelle associated ATP remaining even after extensive sample pretreatment. This effectively restricts the application to dairy products containing levels of bacteria in excess of 10^4 cfu/ml. At this level of contamination, the ATP from microbial cells becomes consistently greater than the ATP from residual non-microbial sources and, therefore, reliable detection and enumeration can occur. The majority of pasteurised milk and cream has a microbiological loading below the threshold of the ATP assay and, therefore, ATP bioluminescence is not directly applicable for the assessment of the microbiological status of such products. The technique has, however, been used, in conjunction with preincubation conditions, to assess post process contamination and to predict the shelf-life or keeping-quality of milk and cream. Psychrotrophic Gram-negative bacteria arising as a consequence of post-pasteurisation contamination are the predominant species associated with spoilage of pasteurised milk and cream products, due to their ability to grow at refrigeration temperatures. Rapid tests to determine post process contamination by these Gram-negative microorganisms have been developed (Waes and Bossuyt, 1981, 1982). Preincubation of milk in the presence of a cocktail of Gram-positive bacterial inhibitors (Benzalkon and crystal violet) for 24 h at 30°C allowed Gram-negative organisms to proliferate and enabled the ATP assay, as described by Bossuyt (1981), to be applied directly to a 0.5 ml sample of the milk suspension. Waes (1982) found this technique to be as sensitive as other equivalent methods for assessing post-process Gram-negative contamination, but with the advantage of achieving results within 24 hours.

A 26 h method has also been reported for assessing post-process Gram-negative contamination and estimation of the keeping-quality of milk (Griffiths, 1989) and cream (Griffiths *et al.*, 1984), termed the P-Inc test. Milk and cream were incubated at 21°C for 25 h in the presence of penicillin, nisin and crystal violet to inhibit Gram-positive microorganisms and select for Gram-negative psychrotrophs. Samples of milk were treated with a somatic cell extractant and apyrase prior to extraction and assay of bacterial ATP (Bossuyt, 1981). They found a 94% agreement and a correlation ($r=0.83$) between the P-Inc ATP assay and the P-Inc plate counts (21°C for 25 h) (Figure 7.7). In addition, the criteria set for a failure by the P-Inc ATP test (>log 3.5 RLU) was shown to be capable of predicting, with 94.5% accuracy, creams that would be considered to be microbiologically spoiled (plate count >3.2×10^7 cfu/ml) after a shelf-life of 7 days at 6°C. Additional work reported by Bautista (1992), using Scottish and Canadian pasteurised milk, indicated that the best correlation between ATP and plate count for the estimation of shelf-life was achieved by preincubation of milk at 15°C for 25 h without inhibitors or at 21°C for 25 h with the standard inhibitors, for UK milk. Canadian milk, however, showed a relatively poor relationship in all conditions except for preincubation at 18°C for 25 h. This was attributed to the limited range of bacterial cell numbers in the Canadian milk

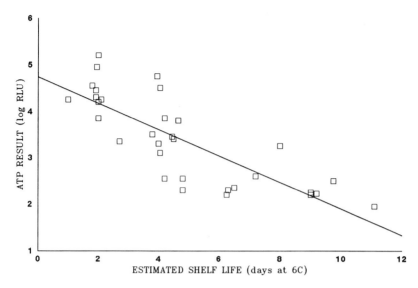

Figure 7.7 Shelf-life of pasteurised milk using the P-Inc ATP assay. ATP assay after incubation of milk at 21°C for 25 h in the presence of Gram-positive inhibitors. (Adapted from Griffiths, 1989.)

after the preincubation stage $(1 \times 10^1 - 1.1 \times 10^3$ cfu/ml). Again, the automated Bactofoss system has been advocated for use in the P-Inc ATP assay.

7.2.3.5 UHT products. UHT dairy products are susceptible to spoilage by very low levels of post-process contaminants due to the long shelf-lives, lack of any inhibitory conditions and storage at ambient temperatures. Manufacturers of such products (UHT milk, cream, etc.) conduct extensive end-product testing as part of their quality-control procedures, in order to detect these low levels of contaminants. Methods traditionally employed for the detection of these contaminants involve preincubation of the samples at elevated temperatures for up to 7 days or more to allow microbial contaminants to grow to sufficient levels to be detectable by streaking the sample on an agar plate. Agar plating adds an additional delay and is labour-intensive. Workers have used alternative techniques to assess microbial contamination of UHT products, such as organoleptic assessment or pH monitoring, again after 7 days preincubation. Such tests are reliant on the growth of microorganisms to high levels and on the subsequent production, by these bacteria, of sufficient volatile compounds to be detectable by taste or smell, or sufficient acidity to decrease the pH.

ATP bioluminescence is being used more frequently for the assessment of the sterility of UHT products, due to the ability to generate faster results allowing product to be released earlier. Sterility testing of UHT products, always involves testing high numbers of samples and would be impractical with manual ATP bioluminescence systems. Systems have, therefore, been developed that automate ATP assays for these products. ATP bioluminescence procedures for UHT products involve incubation of samples at elevated temperatures for several days to increase the microbial levels from a theoretical minimum contamination level of 1 cell per pack to levels in excess of 10^5 cfu/ml. Automated ATP bioluminescence instruments are then able to detect bacterial contamination by assay of microbial ATP (Waes, 1984). The principal commercially available systems for UHT sterility testing are the M4000 (previously M3000) from Lumac BV and the Auto-Lite from Biotrace Ltd (Figure 7.8). Both systems are computer controlled and automate all stages of reagent addition and ATP analysis. Cuvettes containing the preincubated samples are introduced into continuous belt assembly mechanisms, exploited by both systems, which carry

(a)

(b)

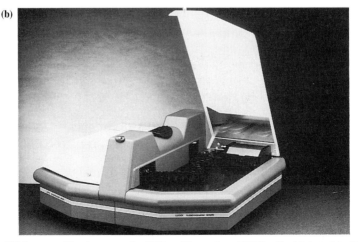

Figure 7.8 Automated luminometers for ATP sterility testing of UHT milk. (a) Auto-Lite (Biotrace Ltd); (b) M4000 (Lumac BV).

the samples to an injector head. Non-microbial ATP is degraded by the automatic injection of apyrase, whilst somatic cells no longer pose a problem since they are lysed by the UHT treatment. In the Lumac M4000 system, a bacterial extractant is then injected into the sample to release bacterial ATP which is detected following the addition of luciferin/luficerase. The amount of light quantified by the system is then related by the computer to defined criteria for the determination of bacterial contamination. The Biotrace system differs slightly from the previous assay in the final stage. Due to the presence of special reagents that protect the luficerase enzyme, the apyrase stage is followed by addition of the luciferin/luciferase to the cuvette. Light detected at this stage reflects the level of non-microbial ATP remaining in the sample and effectively represents a background control for each assay. This ensures that the apyrase enzyme is always active. Bacterial extractant is then injected to release the bacterial ATP and the light emitted is used to determine microbial contamination. Automated systems do necessitate the expenditure of large capital sums (£25k–£30k), but these systems offer large cost saving advantages associated with reduced labour requirement and warehouse costs due to earlier release of product. In addition, automating the assay provides for precision accuracies not achievable with traditional methodologies.

7.2.4 Other applications

7.2.4.1 Starter culture activity.
The activity of starter cultures is essential to the manufacture of a variety of fermented foods. Starter cultures are responsible for both the production of compounds (acids, alcohols, etc.) which are inhibitory to the growth and survival of pathogens and spoilage microorganisms, and for the production of flavour compounds which add to the organoleptic qualities of the end product. Documented outbreaks of food poisoning have occurred due to the poor activity of starter cultures. In the dairy industry, starter bacteria are used for the production of cheese and yoghurts and poor activity generally results from the presence of veterinary residues (antibiotics) in milk, which are capable of suppressing starter growth, together with bacteriophages (bacterial viruses) which infect and lyse starter bacteria. Milk used for fermentation is, therefore, routinely screened for the presence of antibacterial substances and starter cultures are assessed for bacteriophage infection. These tests generally involve incubation of a microorganism (or starter culture) in an nutrient medium (or milk) at a controlled temperature for a defined period of time, usually in excess of 2.5 h. The activity of the microorganism is determined by monitoring changes in pH or development of acidity. The presence of antibiotics in the milk or bacteriophages in the starter culture will be evident by a decrease in the rate of change of pH or acidity.

ATP can be used as an indication of microbial metabolic activity and, therefore, bioluminescence assays can be used in place of pH/acidity for the above applications. Inhibitory effects on the activity of a microorganism will rapidly manifest in a reduction in the rate of intracellular ATP increase and, measured over a period of time, can provide information on the presence of inhibitory compounds. Hawronskyj *et al.* (1993) demonstrated the potential to detect a number of antibiotic residues in milk within 1.5 h, using the Biotrace Milk Microbial ATP Kit (Biotrace Ltd). *Streptococcus thermophilus*, a microorganism susceptible to a variety of antimicrobial substances, was inoculated into a mixture of milk and M17 broth and incubated at 37°C. ATP assays were performed at the beginning of incubation and after 1.5 h incubation. Comparison of ATP assays in the presence of a number of antibiotics with an antibiotic-free control showed the ability of the assay to detect levels as low as 0.0048 IU penicillin/ml. Other bioluminescence assays for the determination of antimicrobial compounds have also been reported (Westhoff and Engler, 1975; Williams, 1984; Anon., 1987). The major drawback of using such assays is the material cost, which can be very high due to the need to perform a number of ATP assays on each sample, and also the labour requirement in comparison with the standard dairy industry assay, Delvotest-P (Anon., 1985c). The recent introduction of more specific immunoassays for the detection of certain antibiotics, such as penicillin, within a shorter period of time, may preclude the adoption of this ATP approach.

1.2.4.2 Microbial biomass. Biomass assays are probably the most simple assays for use with ATP systems, since they are principally for applications where discrimination between non-microbial and microbial ATP is not essential. This is where microbial ATP in a sample is consistently and significantly higher than non-microbial ATP, as a result of either high microbial populations or low non-microbial ATP levels. Protocols for such assays involve addition of microbial extractant directly to samples, followed by luciferin/luciferase and light detection. Miller *et al.* (1978) described an ATP biomass assay for the quantification of viable yeast for brewery fermenter pitching. Samples of yeast were diluted and their ATP extracted using acetone. ATP was quantified using crude luciferase preparation and results were used to estimate the amount of yeast slurry necessary for pitching into the wort for efficient fermentation. The assay was used in a number of production trials and enabled optimum utilisation of yeast in comparison with standard approaches of pitching based on percent solids. This resulted in increased final beer filter life and improved fermentation performance, even when lower volumes of yeast were added. ATP bioluminescence may also be applied for biomass estimation in cooling tower or other process water where biocide dosing is used to control microbial contaminants and where high counts are undesirable. Rapid on-site determination of the biomass is possible and high microbial ATP results would indicate high microbial populations and potential biocide inefficiency. Biomass estimation using ATP assays may also be used to monitor growth of microorganisms in pure culture or during fermentation processes (Stannard, 1989) and, indeed, biomass estimation has also been proposed for the determination of extraterrestrial life (MacLeod *et al.*, 1969).

7.3 Bacteriophage bioluminescence

Bioluminescence reactions require an enzyme, an enzyme substrate and an energy source. ATP bioluminescence, based on the firefly system, uses firefly luciferase, D-luciferin and adenosine triphosphate. Like the firefly, a number of bacteria possess the ability to produce light and are, therefore, bioluminescent. Bacterial bioluminescence differs from firefly luminescence in that it uses energy from reduced flavin mononucleotide ($FMNH_2$) to drive the oxidation of a long-chain aldehyde (do- or tetradecanal) in a reaction catalysed by bacterial luciferase (Meighen, 1991a) (Figure 7.9). It is the conversion of the aldehyde

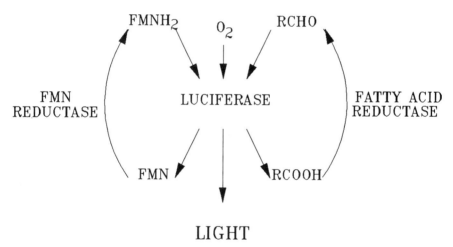

Figure 7.9 Schematic drawing of bacterial bioluminescence. FMN, flavin mononucleotide; $FMNH_2$, reduced flavin mononucleotide; RCHO, long-chain aldehyde; RCOOH, long-chain fatty acid (Adapted from Baker *et al.*, 1992.)

into the corresponding fatty acid that provides the energy for light emission. Bacterial luciferase has a high specificity for its primary energy source $FMNH_2$. Reduced flavin mononucleotide is readily produced in microorganisms via the electron transport chain (Hastings *et al*, 1985). Bioluminescent bacteria belong to four main genera *Vibrio, Photobacterium, Xenorhabdus* and *Alteromonas* (Baker *et al.*, 1992). The phenotypic characteristic of bioluminescence is linked to the possession of genes coding for the production of bacterial luciferase and for a long-chain fatty acid reductase complex. The fatty acid reductase is responsible for providing a source of long-chain aldehydes necessary for the reaction.

The genes responsible for bacterial bioluminescence have been extensively studied (Meighen, 1991b) and are collectively known as *lux* genes. In *Vibrio fischeri*, a bioluminescent marine bacterium, genes exist for structural components of the luciferase enzyme (*lux* A and *lux* B), for structural components of the enzymes responsible for aldehyde biosynthesis (*lux* C, *lux* D and *lux* E) and for regulation of the gene expression (*lux*I and *lux* R), together with other genes of speculative roles (*lux* F, *lux* G and *lux* H) (Meighen, 1991b). The advent of molecular techniques allowing gene manipulation has provided both for the study of the mechanisms of bacterial bioluminescence and for the development of highly specific assays for microbial detection.

7.3.1 Detection of pathogens and indicator microorganisms

Bacteriophages are bacterial viruses which have a high specificity to their targets (Horne, 1978). Specificity may be to a wide host range (several bacterial species) or to a very narrow host range (single species or sub-species). Bacteriophages have no metabolic function, as this is provided entirely by the host bacterium. Bacteriophage infection begins with recognition and attachment to the specific host cell. This is rapidly followed by injection of bacteriophage DNA into the host cell and subsequent incorporation (transduction) of this DNA into the host cell genome. The principal role of the bacterial host is to provide a metabolic facility for the replication of bacteriophage, which is controlled by bacterial expression of the bacteriophage genes. Bacteriophages can be used as specific vectors of the *lux* genes by genetically inserting these genes into the DNA of a phage of predetermined specificity (i.e phage p22 to *Salmonella typhimurium* LT2). Subsequent exposure of the genetically modified bacteriophage to its target bacterium results in rapid infection and expression of the *lux* genes, conferring a bioluminescent property to the host cell. Non-bioluminescent bacteria can therefore be specifically detected using this rapid bacteriophage luminescence assay, with the entire procedure, ending in light production, taking approximately 1 h. Since the end-point of the assay is the generation of light, detection can be achieved using a standard luminometer. The principal problems of these techniques are derived from the need to select bacteriophages of sufficient specificity to ensure minimisation of cross-reactions with non-target bacteria (false-positive reactions) or lack of recognition of the desired bacteria (false-negative reactions) (Baker *et al.*, 1992). In addition, insertion of the *lux* genes into the selected bacteriophage is quite complex and necessitates deletion of bacteriophage DNA and substitution of the *lux* genes. This must be achieved without compromising the integrity of the bacteriophage in such a way as to affect subsequent infection and gene expression. Interestingly, it is not essential to insert all lux genes into the bacteriophage for exploitation of this technology (Meighen, 1991a) and in many cases it is sufficient to insert only the genes coding for the production of luciferase, since the long-chain aldehyde can be supplemented to the reaction.

Stewart *et al.* (1989) inserted *lux* genes into a bacteriophage (p22) with a very narrow host range (*Salmonella typhimurium* LT2) and demonstrated a 60 min assay for the detection of 100 cells of *S. typhimurium*. Other workers have reported similar sensitivity (Meighen, 1991b). Detection of low levels of microorganisms with bacteriophage systems is reliant on a number of factors, which include the sensitivity of the luminometer used for the detection of light, together with the activity of the promoter genes controlling the expression of the luciferase complex (Meighen, 1991a). The current need to detect pathogenic microorganisms at levels as low as 1 cell per sample weight (25 g) necessitates the use of pre-enrichment stages to resuscitate and elevate numbers of the target cell to detectable levels. Turpin *et al.* (1993) developed a most probable number procedure for the detection and enumeration of *Salmonella typhimurium* in environmental samples (soils, water, sewage sludge). Using artificially inoculated samples containing *S. typhimurium* LT2, they were able to demonstrate a 24 h assay with a sensitivity of 1 cell per 100 ml, giving no false-positive or false-negative reactions. Potentially, with improvements in resuscitation and pre-enrichment, it may be possible to elevate microbial populations to sufficient numbers (100 per ml) in shorter periods, which allow these bacteriophage bioluminescence systems to be applied as preliminary screening assays (4–6 h) for pathogen detection (Stewart, 1990). Since the specificity of the assay can be dictated by the chosen bacteriophage, assays can be developed for the detection and enumeration of other pathogens, such as *Listeria monocytogenes,* and, indeed, indicator microorganisms, such as Enterobacteriaceae.

Tests for indicator organisms are used to reflect the degree of hygiene operated in the factory environment and are applied to raw materials, process samples, end products and environmental swabbing samples. Bacteriophage bioluminescence could potentially provide a mechanism for the rapid (<1 h) monitoring of these indicator organisms, with a sensitivity of 100–1000 cells per ml or gram (Stewart *et al.*, 1989). Like ATP bioluminescence hygiene monitoring, the generation of rapid results provides for on-line monitoring and enables corrective action to be taken quickly. Development of tests for these indicator organisms has been reported, but lack of funding by industrial groups has prevented its rapid transition to the market.

7.3.2 Detection of inhibitory substances

Bacterial bioluminescence provides an excellent mechanism to detect the presence of inhibitory substances, such as antibiotics, bacteriophages or other antimicrobial compounds. Starter cultures (*Lactobacillus* spp. or *Streptococcus* spp.) can be transformed to contain the *lux* genes using plasmid vectors instead of bacteriophages. The transformed bacteria are capable of producing bacterial luciferase and, since both the luciferase production and the production of $FMNH_2$ are linked to bacterial metabolic activity, any compound inhibiting bacterial activity can be detected by decreases in the light output of these transformed organisms (Stewart, 1990; Meighen, 1991b). Stewart *et al.* (1989) developed a plasmid vector (pSB154) to introduce *lux* genes into *Streptococcus lactis* (now *Lactococcus lactis*), *Lactobacillus casei* and *Lactobacillus plantarum,* and proposed that cultures actually used by the industry could be transformed and used for monitoring inhibitory substances to accurately reflect the true effects on the cultures in use. Using genetically transformed *Lactobacillus casei* and *Lactococcus lactis*, Ahmad and Stewart (1991) were able to demonstrate a 30–60 min assay for tetracycline (0.2 μg/ml) and penicillin G (0.05 units/ml), respectively. Similar findings were evident for nine other antibiotics. In the same report, a transformed strain of *Lactococcus lactis* subsp. *diacetylactis* was used to detect bacteriophage infection within 50 min (10^7 ml) or 100 min (10^5 ml).

In a similar way, transformed bacteria can be used to rapidly determine the antimicrobial activity of biocides. The effective concentration of an antimicrobial compound can be quickly determined by exposing serial dilutions of a sample of biocide to a transformed target microorganism and then monitoring light output. This approach has been used with *Listeria monocytogenes* (Walker *et al.*, 1992) and *Escherichia coli* (Stewart *et al.*, 1991).

7.4 Future trends

It is evident that the ATP bioluminescence assays which have been utilized with the greatest success are those where the assay format is either automated (UHT sterility testing) or where it is highly simplified (hygiene monitoring). The need of the industry has been satisfied by provision of rapid results with assay kits requiring the minimum of technical expertise. Other assays, such as the Milk Microbial Kit (Biotrace Ltd), are on the verge of use, but require, perhaps, even more simplicity for acceptance, and automated systems, such as the Bactofoss, originally over-priced, may make an impact at a more competitive capital and reagent cost. Few commercial assays exist based on bacterial bioluminescence, even though the technology appears to offer much to a number of industries. It is probable that assays to determine biocide efficacy and, perhaps, specific detection of pathogenic bacteria and indicator organisms may be available within a few years and the familiarity of workers with ATP bioluminescence is likely to play a significant role in the uptake of bacterial bioluminescence systems.

ATP bioluminescence and bacterial bioluminescence assays could benefit from becoming more sensitive for certain applications, but, in the majority of cases, the sensitivity of the instruments and reagents is already sufficient. It is the ability to remove interfering compounds, such as food particles, that compromises the sensitivity of the assays. In fact, this is a phenomenon common to all but a few microbiological methods, no matter how sophisticated. The development of many powerful techniques for the detection of microorganisms is being seen continually, but, almost without exception, it is the presence of food debris that compromises the sensitivity and specificity of these assays. In order to compensate for this, most end detection systems assay only a minute portion of the original sample or enrichment broth to avoid product interference. The vast majority of any sample or broth is discarded, even though it may contain >99.9% of the microorganisms. Future efforts must address the issue of separating and concentrating microorganisms from foods or enrichment broths prior to detection, to harness this 'hidden' population and significantly improve sensitivity. The value of immunomagnetic technologies has already been demonstrated for *Salmonella* and food-spoilage yeasts (see chapter 4) and several are now commercially available for *Salmonella* and *Listeria* (see chapter 3), with others to follow (e.g. *E. coli* O157:H7). In addition, the potential of solid and liquid phase ion exchangers as aids to the rapid detection of food spoilage microorganisms has also been demonstrated in this chapter.

Bioluminescence and chemiluminescence based immunoassays offer highly sensitive detection systems, and these are likely to be applied in food microbiology (e.g. pathogen detection) where extreme sensitivities are needed to significantly reduce the total analysis times, perhaps in combination with separation systems already considered previously. These detection technologies have been reviewed recently by Geiger *et al.* (1989). To cater for the needs of industry, the problem of robustness and reliability of these techniques, particularly in combination with simple sample preparation and sensitive detection formats, needs to be addressed.

Undoubtedly the next major step forward will come by combining advances from a variety of technologies.

References

Ahmad, K.A. and Stewart, G.S.A.B. (1991) The production of bioluminescent lactic bacteria suitable for the rapid assessment of starter culture activity in milk. *J. Appl. Bacteriol.*, **70** (2) 113–20.

Andersen, T.H. (1989) Rapid hygienic testing of milk tankers. *Scand. Dairy Ind.*, **1**, 36–37.

Anon. (1985a) *Code of Practice for the Assessment of Milk Quality*, 5th edn. The Joint Committee of the Milk Marketing Board and Dairy Trade Federation, Thames Ditton, UK.

Anon. (1985b) Resazurin dye reduction tests, in *Code of Practice for the Assessment of Milk Quality*, 5th edn. The Joint Committee of the Milk Marketing Board and Dairy Trade Federation, Thames Ditton, UK.

Anon. (1985c) Delvotest P method for the detection of antibiotics and other inhibitory substances in milk (ampoule method), in *Code of Practice for the Assessment of Milk Quality*, 5th edn. The Joint Committee of the Milk Marketing Board and Dairy Trade Federation, Thames Ditton, UK. (Revised March 1992).

Anon. (1987) Inhibitor tests with *Streptococcus thermophilus*. Bioluminescence (ATP) method. *Bull. Int. Dairy Fed.*, **220**, 27–32.

Avis, J.W. (1988) *Microbiologia Applicata. Metodi Rapidi ed Automatizzati*. Societa Editorial Farmaceutica Milano, Milan, p. 39.

Baker, J.M., Griffiths, M.W. and Collins-Thompson, D.L. (1992) Bacterial bioluminescence: applications in food microbiology. *J. Food Prot.*, **55** (1) 62–70.

Bautista, D.A., McIntyre, L., Laleye, L. and Griffiths, M.W. (1992) The application of ATP bioluminescence for the assessment of milk quality and factory hygiene. *J. Rapid Methods Automat. Microbiol.*, **1**, 179–93.

Bossuyt, R. (1981) Determination of bacteriological quality of raw milk by an ATP assay technique. *Milchwissenschaft*, **36** (5) 257–262.

Bossuyt, R. (1982) A 5-minute ATP platform test for judging the bacteriological quality of raw milk. *Neth. Milk Dairy J.*, **36**, 355–364.

Bramley, A.J. and McKinnon, C.H. (1990) The microbiology of raw milk, in *Dairy Microbiology, Volume 1, The Microbiology of Milk* (ed. Robinson, R.K.) Elsevier Applied Science, London, pp. 163–208.

Campbell, I. (1987) Wild yeasts in brewing and distilling, in *Brewing Microbiology* (eds. Priest, F.G. and Campbell, I.) Elsevier Applied Science London, pp. 187–205.

Coleman, E. (1992) Selection of appropriate ATP systems for the food and beverage industries. *J. Biolumin. Chemilumin.* 7, 256.

DeLuca, M. and McElroy, W.D. (1978) Purification and properties of firefly luciferase, in *Methods in Enzymology* (ed. DeLuca, M.), Academic Press, London, pp. 3–15.

Eriksen, B. and Olsen, O. (1989) Rapid assessment of the microbial status of bulk milk and raw meat with the new instrument: Bactofoss, in *ATP luminescence. Rapid Methods in Microbiology* (eds. Stanley, P.E. McCarthey, B.J. and Smither, R.). Blackwell Scientific, Oxford, pp. 175–181.

Geiger, R., Hauber, R. and Miska, W. (1989) New, bioluminescence - enhanced detection systems for use in enzyme activity tests, enzyme immunoassays, protein blotting and nucleic acid hybridisation. *Molec. Cell. Probes*, **3**, 309–328.

Gilmour, A. and Rowe, M.T. (1990) Microorganisms associated with milk, in *Dairy Microbiology, Volume 1, The Microbiology of Milk* (ed. Robinson, R.K.). Elsevier Applied Science, London, pp. 37–75.

Graumlich, T.R. (1985) Estimation of microbial populations in orange juice by bioluminescence. *J. Food Sci.*, **50**, 116–17.

Griffiths, M.W. (1989) ATP to detect bacteria in dairy products *J. Soc. Dairy Technol.*, **42** (3) 61–62.

Griffiths, M.W., Phillips, J.D. and Muir, D.D. (1981) Thermostability of proteases and lipases for a number of species of psychrotrophic bacteria of dairy origin. *J. Appl. Bacteriol.* **30** (2) 209–303.

Griffiths, M.W., Phillips, J.D. and Muir, D.D. (1984) Methods for rapid detection of post-pasteurisation contamination in cream. *J. Soc. Dairy Technol.*, **(37)** (1) 22–26.

Griffiths, M.W., McIntyre, L., Sully, M. and Johnson I. (1991) Enumeration of bacteria in milk, in *Bioluminescence and Chemiluminescence: Current Status* (eds. Stanley, P.E. and Kricka, L.J.). John Wiley and Sons, Chichester, U.K pp. 479–481.

Hastings, J.W., Poitrikus, C.J., Gupta, S.C., *et al* (1985) Biochemistry and physiology of bioluminescent bacteria, in *Advance in Microbial Physiology*. vol. 26, (eds. Rose, A.H. and Tempest, D.W.). Academic Press, London, pp.236–291.

Hawronskyj, J.-M., Adams, M.R. and Kyriakides, A.L. (1993) Rapid detection of antibiotics in raw milk by ATP bioluminescence. *J. Soc. Dairy Technol.*, **46** (1) 31–33.

Hicks, C.L., Onuorah, C., O'Leary, J. and Langlois, B.E. (1986) Effect of milk quality and low temperature storage on cheese yield - a summation. *J. Dairy Sci.*, **69**, 649–657.

Horne, R.W. (1978) The structure and function of viruses, in *The Institute of Biology's Studies in Biology*. Edward Arnold, London.

Hysert, D.W., Kovecses, F. and Morrison, N.M. (1976) A firefly bioluminescence ATP assay method for rapid detection and enumeration of brewery microorganisms. *J. Am. Soc. Brewing Chemists*, **34**, 145–150.

Jago, P.H., Simpson, W.J., Denyer, S.P., *et al.* (1989) An evaluation of the performance of ten commercial luminometers. *J. Biolumin. Chemilumin.*, **3**, 131–145.

Jassim, S.A.A., Ellison, A., Denyer, S.P. and Stewart, G.S.A.B. (1990) *In vivo* bioluminescence: a cellular reporter for research and industry. *J. Biolumin. Chemilumin.*, **5**, 115–122.

Kennedy, J.E. and Oblinger, J.L. (1985) Application of bioluminescence to rapid determination of microbial levels in ground beef. *J. Food Prot.*, **48** (4) 334–340.

Kilgour, W.J. and Day, A. (1983) The application of new techniques for the rapid determination of microbial contamination in brewing, in *Proceedings of the European Brewing Convention Congress*. IRL Press, Oxford, pp. 177–184.

Kyriakides, A.L., Costello, S.M., Easter, M.C. and Johnson, I. (1991) Rapid hygiene monitoring using ATP bioluminescence, in *Bioluminescence and Chemiluminescence : Current Status* (eds. Stanley, P.E. and Kricka, L.J.). John Wiley and Sons, Chichester, pp. 519–522.

LaRocco, K.A., Galligan, P., Littel, K.J. and Spurgash, A. (1985) A rapid bioluminescent ATP method for determining yeast contamination in a carbonated beverage. *Food Technol.*, **39** 49–52.

LaRocco, K.A., Littel, K.J. and Pierson, M.D. (1986) The bioluminescent ATP assay for determining the microbial quality of foods, in *Foodborne Microorganisms and their Toxins:Developing Methodology* (eds. Pierson, M.D. and Stern, N.J.). Marcel Dekker New York, pp. 145–174.

Learoyd, S.A., Kroll, R.G. and Thurston, C.F. (1992) An investigation of dye reduction by foodborne bacteria. *J. Appl. Bacteriol.*, **72** (6) 479–485.

Littel, K.J. and LaRocco, K.A. (1986) ATP screening method for presumptive detection of microbiologically contaminated carbonated beverages. *J. Food Sci.*, **51** (2) 474–476.

Littel, K.J., Pikelis, S. and Spurgash, A. (1986) Bioluminescent ATP assay for rapid estimation of microbial numbers in fresh meat. *J. Food Prot.*, **49** (1) 18–22.

Lundin, A. (1982) Analytical applications of bioluminescence: the firefly system. *Clin. Biochem. Anal.*, *42*, 43–74.

MacLeod, N.H., Chappelle, E.W. and Crawford, A.M. (1969) ATP assay of terrestrial soils: a test of an exobiological experiment. *Nature*, **223**, 267–268.

Martin, L. (1991) A rapid concentration procedure for microorganisms in raw milk. Presented at the International Association of Milk, Food and Environmental Sanitarians Annual Meeting, 21–24 July, Ames, Iowa.

McElroy, W.D. (1947) The energy source for bioluminescence in an isolated system. *Zoology*, *33*, 342–345.

McElroy, W.D. and Strehler, B.L. (1949) Factors influencing the response of the bioluminescent reactions to adenosine triphosphate.. *Arch. Biochem. Biophys.*, *22*, 420–433.

Meighen, E.A. (1991a) Molecular biology of bioluminescence, in *Bioluminescence and Chemiluminescence: Current Status*, (eds. Stanley, P.E. and Kricka, L.J.). John Wiley and Sons, Chichester, pp. 3–10.

Meighen, E.A. (1991b) Molecular biology of bacterial bioluminescence. *Microbiol. Rev.*, **55**, 123–142.

Miller, L.F., Mabee., M.S., Gress, H.S. and Jangaard, N.O. (1978) An ATP bioluminescence method for the quantification of viable yeast for fermenter pitching. *J. Am. Soc. Brewing Chemists*, **36**, 59–62.

Ono, M., Yamamoto, Y., Yamamoto, R. *et al* (1987) Application of ATP photometry in brewing. The *Brewers Digest*, **62 (5)**.

Patel, P.D. (1984) The potential of chromatographic techniques for the manipulation of viable microorganisms. *PhD thesis*, University of Surrey.

Patel, P.D. (1991) Enzymes as diagnostic tools, in *Enzymes in Food Processing* (eds. Tucker, G.A. and Woods, L.F.J.). Blackie, Glasgow, pp. 262–280.

Patel, P.D. and Williams, A.P. (1985) A note on estimation of food spoilage yeasts by measurements of adenosine triphosphate (ATP) after growth at various temperatures. *J. Appl. Bacteriol.*, **59**, 133–136.

Pettipher, G.L. (1983) *The Direct Epifluorescent Filter Technique for the Rapid Enumeration of Micro-organisms*. Research Studies Press, Letchworth Priest, F.G. (1987) Gram-positive brewery bacteria. In *Brewing Microbiology* (eds. Priest, F.G. and Campbell, I.). Elsevier Applied Science, London, pp. 121–154.

Priest, F.G. (1987) Gram-positive brewery bacteria, in *Brewing Microbiology* (eds Priest, F.G. and Campbell, I.). Elsevier Applied Science, London, pp. 121–154.

Reybroeck, W. and Schram, E. (1992) Study of parameters involved in the assessment of the bacteriological quality of raw milk by two bioluminescent ATP assays. *J. Biolumin. Chemilumin.*, 7, 259.

Reybroeck, W. and Waes, G. (1992) Comparison of two ATP test kits for the assessment of the bacteriological quality of raw milk. *J. Biolumin. Chemilumin.*, 7, 260.

Rigarlsford, J. (1992) The pros and cons of using ATP to monitor hygiene. *J. Biolumin. Chemilumin.*, 7, 258–259.

Robinson, R.K. (1990) *Dairy Microbiology, Volume 1, The Microbiology of Milk*, Elsevier Applied Science, London.

Rowe, M.T., Pearce, J., Crone, L., *et al* (1991) Bioluminescence assay for psychrotroph proteases, in *Bioluminescence and Chemiluminescence : Current Status* (eds. Stanley, P.E. and Kricka, L.J.). John Wiley and Sons, Chichester, pp. 515–518.

Schram, E. (1991) Evolution of bioluminescent ATP assays, in *Bioluminescence and*

Chemiluminescence : Current Status (eds. Stanley, P.E. and Kricka, L.J.) John Wiley and Sons, Chichester, pp. 407–412.

Sharpe, A.N., Woodrow, M.N. and Jackson, A.K. (1970) Adenosine triphosphate (ATP) levels in foods contaminated by bacteria. *J. Appl. Bacteriol.*, **33** 758–767.

Simpson, W.J. and Hammond, J.R.M. (1989) Cold ATP extractants compatible with constant light signal firefly luciferase reagents, in *ATP Luminescence : Rapid Methods in Microbiology* (eds. Stanley, P.E., McCarthey, B.J. and Smither, R.). Blackwell Scientific, Oxford, pp. 45–52.

Simpson, W.J., Hammond, J.R.M., Thurston, P.A. and Kyriakides, A.L. (1989) Brewery process control and the role of `instant' microbiological techniques, in *Proceedings of the European Brewing Convention Congress*. IRL Press, Oxford, pp. 663–674.

Slater, K. (1993) Plant hygiene – the new dimension. *Int. Food Hyg.*, **3** (5) 11–15.

Stanley, P.E. (1982) Rapid measurements of bacteria by ATP assay. *Lab. Equipment Digest*, **February**, 62–67.

Stanley, P.E. (1989) A concise beginner's guide to rapid microbiology using adenosine triphosphate (ATP) and luminescence, in *ATP Luminescence: Rapid Methods in Microbiology* (eds. Stanley, P.E., McCarthey, B.J. and Smither, R.). Blackwell Scientific Oxford, pp. 1–11.

Stanley, P.E. (1992) A survey of more than 90 commercially available luminometers and imaging devices for low-light measurements of chemiluminescence and bioluminescence, including instruments for manual, automatic and specialised operation, for HPLC, LC, GLC and microtitre plates. Part1. Descriptions. *J. Biolumin. Chemilumin.*, **7**, 77–108.

Stannard, C.J. (1989) ATP estimation, in *Rapid Methods in Food Microbiology* (eds. Adams, M.R. and Hope, C.F.A.). Elsevier Science Publishers, Amsterdam, pp. 1–18.

Stannard, C.J. and Williams, A.P. (1987) ATP assay for shelf-life prediction of food, in *Bioluminescence and Chemiluminescence, New Perspectives* (eds. J. Scholmerich, R. Andresen, A. Kapp, M. Ernst and W.G. Woods) John Wiley, Chichester, pp. 481–490.

Stannard, C.J. and Wood, J.M. (1983a) The rapid estimation of microbial contamination of raw meat by measurement of adenosine triphosphate (ATP). *J. Appl. Bacteriol.*, **55**, 429–438.

Stannard, C.J. and Wood, J.M. (1983b) *Rapid Estimation of Yeast in Fruit Juices by ATP Measurements* (Leatherhead F.R.A. Research Report No. 443). Leatherhead Food Research Association, Leatherhead, Surrey.

Stewart, G.S.A.B. (1990) *In vivo* bioluminescence: new potentials for microbiology. *Letters Appl. Microbiol.*, **10**, 1–8.

Stewart, G., Smith, T. and Denyer, S. (1989) Genetic engineering for bioluminescent bacter. *Food Sci. Technol. Today*, **3** (1) 19–22.

Stewart, G.S.A.B., Jassim, S.A.A. and Denyer, S.P. (1991) *Mechanisms of Action of Chemical Biocides: Their Study and Exploitation* (eds. Denyer, S.P. and Hugo, W.B.). Blackwell Scientific, Oxford, pp. 319–329.

Turpin, P.E., Maycroft, K.A., Bedford, J., *et al* (1993) A rapid luminescent-phage based MPN method for the enumeration of *Salmonella typhimurium* in environmental samples. *Lett. Appl. Microbiol.*, **16** (1) 24–27.

Van Crombrugge, J. (1989) The ATP-F test for estimation of bacteriological quality of raw milk. *Neth. Milk Dairy J.*, **43**, 347–354.

Waes, G. (1982) A rapid method to detect postcontamination in pasteurised milk. *Antonie van Leuwenhoek*, **48**, 407–408.

Waes, G. (1984) A rapid method for the detection of non-sterile UHT milk by the determination of the bacterial ATP. *Milchwissenschaft*, **39** (12) 707–711.

Waes, G. and Bossuyt, R.G. (1981) A rapid method to detect postcontamination in pasteurised milk. *Milchwissenschaft*, **36**, 548–552.

Waes, G.M. and Bossuyt, R.G. (1982) Usefulness of the benzalkon-crystal violet-ATP method for predicting the keeping quality of pasteurised milk. *J. Food Prot.*, **45**, 928–931.

Waes, G., Van Crombrugge, J. and Reybroeck, W. (1989) The ATP-F test for estimation of bacteriological quality of raw milk, in *Modern Microbiological Methods for Dairy Products*. IDF, Brussels, pp. 279–286.

Walker, A.J., Jassim, S.A.A., Holah, J.T., *et al* (1992) Bioluminescent *Listeria monocytogenes* provide a rapid assay for measuring biocide efficacy. *FEMS Microbiol. Lett.*, **91**, 251–256.

Ward, D.R., LaRocco, K.A. and Hopson, D.J. (1986) Adenosine triphosphate assay to enumerate bacterial numbers on fresh fish. *J. Food Prot.*, **49** (8) 647–650.

Webster, J.J., Hall, M.S., Rich, C.N., *et al* (1988) Improved sensitivity of the bioluminescent determination of numbers of bacteria in milk samples. *J. Food Prot.*, **51** (12) 949–954.

Weeks, I. and Woodhead, J.S. (1988) Chemiluminescence immunoassays. *Trends Analyt. Chem.*, **7**, 55–58.

Westhoff, D.C. and Engler, T. (1975) Detection of penicillin in milk by bioluminescence. *J. Milk Food Technol.*, **38**, 537–539.

Williams, G.R. (1984) Use of bioluminescence in the determination of antibiotic content of milk. *J. Soc Dairy Technol.*, **37**, 40–41.

8 Modern methods for detecting and enumerating foodborne fungi

J. I. PITT and A. D. HOCKING

8.1 Introduction

Quantification of the growth of filamentous fungi is more difficult than for bacteria or yeasts. Vegetative growth consists of hyphae that are not readily detached from the substrate and which survive blending poorly. When sporulation occurs, very high numbers of spores may be produced, causing sharp rises in viable counts, often without any great increase in biomass (Pitt, 1984).

The estimation of biomass itself is not easy, because no primary standard exists (such as cell numbers used for yeasts and bacteria). Although progress has been made in recent years in techniques for quantifying biomass, most food laboratories continue to rely on viable counting for detecting and quantifying fungal growth in foods. Two main microbiological methods exist for estimating fungal numbers: direct and dilution plating. These are described in detail below. Techniques for biomass estimation will be discussed later.

8.2 Cultural methods

8.2.1 Direct plating

Direct plating is the preferred method for detecting, enumerating and isolating fungi from particulate foods such as grains and nuts (Samson et al., 1992). In direct plating, food particles are placed directly on solidified agar media. In most situations, particles should be surface disinfected before plating, as this removes the inevitable surface contamination arising from dust and other sources, and permits recovery of the fungi actually growing in the particles. This process provides an effective measure of inherent mycological quality and permits assessment of the potential presence of mycotoxins as well.

Surface disinfection should be omitted only where surface contaminants become part of the downstream mycoflora, for example in grain intended for flour manufacture. Even here, surface disinfection before direct plating provides the most realistic appraisal of actual grain quality.

Results from direct plating analysis are expressed as percentage infection of particles. The technique provides no direct indication of the *extent* of fungal invasion in individual particles. However, it is reasonable to assume that a high percentage infection is correlated with extensive invasion in the particles and a higher risk of mycotoxin occurrence.

Standard protocols for direct plating now exist (Pitt *et al.*, 1992), and are described below.

8.2.1.1 Surface disinfection.

Surface disinfection is carried out by immersing particles in household chlorine bleach (nominally 4–5% active chlorine), diluted 1 to 10 with water. Immerse particles for 2 min, stirring occasionally, then drain the chlorine. Fifty or more particles should be treated.

8.2.1.2 Rinsing.

After the chlorine is poured off, particles may be rinsed once with sterile water. Use a 1 min treatment, with stirring. The rinsing step can probably be omitted without loss of efficacy of the treatment, with savings in time and materials and reduced risk of recontamination from the air.

8.2.1.3 Plating.

After disinfection and the optional rinse, particles should be plated onto solidified agar, at the rate of 6–10 particles per plate, depending on particle size.

8.2.1.4 Incubation.

Incubate plates upright, under normal circumstances, for 5 days at 25°C.

8.2.1.5 Examination.

After incubation, examine plates visually, count the numbers of infected particles, and express results as a percentage. Differential counting of various genera is often possible. Correct choice of media, a stereo-microscope and experience will all assist in this process.

8.2.2 Dilution plating

Dilution plating is the appropriate method for mycological analysis of liquid or powdered foods. It is also suitable for grains intended for flour manufacture and other situations where total fungal contamination is relevant.

8.2.2.1 Sample preparation.

The two most common methods of sample preparation for dilution plating are stomaching and blending: stomaching is recommended (King *et al.*, 1986). Treatment time in the stomacher should be 2 min. For harder or particulate foods, such as grains and nuts, comminution in a Waring blender or similar machine may give a more satisfactory homogenate. Blending times should not exceed 60 s.

The sample size used should be as large as possible. If a stomacher 400 is used, a 10–40 g sample is suitable.

8.2.2.2 Diluents.

The recommended diluent is aqueous 0.1% peptone (Kurtzman *et al.*, 1971). This is suitable for both filamentous fungi and yeasts. Saline solutions, phosphate buffer or distilled water may also be used. The addition of a wetting agent such as polysorbitan 80 (Tween 80) may be desirable for some products.

Special diluents may be necessary in some circumstances. If yeasts are to be enumerated from dried products or juice concentrates, the diluent should also contain 20–30% sucrose, glucose or glycerol, as the cells may be injured or be susceptible to osmotic shock.

8.2.2.3 Dilution. Serial dilutions of fungi are carried out by the same procedures as those used in bacteriology, and the recommended dilution rate is 1:10 (=1+9). Fungal spores sediment more quickly than bacteria, so it is important to draw samples for dilution or plating as soon as possible, preferably within 1 min, to avoid inaccurate results due to settling (Beuchat, 1992).

8.2.2.4 Plating. Spread (surface) plating is recommended. When pour plates are used, fungi develop more slowly from beneath the agar surface and may be obscured by faster growing colonies from surface spores. Hence, spread plating allows more uniform colony development, improves the accuracy of enumeration of the colonies and makes subsequent isolation of pure cultures easier.

The optimum inoculum volume for surface plating is 0.1 ml. Best results will be obtained if plates are dried slightly before use. It is usually possible to enumerate plates with up to 150 colonies, but if a high proportion of rapidly growing fungi are present, the maximum number which can be distinguished with any accuracy will be lower. Because of this restriction on maximum numbers, it may be necessary to accept counts from plates with as few as 10–15 colonies. Clearly, such limitations on numbers per plate and the overgrowth of slow colonies will result in counting errors which are higher than those usually achieved with bacteria or yeasts.

Enumerating yeasts is less difficult. In the absence of filamentous fungi, from 30 to 300 colonies per plate can be counted and errors will be comparable with those to be expected in bacterial enumeration.

8.2.2.5 Incubation. The standard incubation conditions are 25°C for 5 days.

8.2.2.6 Reporting results. As in bacteriology, results from dilution plating are expressed as viable counts per gram of sample. Note that such results are not directly comparable with those obtained from direct plating, and again do not offer a direct indication of the extent of fungal growth.

8.2.3 Modern media

Food laboratories often rely on a single all purpose medium to produce a 'yeast and mould' count in everything from raw material to finished product. But just as the food bacteriologist uses selective media for particular purposes, so too food mycologists are developing a range of media suited to specific applications. It is plainly unrealistic to expect a single medium to answer all questions about mould and yeast contamination in foods. The fungi which spoil meats or fresh vegetables are not the same as those which grow on dried fish. Although often used for this purpose, very dilute media such as potato dextrose agar are of little or no value for enumerating fungi from dried foods.

The most important division in types of enumeration media lies between those suitable for high water activity foods such as fruits, vegetables, eggs, meat and dairy products, and those suited to the enumeration of fungi in intermediate moisture and dried foods. The most suitable media for dried foods depend on the type of food, the three major categories being foods low in soluble solids, such as cereals, grains and nuts, high-sugar foods such as confectionery and dried fruit, and salt foods. A second consideration lies in whether the primary interest is in moulds, or yeasts, or both, and a third concerns the presence or absence of preservatives. Finally, media are available for specific mycotoxigenic fungi, notably *Aspergillus flavus* and related species and *Penicillium verrucosum* plus *P. viridicatum*.

8.2.3.1 Media for routine enumeration of fungi.   Modern fungal enumeration media rely on the use of antibiotics at neutral pH for the inhibition of bacteria. Such media allow better recovery of moribund and sensitive fungi than the acidified media commonly used in the past. For many years rose bengal has been added to media to decelerate colony spread, while a more recent development is the use of 2,6-dichloro-4-nitroaniline (dichloran) to inhibit the most rapidly spreading moulds. Many common spoilage fungi, *Aspergillus* and *Penicillium* species in particular, develop better on media with adequate nutrients. Low nutrient media, such as potato dextrose agar, have lost favour because they are selective against some species in these genera.

The formulations given below are considered to be the most satisfactory general purpose enumeration media available at this time (Samson *et al.*, 1992).

8.2.3.1.1 Dichloran rose bengal chloramphenicol agar (DRBC).   DRBC (King *et al.*, 1979; modified by Pitt and Hocking, 1985) is recommended for both moulds (Beuchat and Brackett, 1992) and yeasts (Deak, 1992). It is particularly suited to fresh and high a_w foods (Hocking *et al.*, 1992). This medium contains both rose bengal (25 mg/kg) and dichloran (2 mg/kg), which restrict colony spreading without unduly affecting spore germination. Compact colonies allow crowded plates to be counted more accurately. This combination of inhibitors also effectively restricts the rampant growth of most of the common mucoraceous fungi such as *Rhizopus* and *Mucor*, although it does not completely control some other troublesome genera such as *Trichoderma* and *Chrysonilia*.

DRBC consists of

Glucose	10 g
Peptone, bacteriological	5 g

KH_2PO_4	1.0 g
$MgSO_4.7H_2O$	0.5 g
Agar	15 g
Water, distilled	1 litre
Rose bengal (5% in water)	0.5 ml
Dichloran (0.2% in ethanol)	1.0 ml
Chloramphenicol	100 mg

After addition of all ingredients, sterilise by autoclaving at 121°C for 15 min. Store prepared media away from light; photoproducts of rose bengal are highly inhibitory to some fungi, especially yeasts. In the dark, the medium will keep for months. The stock solutions of rose bengal and dichloran need no sterilisation, and are also stable for very long periods, provided they are protected from light.

In routine use, it is recommended that DRBC plates be incubated at 25°C for 5 days.

8.2.3.1.2 Dichloran 18% glycerol agar (DG18). Hocking and Pitt (1980) developed DG18 for enumeration of xerophilic fungi from low moisture foods such as stored grains, nuts, flour and spices. DG18 was designed for enumeration of a range of non-fastidious xerophilic fungi and yeasts. However, DG18 can now be described as a general-purpose medium, with emphasis on enumeration of fungi from intermediate moisture and dried foods.

DG18 consists of

Glucose	10 g
Peptone	5 g
KH_2PO_4	1.0 g
$MgSO4.7H_2O$	0.5 g
Glycerol, AR	220 g
Agar	15 g
Water, distilled	1 litre
Dichloran (0.2% in ethanol)	1.0 ml
Chloramphenicol	100 mg

To produce this medium, add minor ingredients and agar to *c.* 800 ml distilled water. Steam to dissolve agar, then make up to 1 litre with distilled water. Add glycerol: note that the final concentration is 18% weight in weight, not weight in volume. Sterilise by autoclaving at 121°C for 15 min. The final a_w of this medium is 0.955.

8.2.3.1.3 Alternative media. Under circumstances where rapidly spreading moulds do not cause problems, two alternative general purpose enumeration media are satisfactory. These are rose bengal chloramphenicol agar (RBC; Jarvis, 1973), from which DRBC was developed, and oxytetracycline glucose yeast extract agar (OGY; Mossel *et al.*, 1970). OGY has been found to be very suitable for yeasts in the absence of moulds (Andrews, 1992).

RBC consists of

Glucose	10 g
Peptone	5 g
KH_2PO_4	1.0 g
$MgSO_4.7H_2O$	0.5 g
Agar	15 g
Water, distilled	1 litre
Rose bengal	50 mg
Chloramphenicol	100 mg

After addition of all ingredients, sterilise by autoclaving at 121°C for 15 min. Store away from light.

OGY consists of

Glucose	20 g
Yeast extract	5 g
Agar	15 g
Water, distilled	1 litre

Sterilise by autoclaving at 121°C for 15 min. After tempering to 50°C, add 10 ml of filtered sterilised oxytetracycline (Terramycin, Pfizer; 0.1% aqueous) per 100 ml of medium.

8.2.3.2 Media for xerophilic fungi. Xerophilic fungi are of great importance in food spoilage, and hence media and techniques for their enumeration and isolation have received much attention in recent years. Xerophiles range from those which grow readily on normal media, such as many *Aspergillus* and *Penicillium* species, which are only marginally xerophilic, to those, such as *Xeromyces bisporus*, which will not grow at all on normal media. It is not surprising that no single medium is suitable for quantitative estimation of all xerophilic fungi causing food spoilage. As noted earlier, DG18 was developed as a general medium for xerophiles, and remains the medium of choice for this purpose. DG18 should be used in any general examination of the mycoflora of dried foods. However, the user should be aware that more specialised media will be necessary to deal with extreme xerophiles (see below).

8.2.3.3 Media for yeasts. The simplest enumeration and growth medium for most food spoilage yeasts is malt extract agar (MEA) (Raper and Thom, 1949). Although originally introduced as a growth medium for moulds, its rich nutritional status makes it very suitable for yeasts, and its relatively low pH (usually near 5.0) reduces problems with bacterial contamination.

MEA consists of

Malt extract	20 g
Peptone	1.0 g
Glucose	20 g

| Agar | 20 g |
| Water, distilled | 1 litre |

Sterilise by autoclaving at 121°C for 15 min. Do not sterilise for longer, as this medium will become soft on prolonged or repeated heating.

MEA is suitable for enumeration of yeasts in liquid products such as fruit juices and yoghurt, where moulds are usually present only in low numbers. If large numbers of moulds are present, which is often the case with solid products such as cheese, a general-purpose enumeration medium such as DRBC should be used.

Enrichment for yeasts in liquid products. In liquid food products, low numbers of yeasts may be difficult to detect, but may have a serious potential to cause spoilage. Enrichment techniques are the only satisfactory way of monitoring products in these circumstances.

For products or raw materials free from suspended solids and of low viscosity, standard membrane filtration techniques are a satisfactory method for detecting low numbers of yeasts. The filter can be placed directly onto a suitable medium such as MEA and staining can be carried out subsequently. Centrifugation can also be used, but has the disadvantage that only relatively small volumes of product can be screened.

If, as is often the case, products or raw materials are viscous, of a low a_w, or contain pulps and cannot be filtered efficiently, other enrichment techniques are needed. In many cases, the best enrichment medium is the product itself, usually diluted 1:1 with sterile water. A 1:1 dilution increases the a_w of juice concentrates or honey to a level which will allow growth of potential spoilage yeasts, without causing a lethal osmotic shock to the cells. If the product contains a preservative, dilution will lower the concentration and allow cells to grow.

To detect low numbers of spoilage yeasts in cordials, fruit juice concentrates and similar materials, simply decant half the product from the container and replace it with sterile water. Leave the cap loose, incubate at room temperature or 25°C, and watch for evidence of fermentation. Shaking the container daily will help to detect gases resulting from fermentation.

8.2.3.4 Selective media for mycotoxigenic fungi.

8.2.3.4.1 Aspergillus flavus *and related species.*   In the presence of appropriate nitrogen sources and ferric salts, *Aspergillus flavus* and *A. parasiticus* produce conspicuous, diagnostic orange–yellow colours in the colony reverses (Bothast and Fennell, 1974). Few other fungi produced a similar coloration. Based on this finding, Pitt *et al.* (1983) developed Aspergillus flavus and parasiticus agar (AFPA).

AFPA consists of

Peptone, bacteriological	10 g
Yeast extract	20 g
Ferric ammonium citrate	0.5 g

Chloramphenicol	100 mg
Agar	15 g
Water, distilled	1 litre
Dichloran (0.2% in ethanol)	1.0 ml

After addition of all ingredients, sterilise by autocaving at 121°C for 15 min. The final pH of this medium is c. 6.2.

When incubated at 30°C for 42–48 h, colonies of *Aspergillus flavus*, *A. parasiticus* and *A. nomius* are distinguished by bright orange–yellow reverse colours. Only *A. niger* can be a source of error: it grows as rapidly as *A. flavus* and sometimes produces a yellow, but not orange, reverse colour. After 48 h, *A. niger* colonies begin production of their diagnostic black or dark brown heads, which provide a ready distinction from *A. flavus*. Prolonged incubation of AFPA, beyond four days, is not recommended, because *Aspergillus ochraceus* and closely related species may also produce a yellow reverse after this time.

AFPA is recommended for the detection and enumeration of potentially aflatoxigenic fungi in nuts, corn, spices and other commodities (Hocking, 1982). Its advantages include rapidity, as 42 h incubation is usually sufficient; specificity; and simplicity, as little skill is required in interpreting results. In consequence, it can be a simple, routine guide to possible aflatoxin contamination (Pitt, 1984).

8.2.3.4.2 Media for fungi producing ochratoxin A. *Aspergillus ochraceus* was the original species found to produce ochratoxin A, but it appears to be unimportant as a source of this potent mycotoxin in foods (Pitt *et al.*, 1993). Most ochratoxin A entering the food and feed chain is produced by *Penicillium verrucosum* (Pitt, 1987). Frisvad (1983) developed dichloran rose bengal yeast extract sucrose agar (DRYS) for selective enumeration of *Penicillium verrucosum*. *Penicillium viridicatum* and *P. aurantiogriseum* are also selected by DRYS. *Penicillium verrucosum* produces ochratoxin A and citrinin, while the latter two species produce xanthomegnin and viomellein. According to Frisvad (1983), *P. verrucosum* colonies on DRYS have a violet–brown reverse, and the latter two species produce yellow colonies with a yellow reverse.

DRYS consists of

Yeast extract	20 g
Sucrose	150 g
Dichloran (0.2% in ethanol)	1 ml
Rose bengal (5% in water)	0.5 ml
Chloramphenicol	50 mg
Agar	20 g
Water, distilled	1 litre
Chlortetracycline (1% in water, filter sterilised)	5 ml

Sterilise all ingredients except chlortetracycline by autoclaving at 121°C for 15 min. Add chlortetracycline after tempering to 50°C. In the authors' experience, chloramphenicol at twice the concentration specified (i.e. 100 mg/litre) adequately controls bacteria in most situations, avoiding the need for a second antibiotic which must be filter sterilised. The incubation regime recommended by Frisvad (1983) is 7–8 days at 20°C.

8.2.3.4.3 Media for Fusarium *species.* A medium effective for isolation of most *Fusarium* species occurring in foods is Czapek Iprodione Dichloran Agar (CZID; Abildgren *et al.*, 1987). As well as dichloran, this medium contains the fungicide iprodione.

CZID consists of

Czapek–Dox broth (Difco)	35 g
CuSO₄.5H₂O	0.005 g
ZnSO₄.5H₂O	0.01 g
Chloramphenico	0.05 g
Dichloran (0.2% in ethanol)	1 ml
Agar	20 g
Chlortetracycline (1% in water, filter sterilised)	1.0 ml

Chlortetracycline is filter sterilised and added after the medium has been autoclaved and cooled to 50°C; iprodione suspension is 0.3 g Roval 50WP (Rhône–Poulenc Agro Chemie, Lyon, France) in 50 ml sterile water, shaken before addition to the medium. In the authors' experience, the use of chloramphenicol at twice the concentration specified (i.e. 100 mg/litre) is a satisfactory replacement for the chlortetracycline.

CZID is highly selective for *Fusarium* species, and is undoubtedly the best currently available medium for *Fusarium* enumeration and isolation. Some questions remain concerning whether it may indeed be too selective and not support growth of all foodborne *Fusarium* species. However, the common species are well supported.

8.2.4 Special groups

8.2.4.1 Heat-resistant fungi. Heat-resistant spoilage fungi, such as *Byssochlamys*, *Talaromyces*, *Neosartorya* and *Eupenicillium* species can be selectively isolated from fruit juices, pulps and concentrates by laboratory pasteurisation. Two methods are described here: the first is the plating method of Murdoch and Hatcher (1978), adapted for larger samples, and the second the direct incubation method. For further details see Beuchat and Rice (1979), Pitt and Hocking (1985) or Beuchat and Pitt (1992).

8.2.4.1.1 Plating method. If the sample to be tested is greater than 35° Brix, it should first be diluted 1:1 with 0.1% peptone or similar diluent. For very acid juices, such as passionfruit, normally about pH 2.0, the pH should be adjusted to 3.5–4.0. Two 50 ml samples are taken for examination. The two samples are heated in 200×30 mm test tubes in a closed water bath at 80°C for 30 min, then rapidly cooled. Each 50 ml sample is then distributed over four 150 mm Petri dishes and mixed with double-strength potato dextrose agar or malt extract agar. The Petri dishes are loosely sealed in a plastic bag to prevent drying, and incubated at 30°C for up to 30 days. Plates are examined weekly for growth. Most moulds will produce visible colonies within 10 days, but incubation for up to 30 days allows for the possible presence of badly heat damaged spores, which may germinate very slowly. This long incubation time also allows most moulds to mature and sporulate, aiding their identification.

The main problem associated with this technique is the possibility of aerial contamination of the plates with common mould spores, which will give false-positive results. The growth of green *Penicillium* colonies, or colonies of common *Aspergillus* species such as *A. flavus* and *A. niger*, is a clear indication of contamination, as these fungi are not heat-resistant. To minimise contamination, plates should be poured in clean, still air or in a laminar-flow cabinet if possible. If a product contains large numbers of heat resistant bacterial spores (e.g. *Bacillus* species), antibiotics can be added to the potato dextrose agar. The addition of chloramphenicol (100 mg/l of medium) will prevent the growth of these bacteria.

8.2.4.1.2 Direct incubation method   A more direct method used for screening fruit pulps and other semisolid materials avoids the problems of aerial contamination. Place approximately 30 ml of pulp in each of three or more flat bottles, such as 100 ml medicine flats. Heat the bottles in the upright position for 30 min at 80°C and cool, as described previously. The bottles of pulp can then be incubated directly, without opening and without the addition of agar. They should be incubated flat, allowing as large a surface area as possible, for up to 30 days at 30°C. Any mould colonies which develop will need to be subcultured onto a suitable medium for identification. If containers such as Roux bottles are available, larger samples can be examined by this technique, but heating times must be increased. Bottle contents should reach at least 75°C for 20 min when checked by a thermometer suspended near the centre of the pulp.

8.2.4.2 Preservative-resistant yeasts.   A few species of yeasts are able to grow in products containing preservatives such as sorbic, benzoic and acetic acids, or sulphur dioxide, or combinations of these preservatives. The most important of these is *Zygosaccharomyces bailii*. The simplest and most effective way to screen for preservative resistant yeasts is to spread or streak product onto plates of malt acetic agar, which is MEA with 0.5% acetic acid added (Pitt and Richardson, 1973).

Malt acetic agar (MAA) is made by adding glacial (16M) acetic acid to melted and tempered MEA, to give a final concentration of 0.5%. Mix and pour immediately. This medium cannot be held molten for long periods or remelted, because of its low pH. The acetic acid does not need sterilisation before use.

MAA is a suitable medium for monitoring raw materials, process lines and products containing preservatives for resistant yeasts. It is also effective for testing previously isolated yeasts for preservative resistance.

8.2.4.3 Extreme xerophiles. Extremely xerophilic fungi grow slowly if at all on DG18, and are usually quickly overgrown by rapidly spreading xerophiles such as *Eurotium* species. Extreme xerophiles include *Xeromyces bisporus*, xerophilic *Chrysosporium* species, *Eremascus* species and the halophilic xerophiles *Polypaecilum pisce* and *Basipetospora halophila*. Such species require special media and techniques.

The most effective medium, suitable for all except the halophilic xerophiles, is malt extract yeast extract 50% glucose agar (MY50G; Pitt and Hocking, 1985).

MY50G consists of

Malt extract	10 g
Yeast extract	2.5 g
Agar	10 g
Water, distilled	500 g
Glucose, AR	500 g

Add the minor constituents and agar to *c*. 450 ml distilled water and steam to dissolve the agar. Immediately make up to 500 g with distilled water. While the solution is still hot, add the glucose while stirring rapidly to prevent the formation of hard lumps of glucose monohydrate. If lumps do form, dissolve them by steaming for a few minutes. Sterilise by steaming for 30 min; note that this medium is of a sufficiently low a_w not to require autoclaving. Glucose monohydrate (food grade) may be used in this medium instead of analytical reagent grade glucose, but allowance must be made for the additional water present. Use 550 g of $C_6H_{12}O_6.H_2O$, and 450 g of the basal medium. As the final concentration of water is unaltered, the concentrations of the minor ingredients are unaffected. The final a_w of this medium is 0.89.

8.2.4.3.1 Media for halophilic xerophiles. Some xerophilic fungi from salted foods, such as salt fish, grow more rapidly on media containing NaCl and hence are correctly termed halophilic xerophiles. The following media, malt extract; yeast extract; 5% salt; 12% glucose agar (MY5–12), and malt extract; yeast extract; 10% salt; 12% glucose agar (MY10–12) are suitable for these fungi.

MY5-12 (or MY10-12) consists of

Malt extract	20 g
Yeast extract	5 g
NaCl	50 g (100 g for MY10-12)
Glucose	120 g
Agar	20 g
Water, distilled	1 litre

Sterilise MY5–12 by autoclaving at 121°C for 10 min, and MY10–12 by steaming for 30 min. Overheating of these media will cause softening. The final a_w of MY5–12 is 0.93 and of MY10–12 is 0.88.

8.2.4.3.2 Isolation techniques for extreme xerophiles. Extreme xerophiles are usually sensitive to diluents of high a_w, and hence cannot be isolated by dilution plating. Direct plating is the method of choice: a convenient technique is to place small pieces of sample, without surface sterilisation, onto a rich, low a_w medium, such as MY50G. For details see Pitt and Hocking (1985).

8.2.4.4 Xerophilic yeasts. Xerophilic yeasts such as *Zygosaccharomyces rouxii* cause fermentative spoilage in glucose syrups, malt extract, fruit juice concentrates and similar products. If fermentation is evident (gas production in the product, or swelling of the container), then the yeasts will often be present in sufficiently high numbers to be detected by dilution plating. To minimise osmotic shock, the a_w of the diluent should be reduced by the addition of 20% sucrose, glucose or glycerol to 0.1% peptone water. A suitable plating medium is MY50G.

If the presence of xerophilic yeasts is suspected, but there are no obvious signs of growth, then the yeasts may be effectively enriched in the product itself, by decanting 30–50% of the contents, replacing with sterile distilled water and incubating at 25°C until signs of fermentation occur. A simple presence–absence test for small numbers of xerophilic yeasts in foods was described by Jermini *et al.* (1987), using an enrichment medium of 0.5% yeast extract and 50% (w/w) glucose incubated with agitation at 30°C. However, neither of these methods is quantitative.

8.2.4.5 A culture film method: Petrifilm YM. Petrifilm YM (3M Company, St Paul, Minnesota) is a proprietary system for enumerating fungi on a layer of medium enclosed in a plastic film, which eliminates the use of Petri dishes. Beuchat *et al.* (1990) compared counts obtained for a wide variety of high a_w foods on Petrifilm YM with standard plate counts on plate count agar and acidified potato dextrose agar. They reported that Petrifilm YM was at least as effective for recovering fungi from high a_w products, and was an acceptable alternative technology. However, sub-culturing colonies from Petrifilm YM plates for subsequent identification was more difficult than for traditional Petri dishes.

8.3 Rapid methods

8.3.1 Detection of secondary metabolites

8.3.1.1 Impedimetry and conductimetry. Metabolites produced by growth of microorganisms in liquid media alter the medium's impedance and conductance.

The use of changes in these properties as a measure of bacterial growth was suggested by Hadley and Senyk (1975). These techniques were first applied to yeasts by Evans (1982) and to moulds by Jarvis *et al.* (1983). Most of the subsequent work on fungi has been carried out with yeasts, but the methodology is often applicable to moulds also.

A major problem with these techniques involves the selection of suitable media, which will induce detectable and reproducible changes in either conductance or capacitance during fungal growth. Early success in impedance measurements was reported with malt extract glucose (Shapton and Cooper, 1984) and yeast carbon base ammonium sulphate (Zindulis, 1984). Williams and Wood (1986) used malt extract (2%) glucose (2%) as a medium and reported that most of the 32 fungal species tested could be detected. However, conventional media such as MEA were subsequently reported to be of little value (Schaertel *et al.*, 1987).

After investigating a wide variety of media for detection of yeast growth by conductance, Connolly *et al.* (1988) recommended yeast carbon base ammonium tartrate, a derivative of the yeast carbon base ammonium sulphate medium of Zindulis (1984).

Watson-Craik *et al.* (1989) studied 27 mould species on a wide range of both commercially available and specially prepared media and concluded that conductance and capacitance were both medium and species specific. They reported that although the medium of Williams and Wood (1986) gave consistent capacitance increases, stability and rate of change were poor. A medium containing soya peptone (0.5%) and yeast extract (0.5%) was more effective. However, the addition of fruit juice produced a signal reversal. This was partly overcome by the addition of glucose and $(NH_4)_2SO_4$. The addition of KH_2PO_4 to increase buffering capacity produced an improved performance (Watson-Craik *et al.*, 1990). Product interference with the quality of the capacitance curves was greatly improved. As the optimal pH was about 6.0, antibiotic addition was necessary to prevent bacterial growth.

To decrease the influence of product variability on media used in conductance estimations, Owens *et al.* (1992) recommended use of media high in ammonium ions and glucose, with added yeast extract and peptone. The high ammonium concentration induced high conductance changes, while the other ingredients minimised the effects of high carbohydrate and low organic nitrogen levels in foods.

An impedimetric method for detection of heat resistant fungi in fruit juices has recently been descried. Nielsen (1992) reported that the addition of yeast extract (0.75%), potassium dihydrogen phosphate (0.6%) and ammonium sulphate (0.1%) to the fruit juices markedly improved curve quality, resulting in earlier and more reproducible detection. The detection limit in artificially contaminated juices was 1 ascospore per ml, detectable in 100 h.

In summary, impedimetry and conductimetry appear to be effective rapid methods when used under well defined conditions for a specific purpose with a particular kind of food. Their utility therefore lies in production quality control. They cannot be applied on a broader scale without considerable developmental studies.

8.3.1.2 Fungal volatiles. Methods described above have all been concerned with detecting fungal growth or reproduction. Methods described hereunder have taken the opposite approach, that of measuring the effects of fungi on foods, rather than fungi *per se.* Deterioration of stored grain results from a combination of chemical and biological changes, with changes due to fungal growth often predominant. Deterioration is marked by off-odour development, loss of germinability, caking, rancidity and sometimes mycotoxin development (Abramson *et al.*, 1980, 1983). Fungi produce volatile chemicals during growth and particular chemicals may be associated with grain deterioration (Kaminski *et al.*, 1972, 1974, 1975). Fungal growth in grain storage was studied by Sinha *et al.* (1988), who monitored production of 3-methyl-l-butanol, 1-octen-3-ol and 3-octanone. The presence of these volatiles usually correlated with seed infection by *Alternaria alternata, Eurotium repens, Aspergillus versicolor* and several *Penicillium* species. The latter two compounds in particular seemed to be associated with deterioration in grain due to fungal growth. Adamek *et al.* (1992) identified methylfuran, 2-methylpropanol and 3-methylbutanol as the most important metabolites from *E. amstelodami, A. flavus, P. cyclopium* and *Fusarium culmorum* growing on wheat. The volatiles produced by several *Penicillium* species were also studied. The use of volatile compound production as an indicator of mould deterioration in grains has been reviewed by Kaminski and Wasowicz (1991).

To clarify the precise role of these compounds and to enable species-specific fungal detection, Börjesson *et al.* (1989) grew several food-spoilage fungi in pure culture on wheat grain. Volatiles were collected on a chemical adsorbent and analysed by gas chromatography. Some compounds, especially 3-methyl-1-butanol, were produced early in fungal growth and could be used as an early warning of potential deterioration, before fungal growth became visible. The production of volatile metabolites of *Penicillium* species in grain correlated well with carbon dioxide production and ergosterol formation (Börjesson *et al.*, 1990, 1992).

8.3.2 Estimation of fungal biomass

A deficiency in all of the enumeration techniques which rely on culturing fungi is that the result is at best poorly correlated with growth or *biomass.* Biomass is usually regarded as the fundamental measure of fungal growth in biotechnology, but it is not easy to quantify under the conditions existing in foods. Mycelial dry weight is most commonly used as a biomass estimate, but its relationship to mycelial wet weight and to metabolism varies widely in foods, due to the great influence of a_w on both of the latter parameters. Fungi growing at reduced a_w can be expected to be more dense than at high a_w, due to high concentrations of internal solutes, though this is exceptionally difficult to measure experimentally. The question of a satisfactory fundamental measure of fungal biomass remains unanswered.

Despite these basic problems, several chemical and biochemical techniques have been developed to estimate the extent of fungal growth in commodities. These techniques rely either on some unique component of the fungus, which is not found in other microorganisms or foods, or on immunological or molecular techniques. Most are still in the developmental phase: the most important ones are described briefly here.

8.3.2.1 Chitin.   Chitin is a polymer of N-acetyl-D-glucosamine, and is a major constituent of the walls of fungal spores and mycelium. It also occurs in the exoskeleton of insects, but is not present in bacteria or in foods. Hence, the chitin content of a food or raw material can provide an estimate of fungal contamination.

Chitin is most effectively assayed by the method of Ride and Drysdale (1972). Alkaline hydrolysis of the food sample at 130°C causes partial depolymerisation of chitin to produce chitosan. Treatment with nitrous acid then causes partial solubilisation and deamination of glucosamine residues to produce 2, 5-anhydromannose, which is estimated colorimetrically using 3 methyl-2-benzothiazolone hydrazone hydrochloride as the principal reagent. Alkaline hydrolysis is more readily accomplished at 121°C in an autoclave (Jarvis, 1977). An improved assay sensitivity was achieved by derivatisation of glucosamine and other products with *o*-phthalaldehyde, separation by high-performance liquid chromatography and detection of fluorescent compounds with a spectro-fluorimeter (Lin and Cousin, 1985). The chitin assay remains rather complex and slow, usually requiring about 5 h.

A number of studies have indicated that the chitin assay is a valuable technique for estimating the extent of fungal invasion in foods such as corn and soy beans (Donald and Mirocha, 1977), wheat (Nandi, 1978) and barley (Whipps and Lewis, 1980). Particular attention has been paid to the possibility of developing the chitin assay as a replacement for the Howard mould count for tomato products (Jarvis, 1977; Bishop *et al.*, 1982; Cousin *et al.*, 1984). Recently, Patel *et al.* (1993) have demonstrated the potential of N-acetyl-D-glucosamine-specific lectins for detecting a range of food-spoilage moulds.

The chitin assay has some shortcomings, and has been severely criticised by some authors (e.g. Sharma *et al.*, 1977). The relationship between dry weight and chitin content varies at least two-fold for different food-spoilage fungi (Cousin *et al.*, 1984; Lin and Cousin, 1985). Some foods contain naturally occurring amino sugars such as glucosamine and galactosamine, which should be removed by acetone extraction prior to hydrolysis (Whipps and Lewis, 1980). Products from rot-free tomatoes gave positive glucosamine assays even after acetone extraction (Cousin *et al.*, 1984) and chitin content does not increase proportionally with fungal growth (Sharma *et al.*, 1977). Insect contamination of grain samples has been reported to produce grossly misleading results (Sharma *et al.*, 1977), but the presence of fruit flies in tomato-based products was less serious (Lin and Cousin, 1985). Materials such as stored grains frequently contain insect fragments and need to be checked before chitin assays are attempted. Because of these difficulties, the use of chitin as a chemical assay for fungi in foods has largely been superseded by the ergosterol assay in recent years.

8.3.2.2 Ergosterol. Ergosterol is the major steroid produced by fungi, but at most is a minor component of plant sterols (Weete, 1974). Ergosterol occurs as a component of fungal cell membranes, so is inherently likely to be correlated with hyphal growth and biomass. It is, therefore, a good candidate as a chemical for measuring fungal growth in foods and raw materials. Methodology for estimating ergosterol in cereals has been developed by Seitz *et al.* (1977, 1979). Samples are blended with methanol, saponified with strong alkali, extracted with petroleum ether, and fractionated by high-pressure liquid chromatography. Ergosterol is detected by ultraviolet absorption, optimally at 282 nm, a wavelength at which other sterols exhibit little or no absorbance. For refinements to this methodology see Newell *et al.* (1988).

The ergosterol assay promises to be widely useful for quantifying fungal growth. Therefore, several studies have been carried out to assess the relationship between fungal growth and ergosterol production, using both liquid and solid substrates.

Using liquid cultures, Zill *et al.* (1988) showed a correlation between ergosterol production, mycelial wet weight and mycelial protein in *Fusarium graminearum*. Matcham *et al.* (1985) reported that ergosterol correlated better with mycelial dry weight than chitin or laccase, a polyphenol oxidase. Variation exists, however. Torres *et al.* (1992) reported that *Aspergillus ochraceus* grown in liquid culture showed a three-fold increase in ergosterol concentration in relation to mycelial dry weight as the culture aged from 2 to 26 days. Other reports (Seitz *et al.*, 1979; Matcham *et al.*, 1985; Newell *et al.*, 1987) indicated smaller variations, only one to two-fold. Marfleet *et al.* (1991) showed that fungal biomass and ergosterol levels were correlated for three representative fungal species over a range of a_w on solid substrates, but not in liquid media. Nout *et al.* (1987) showed that the ergosterol content of *Rhizopus oligosporus* varied widely, from 2 to 24 μg/mg of mycelial mass, and varied with substrate, aeration and growth phase. The ergosterol content was low during the rapid growth phase, but tended to increase, at times sharply, as growth slowed.

Methods for estimating fungal biomass during growth on agar media were compared by Schnürer (1993). Changes in ergosterol content correlated with hyphal length, while ergosterol was 0.04–0.14% of the calculated fungal dry weight. These levels were lower (0.2 to 6%) than those reported by Seitz *et al.* (1979).

Quantifying ergosterol production in foods has proved difficult. Seitz *et al.* (1977) showed a good correlation between damage in rice grains and their ergosterol content, between ergosterol in wheat and rainfall during the growing season, and between ergosterol content and fungal invasion in several sorghum hybrids. Matcham *et al.* (1985) reported good correlations between linear extension of *Agaricus bisporus* grown on rice grains and chitin, ergosterol and laccase production. Ergosterol content correlated with colony counts of fungi on wheat grains at 0.95 a_w, but not at 0.85 a_w (Tothill *et al.*, 1992). Using a stereo-microscope for visual examination, these authors concluded that sound grain contained up to 6 μg ergosterol/g, microscopically mouldy grain 7.5–10 μg/g, and visibly mouldy grain more than 10 μg/g. From studies on ergosterol levels, colony counts and mould growth in a variety of grain samples, Schnürer and Jonsson (1992) concluded that ergosterol correlated with colony counts better on DG18 (r=0.77) than on MEA (r=0.69). Ergosterol levels of food grade wheat ranged from 2.4 to 2.8 μg/g dry weight, samples from field trials (of unspecified quality) from 3.0 to 5.6 μg/g and feed grains from 8 to 15 μg/g dry weight.

After an extensive survey of ergosterol levels in Danish crops, Hansen and Pedersen (1991) concluded that the normal levels of ergosterol in barley were 7.6 ± 2.8, in wheat for breadmaking 5.0 ± 1.5, rye for breadmaking 6.8 ± 2.2, peas 2.2 ± 2.7 and rapeseed 2.4 ± 1.3 μg/g dry weight, respectively. ochratoxin A in barley correlated well with ergosterol content and reached significant levels when ergosterol increased to 25 μg/g dry weight. However, aflatoxin B_1 became detectable in cottonseed meal when ergosterol reached only 4 μg/g. 'Burned' rapeseed, a measure of quality, became significant when ergosterol reached 1.4 μg/g dry weight.

The ergosterol assay is reported to have a high sensitivity and, in contrast to the chitin assay, requires only 1 h for completion (Seitz *et al.*, 1979). It appears to be a useful indicator of fungal invasion of foods and to hold promise as a routine technique for quality control purposes.

8.3.2.3 ATP   ATP has also been suggested as a measure of microbial biomass, because bioluminescence techniques provide a very sensitive assay (Jarvis *et al.*, 1983). Provided that background levels of ATP in plant or other cells are very low, or that microorganisms can be effectively separated from such other materials, the method has some potential as a microbial assay. A good correlation was shown between viable counts of six species of psychrotrophic yeasts grown in pure culture and ATP production (Patel and Williams, 1985). The effective detection of low levels of yeasts in carbonated beverages by ATP has also been reported (LaRocco *et al.*, 1985). However, living plant cells contain high levels of ATP and fungi are often very difficult to separate from food materials. The potential of ferrofluids (in combination with lectins) for the separation of food-spoilage yeasts has been shown in chapter 4. Moreover, extraction of molecules from fungal cells is notoriously difficult, so the potential of this method may be difficult to realise in food mycology.

8.3.2.4 Immunological techniques. Cell wall proteins of fungi produce antigens, which can be detected by immunological methods. Some antigens are derived from components common to a wide range of fungi, and hence antigens are indicative of general fungal growth, while others are genus or even species specific. A variety of methods has been developed to take advantage of the antigenic properties of fungi.

8.3.2.4.1 Enzyme-linked immunosorbent assay (ELISA). The preparation of antigens from three common foodborne fungi *(Penicillium aurantiogriseum, Mucor racemosus* and *Fusarium oxysporum)* was described by Notermans and Heuvelman (1985). Preparation of immunoglobulin antibodies against these antigens was followed by development of an ELISA. Fungi were detected in both unheated and heat-processed foods by this method. Antigens were relatively genus specific: the *Mucor racemosus* antigens reacted with other *Mucor* and *Rhizopus* species, and the *Penicillium* antigen reacted with the *Aspergillus* species tested.

It was subsequently shown that the *Penicillium* antigen reacted with 43 of 45 *Penicillium* species tested, that antigen production correlated with mycelial weight and that it was unaffected by culture conditions, medium, temperature and a_w (Notermans *et al.*, 1986). The *Penicillium* antigen also reacted with *Aspergillus flavus*: the level of antigen correlated with aflatoxin production (Notermans *et al.*, 1986).

ELISA techniques have also been studied as a potential replacement for the Howard mould count. Antigens from tomato moulds (*Alternaria alternata, Geotrichum candidum* and *Rhizopus stolonifer*) were used to produce an ELISA test sensitive to 1 μg of mould/g in tomato. A correlation was observed between antigen formation and mould added to tomato purée, while background interference was very low (Lin *et al.*, 1986). The method was tested against a broader range of foods, with encouraging results (Lin and Cousin, 1987). Robertson and Patel (1989) improved the sensitivity of the method for tomato paste by using a polyclonal antibody against *Botrytis cinerea*, Mucor pyriformis and *Fusarium solani* in addition to the three species used by Lin *et al.* (1986).

ELISA tests for *Botrytis* and *Monascus* in foods (Cousin *et al.*, 1990) and for the detection of *Penicillium islandicum* (Dewey *et al.*, 1990) and *Humicola lanuginosa* (Dewey *et al.*, 1992) in rice have also been described.

8.3.2.4.2 Latex agglutination. A different approach to the immunological detection of fungi in foods has been the coating of latex beads with antibodies and detection of agglutination of the beads in the presence of antigens (Kamphuis *et al.*, 1989; Notermans and Kamphuis, 1992; Stynen *et al.*, 1992). It was found that 0.8 μm latex beads coated with antibodies from the extracellular polysaccharide produced by *Penicillium digitatum* specifically detected *Aspergillus* and *Penicillium* species (Kamphuis *et al.*, 1989). Detection limits were as low as 5–10 ng of the purified antigen/ml. Commercially produced latex agglutination tests were found to be useful for screening the mycological quality of grains and processed foods (Braendlin and Cox, 1992; van der Horst *et al.*, 1992), although one kit performed poorly in detection of mould in tomato products (van der Horst *et al.*, 1992).

Schwabe *et al.* (1992) compared the latex agglutination assay with ergosterol production for detection of *Penicillium*, *Aspergillus* and *Fusarium* species in pure culture. They concluded that the two methods were comparable for *Penicillium* and *Aspergillus*, but that ergosterol was more sensitive for *Fusarium*. In food samples, both the latex agglutination test and ergosterol were effective means of detecting mould growth, but no clear correlation existed between values obtained by the two methods.

8.3.2.4.3 Fluorescent antibody techniques. These have also been used directly for the detection of mould in foods. Warnock (1971) detected *Penicillium aurantiogriseum* in barley by this method, while Robertson *et al.* (1988) used antisera against five fungi to visualise moulds and simplify their detection in the Howard mould count technique.

8.3.3 Molecular methods

Immunological techniques rely on the development of antibodies to specific antigens on molecules expressed by RNA and ultimately genomic DNA. The next step up the chain of detection methods for microorganisms is to use DNA itself. During the past decade, techniques for detecting DNA sequences using specific probes have been developed (Southern, 1975). With the development of gene cloning techniques and oligonucleotide synthesis, DNA sequences can now be prepared in large quantities for use in probes. Depending on its role in the genome, DNA may be specific at almost any taxonomic level. In theory, producing a broadly based or even strain specific DNA probe is now possible for any organism (Walker and Dougan, 1989). In practice, the genome of only a few organisms is well-enough known for this to be true.

To date, probes relevant to food mycology have been confined to recognition of species after isolation and growth in pure culture. One probe is designed to distinguish between the closely related yeasts *Zygosaccharomyces bailii*, *Z. rouxii* and *Fusarium moniliforme* from two closely related species (Lodolo *et al.*, 1992). Rapid developments may be expected in this area.

References

Abildgren, M.P., Lund, F., Thrane, U. and Elmholt, S. (1987) Czapek–Dox agar containing iprodione and dichloran as a selective medium for the isolation of *Fusarium*. *Lett. Appl. Microbiol.* **5**, 83–86.
Abramson, D., Sinha, R.N. and Mills, J.T. (1980) Mycotoxin and odor formation in moist cereal grain during granary storage. *Cereal Chem.* **57**, 346–351.
Abramson, D., Sinha, R.N. and Mills, J. T. (1983) Mycotoxin and odor formation in barley stored at

16 and 20% moisture in Manitoba. *Cereal Chem.* **60** 350–355.

Adamek, P., Bergström, B., Börjesson, T. and Stöllman, U. (1992) Determination of volatile compounds for the detection of moulds, in *Modern Methods in Food Mycology* (eds Samson, R.A., Hocking, A.D., Pitt, J.I. and King, A.D.). Elsevier, Amsterdam, pp. 327–336.

Andrews, S. (1992) Comparative study of WL nutrient agar with DRBC and OGY for yeast enumeration in foods, in *Modern Methods in Food Mycology* (eds. Samson, R.A., Hocking, A.D., Pitt, J. I. and King, A.D.). Elsevier, Amsterdam, pp. 61–65.

Beuchat, L.R. (1992) Enumeration of fungi in grain flours and meals as influenced by settling time in diluent and by the recovery medium. *J. Food Prot.*, **55**, 899–901.

Beuchat, L.R. and Brackett, R.E. (1992) Collaborative study of media and methods for enumerating heat stressed fungi in wheat flour, in *Modern Methods in Food Mycology* (eds. Samson, R.A., Hocking, A.D., Pitt, J.I. and King, A.D.). Elsevier, Amsterdam, pp. 89–91.

Beuchat, L.R. and Pitt, J.I. (1992) Detection and enumeration of heat resistant molds, in *Compendium of Methods for the Microbiological Examination of Foods* (eds Vanderzant, C. and Splittstoesser, D.F.). American Public Health Association, Washington, DC, pp. 251–263.

Beuchat, L.R. and Rice, S.L. (1970) *Byssochlamys* spp. and their importance in processed fruits. *Adv. Food Res.* **25**, 237–288.

Beuchat, L.R., Nail, B.V., Brackett, R.E. and Fox, T.L. (1990) Evaluation of a culture film (Petrifilm™ YM) method for enumerating yeasts and molds in selected dairy and high-acid foods. *J. Food Prot.*, **53**, 864, 869–874.

Bishop, R.H., Duncan, C.L., Evancho, G.M. and Young, H. (1982) Estimation of fungal contamination of tomato products by a chemical assay for chitin. *J. Food Sci.* **47**, 437–439, 444.

Börjesson, T., Stöllman, U., Adamek, P. and Kaspersson, A. (1989) Analysis of volatile compounds for detection of molds in stored cereals. *Cereal Chem.*, **66**, 300–304.

Börjesson, T., Stöllman, U. and Schnürer, J. (1990) Volatile metabolites and other indicators of *Penicillium aurantiogriseum* growth on different substrates. *appl. Environ. Microbiol.* **56**, 3705–3710.

Börjesson, T., Stöllman, U. and Schnürer, J. (1992) Volatile metabolites produced by six fungal species compared with other indicators of fungal growth on cereal grains. *Appl. Environ. Microbiol.* **58**, 2599–2605.

Bothast, R.J. and Fennell, D.I. (1974) A medium for rapid identification and enumeration of *Aspergillus flavus* and related organisms. *Mycologia*, **66**, 365–369.

Braendlin, N. and Cox, L. (1992) Immunoagglutination assay for rapid detection of *Aspergillus* and *Penicillium* contamination in food, in *Modern Methods in Food Mycology* (eds Samson, R.A., Hocking, A.D., Pitt, J.I. and King, A.D.). Elsevier, Amsterdam, pp. 233–240.

Connolly, P., Lewis, S.J. and Corry, J.E.L. (1988) A medium for the detection of yeasts using a conductimetric method. *Int. J. Food Microbiol.* **7**, 31–40.

Cousin, M.A., Zeidler, C.S. and Nelson, P.E. (1984) Chemical detection of mold in processed foods. *j. Food Sci.*, **49**, 439–445.

Cousin, M.A., Dufrenne, J., Rombouts, F.M. and Notermans, S. (1990) Immunological detection of *Botrytis* and *Monascus* species in food. *Food Microbiol.* **7**, 227–235.

Deak, T. (1992) Media for enumerating spoilage yeasts – a collaborative study, in *Modern Methods in Food Mycology* (ed Samson, R.A., Hocking, A.D., Pitt, J.I. and King, A.D.). Elsevier, Amsterdam, pp. 31–38.

Dewey, F.M., MacDonald, M.M., Phillips, S.I. and Preistley, R.A. (1990) Development of monoclonal-antibody-ELISA and dip-stick immunoassays for *Penicillium islandicum* in rice grains. *J. Gen. Microbiol.*, **136**, 753–760.

Dewey, F.M., Twiddy, D.R., Phillips, S.I., Grose, M.J. and Wareing, P.W. (1992) Development of a quantitative monoclonal antibody-based innumoassay for *Humicola lanuginosa* in rice grains and comparison with conventional assays. *Food Agric. Immunol.* **4**, 153–167.

Donald, W.W. and Mirocha, C.J. (1977) Chitin as a measure of fungal growth in stored corn and soybean seed. *Cereal Chem.* **54**, 466–474.

Evans, H.A.V. (1982) A note on two uses for impedimetry in brewing microbiology. *J. Appl. Bacteriol.*, **53**, 423–426.

Frisvad, J.C. (1983) A selective and indicative medium for groups of *Penicillium viridicatum* producing different mycotoxins in cereals. *J. Appl. Bacteriol*, **54**, 409–416.

Hadley, W.K. and Senyk, G. (1975) Early detection of metabolism and growth by measurement of electrical impedance. *Microbiology*, **1975**, 12–21.

Hansen, I.D. and Pedersen, J.G. (1991) *Ergosterol – et Nyttigt Vaerktøj ved Verdering af*

Svampebelastning. Biotechnolosk Institut Report No. 91–2–4, Kolding, Denmark.

Hocking, A.D. (1982) Aflatoxigenic fungi and their detection. *Food Technol. (Aust.)*, **34**, 236–238.

Hocking, A.D. and Pitt, J.I. (1980) Dichloran-glycerol medium for enumeration of xerophilic fungi from low moisture foods. *Appl. Environ. Microbiol.*, **39**, 488–492.

Hocking, A.D., Pitt, J.I., Samson, R.A. and King, A.D. (1992) Recommendations from the closing session of SMMEF II, in *Modern Methods in Food Mycology* (eds Samson, R.A., Hocking, A.D., Pitt, J.I. and King, A.D.). Elsevier, Amsterdam, pp. 359–364.

Jarvis, B. (1973) Comparison of an improved rose bengal–chlortetracycline agar with other media for the selective isolation and enumeration of moulds and yeasts in foods. *J. Appl. Food Technol.* **12**, 581–591.

Jarvis, B., Seiler, D.A.L., Ould, A.J.L. and Williams, A.P. (1983) Observations on the enumeration of moulds in food and feedingstuffs. *J. Appl. Bacteriol.*, **55**, 325–336.

Jermini, M.F., Geiges, O. and Schmidt–Lorenz, W. (1987) Detection, isolation and identification of osmotolerant yeasts from high-sugar products. *J. Food Prot.*, **50**, 468–472.

Kaminski, E. and Wacowicz, E. (1991) The useage of volatile compounds produced by moulds as indicators of grain deterioration, in *Cereal Grain. Mycotoxins, Fungi and Quality in Drying and Storage* (ed. Chelkowski, J.). Elsevier, Amsterdam, pp. 229–258.

Kaminski, E., Libbey, L.M., Stawicki, S. and Wasowicz, E. (1972) Identification of the predominant volatile compounds produced by *Aspergillus flavus. Appl. Microbiol.* **24**, 721–726.

Kaminski, E., Stawicki, S. and Wasowicz, E. (1974) Volatile flavor compounds produced by molds of *Aspergillus, Penicillium* and Fungi Imperfecti. *Appl. Microbiol.* **27**, 1001–1004.

Kaminski. E/. Stawicki. S. and Wasowicz. E. (1975) Volatile flavor substances produced by moulds on wheat grain. *Acta Aliment. Pol.* **1**, 153–164.

Kamphuis, H.J., Notermans, S., Veeneman, G.H., *et al.* (1989) A rapid and reliable method for the detection of molds in foods: using the latex agglutination assay. *J. Food Prot.* **52**, 244–247.

King, A.D., Hocking, A.D. and Pitt, J.I. (1979) Dichloran-rose bengal medium for enumeration and isolation of molds from foods. *Appl. Environ. Microbiol.* **37** 959–964.

King, A.D., Pitt, J.I., Beuchat, L.R. and Corry, J.E.L. (1986) *Methods for the Mycological Examination of Food.* Plenum Press, New York.

Kurtzman, C.P., Rogers R. and Hesseltine, C.W. (1971) Microbiological spoilage of mayonnaise and salad dressings. *Apl. Microbiol.* **21**, 870–874.

LaRocco, K.A., Galligan, P., Littel, K.J. and Spurgash, A. (1985) A rapid bioluminescent ATP method for determining yeast contamination in a carbonated beverage. *Food Technol.* **39 (7)** 49–52.

Lin, H.H. and Cousin, M.A. (1985) Detection of mold in processed foods by high performance liquid chromatography. *J. Food Prot.* **48**, 671–678.

Lin, H.H. and Cousin, M.A. (1987) Evaluation of enzyme-linked immunosorbent assay for detection of molds in foods. *J. Food Sci.* **52**, 1089–1096.

Lin, H.H., Lister, R.M. and Cousin, M.A. (1986) Enzyme-linked immunosorbent assay for detection of mold in tomato puree. *J. Food Sci.*, **51**, 180–183.

Lodolo, E.J., Van Zyl, W.H. and Rabie, C.J. (1992) A rapid molecular technique to distinguish *Fusarium* species. *Mycol. Res.*, **97**, 345–346.

Marfleet, I., Magan, N. and Lacey, J. (1991) The relationship between fungal biomass, ergosterol and grain spoilage, in *Proceedings of the Fifth International Working Conference on Stored Product Protection* (eds. Fleurat–Lesard, F. and Ducom, P.). Institut National de la Recherche Agronomique, Bordeaux, France.

Matcham, S.E., Jordan, B.R. and Wood, D.A. (1985) Estimation of fungal biomass in a solid substrate by three independent methods. *Appl. Microbiol. Biotechnol.* **21** 108–112.

Mossel, D.A.A., Kleynen–Semmeling, A.M.C., Vincentie, H.M., *et al.* (1970) Oxytetracycline-glucose-yeast extract agar for selective enumeration of moulds and yeasts in foods and clinical material. *J. Appl. Bacteriol.* **33**, 454–457.

Murdoch, D.I. and Hatcher, W.S. (1978) A simple method to screen fruit juices and concentrates for heat-resistant mold. *J. Food Prot.*, **41**, 254–256.

Nandi, B. (1978) Glucosamine analysis of fungus-infected wheat as a method to determine the effect of antifungal compounds ingrain preservation. *Cereal Chem.*, **55**, 121–126.

Newell, S.Y., Miller, J.D. and Fallon, R.D. (1987) Ergosterol content of salt-marsh fungi: effect of growth conditions and mycelial age. *Mycologia*, **79**, 688–695.

Newell, S.Y., Arsuffi, T.L. and Fallon, R.D. (1988) Fundamental procedures for determining ergosterol content of decaying plant material by liquid chromatography. *Appl. Environ. Microbiol.* **54**, 1876–1879.

Nielsen, P.V. (1992) Rapid detection of heat resistant fungi in fruit juices by an impedimetric method, in *Modern Methods in Food Mycology* (eds. Samson, R.A., Hocking, A.D., Pitt, J.I. and King, A.D.). Elsevier, Amsterdam, pp. 311–319.

Notermans, S. and Huevelman, C.J. (1985) Immunological detection of moulds in food by using the enzyme-linked immunosorbent assay (ELISA): preparation of antigens. *Int. J. Food Microbiol.*, **2**, 247–258.

Notermans, S. and Kamphuis, H.J. (1992) Detection of fungi in foods by latex agglutination: a collaborative study, in *Modern Methods in Food Mycology* (eds. Samson, R.A., Hocking, A.D., Pitt, J.I. and King, A.D.). Elsevier, Amsterdam, pp. 205–212.

Notermans, S., Huevelman, C.J., Beumer, R.R. and Maas, R. (1986) Immunological detection of moulds in food: relation between antigen production and growth. *Int. J. Food Microbiol.*, **3**, 253–261.

Nout, M.J.R., Bonants–Van Laarhoven, T.G.M., De Jongh, P. and De Koster, P.G. (1987) Ergosterol content of *Rhizopus oligosporus* NRRL 5905 grown in liquid and solid substrates. *Appl. Microbiol. Biotechnol.* **26**, 456–461.

Owens, J.D., Konirova, L. and Thomas, D.S. (1992) Causes of conductance change in yeast cultures. *J. Appl. Bacteriol.* **72**, 32–38.

Patel, P.D. and Williams, A.P. (1985) A note on enumeration of food spoilage yeasts by measurement of adenosine triphosphate (ATP) after growth at various temperatures. *J. Appl. Bacteriol.* **59**, 133–136.

Patel, P.D., Williams, D.W. and Haines, S.D. (1993) Rapid separation and detection of foodborne yeasts and moulds by means of lectins, in *New Techniques in Food and Beverage Microbiology* (eds Kroll, R.G., Gilmour, A. and Sussman, M.). Blackwell Scientific, London, pp. 31–43.

Pearson, B.M. and McKee, R.A. (1992) Rapid identification of *Saccharomyces cerevisiae, Zygosaccharomyces bailii* and *Zygosaccharomyces rouxii. Int. J. Food Microbiol.*, **16**, 63–67.

Pitt, J.I. (1984) The significance of potentially toxigenic fungi in foods. *Food Technol. (Aust.)*, **36**, 218–219.

Pitt, J.I. (1987) *Penicillium viridicatum, Penicillium verrucosum* and production of ochratoxin A. *Appl. Environ. Microbiol.* **53**, 266–269.

Pitt, J.I. and Hocking, A.D. (1985) *Fungi and Food Spoilage.* Academic Press, Sydney.

Pitt, J.I. and Richardson, K.C. (1973) Spoilage by preservative-resistant yeasts. *CSIRO Food Res. Q.*, **33**, 80–85.

Pitt, J.I., Hocking, A.D. and Glenn, D. R. (1983) An improved medium for the detection of *Aspergillus flavus* and *A. parasiticus. J. Appl. Bacteriol.* **54** 109–114.

Pitt, J.I., Hocking, A.D., Samson, R.A. and King, A.D. (1992) Recommended methods for mycological examination of foods, 1992, in *Modern Methods in Food Mycology*, eds R.A. Samson, A.D. Hocking, J.I. Pitt and A.D. King, pp. 365–368. Elsevier, Amsterdam.

Pitt, J.I., Hocking, A.D., Bhudhasamai, K., *et al.* (1993) The normal mycoflora of commodities from Thailand. 1. Nuts and oilseeds. *Int. J. Food Microbiol.*, **20**, 211–226.

Raper, K.B. and Thom, C. (1949) *A Manual of the Penicillia.* Williams and Wilkins, Baltimore, Maryland.

Ride, J.P. and Drysdale, R.B. (1972) A rapid method for the chemical estimation of filamentous fungi in plant tissue. *Physiol. Pl. Pathol.*, **2**, 7–15.

Robertson, A. and Patel, N. (1989) Development of an enzyme immunoassay for the quantification of mould in tomato paste – an alternative to the Howard mould count. *Food Agric. Immunol.* **1**, 73–81.

Robertson, A., Patel, N. and Sargeant, J.G. (1988) Immunofluorescence detection of mould – an aid to the Howard mould counting technique. *Food Microbiol.* **5**, 33–42.

Samson, R.A., Hocking, A.D., Pitt, J.I. and King, A.D. (1992) *Modern Methods in Food Mycology.* Elsevier, Amsterdam.

Schaertel, B.J., Tsang, N. and Firstenberg–Eden, R. (1987) Impedimetric determination of yeast and mould. *Food Microbiol.*, **4**, 155–163.

Schnürer, J. (1993) Comparison of methods for estimating the biomass of three food-borne fungi with different growth patterns. *Appl. Environ. Microbiol.* **59** 552–555.

Schnürer, J. and Jonsson, A. (1992) Ergosterol levels and mould colony forming units in Swedish grains of good and feed grade. *Acta Agric. Scand., Sect. B, Soil and Plant Sci.* **42**, 240–245.

Schwabe, M., Kamphuis, H., Trümner, U., *et al.* (1992) Comparison of the latex agglutination test and the ergosterol assay for the detection of moulds in foods and feedstuffs. *Food Agric. Immunol.*, **4**, 19–25.

Seitz, L.M., Mohr, H.E., Burroughs, R. and Sauer, D.B. (1977) Ergosterol as an indicator of fungal invasion in grains. *Cereal Chem.*, **54**, 1207–1217.

Seitz, L.M., Sauer, D.B., Burroughs, R., Mohr, H.E. and Hubbard, J.D. (1979) Ergosterol as a measure of fungal growth. *Phytopathology*, **69**, 1202–1203.

Shapton, N. and Cooper, P.J. (1984) Rapid determination of yeast numbers by impedance measurements using a Bactometer 32. *J. Soc. Dairy Technol.*, **37**, 60–62.

Sharma, P.D., Fisher, P.J. and Webster, J. (1977) Critique of the chitin assay technique for estimation of fungal biomass. *Trans. Br. Mycol. Soc.*, **69**, 479–483.

Sinha, R.N., Tuma, D., Abramson, D. and Muir, W.E. (1988) Fungal volatiles associated with moldy grain in ventilated and non-ventilated bin-stored wheat. *Mycopathologia*, **101**, 53–60.

Southern, E. (1975) Detection of specific sequences among DNA fragments separated by gel electrophoresis. *J. Molec. Biol.* **98**, 503–517.

Stynen, D., Meulmans, L., Goris, A., *et al.* (1992) Characteristics of a latex agglutination test based on monoclonal antibodies for the detection of fungal antigens in food, in *Modern Methods in Food Mycology* (eds Samson, R.A., Hocking, A.D., Pitt, J.I. and King, A.D.). Elsevier, Amsterdam, pp. 213–219.

Torres, M., Viladrich, R., Sanchis, V. and Canela, R. (1992) Influence of age on ergosterol content in mycelium of *Aspergillus ochraceus. Lett. Appl. Microbiol.*, **15**, 20–22.

Tothill, I.E., Harris, D. and Magan, N. (1992) The relationship between fungal growth and ergosterol content of wheat grain. *Mycol. Res.*, **96**, 965–970.

Van der Horst, M., Samson, R.A. and Karman, H. (1992) Comparison of two commercial kits to detect moulds by latex agglutination, in *Modern Methods of Food Mycology* (eds Samson, R.A., Hocking, A.D., Pitt, J.I. and King, A.D. Elsevier, Amsterdam, pp. 241–245.

Walker, J. and Dougan, G, (1989) DNA probes: a new role in diagnostic microbiology. *J. Appl. Bacteriol.*, **67**, 229–238.

Warnock, D.W. (1971) Assay of fungal mycelium in grains of barley, including the use of the fluorescent antibody technique for individual species. *J. Gen. Microbiol*, **67**, 197–205.

Watson–Craik, I.A., Aidoo, K.E. and Anderson, J.G. (1989) Induction of conductance and capacitance changes by food–borne fungi. *Food Microbiol.*, **6**, 231–244.

Watson-Craik, I.A., Aidoo, K.E. and Anderson, J.G. (1990) Development and evaluation of a medium for the monitoring of food-borne moulds by capacitance changes. *Food. Microbiol.* **7**, 129–145.

Weete, J.D. 1974. *Fungal Lipid Biochemistry: Distribution and Metabolism.* Plenum Press, New York.

Whipps, J.M. and Lewis, D.H. (1980) Methodology of a chitin assay. *Trans. Br. Mycol. Soc.*, **74**, 416–418.

Williams, A.P. and Wood, J.M. (1986) Impedimetric estimation of molds, in *Methods for the Mycological Examination of Food* (eds King, A.D., Pitt, J.I., Beuchat L.R. and Corry J.E.L.). Plenum Press, New York, pp. 230–238.

Zill, G., Engelharst, G. and Wallnöfer, P.R. (1988) Determination of ergosterol as a measure of fungal growth using Si 60 HPLC. *Z. Lebensm. Unters. Forsch.*, **187**, 246–249.

Zindulis, J. (1984) A medium for the impedimetric detection of yeasts in foods. *Food Microbiol.*, **1**, 159–167.

9 Scope for rapid microbiological methods in modern food production

P.A. HALL

9.1 Introduction

It has been estimated that between 5 and 6 million cases of foodborne illness with greater than 9 million deaths occur each year in the United States (Lechowich, 1992). The majority of these outbreaks are linked to consumer mishandling of seafood, red meat and poultry. The greatest number of confirmed outbreaks in the USA from 1977 to 1982 was caused by *Salmonella* followed by *Staphylococcus aureus*, *Clostridium perfringens* and *Clostridium botulinum*. In Canada, during the period of 1975–1984, *Salmonella*, *S. aureus*, *C. perfringens* and *Bacillus cereus* were the most common foodborne pathogens reported (Todd, 1992). Meat and poultry were the predominant food vehicles for illness in Canada. Most of the Canadian incidents were attributed to mishandling of food at foodservice establishments. Grossklaus *et al.* (1991) compiled data on foodborne outbreaks in (former) West Germany and reported that meat, milk and eggs were the major vehicles of disease. As opposed to the mishandling of raw meats, poultry and seafood by the consumer or foodservice establishments, commercially processed foods have a long tradition of being safe and wholesome. Today, however, more so than ever, microbiological control during food production is essential to the manufacture and distribution of safe, wholesome products which meet desired shelf life. In many parts of the world, the food microbiologist is challenged by the advent of the 'new generation' of consumer-preferred products having convenience and 'closer to fresh' characteristics. In general, these products are minimally processed, contain little or no preservative, and may be packaged under modified atmosphere toe extend shelf life. It should also be recognized that food manufacturers are not only concerned with manufacturing product that is free from pathogens, but are also concerned with manufacturing product which is resistant to microbiological spoilage. In the USA alone, economic loss due to microbiological spoilage has been estimated to be as high as $30 billion per year (Banwart, 1989). Overall, microbiological analysis of foods and their production plays a key role in assessing safety, quality, shelf life and compliance with regulatory specifications. Over the past 20 years, much effort has been focused on developing rapid microbiological methods for monitoring finished product, ingredients, proces equipment and environment. Traditional microbiological testing methods are slow, laborious, and often provide only retrospective data on manufactured product.

9.2 Traditional approach to microbiological control

With the exceptions of retort processing for canned foods and pasteurization of milk and dairy products, the traditional approach prior to the early 1970s for microbiological control of foods was very heavily focused on end product testing. Food products were manufactured and then often tested for compliance with pre-determined microbiological specifications. Typically, food laws are enforced by an inspectional approach. The drawbacks to this approach are discussed in depth by the International Commission on Microbiological Specifications for Foods (1988). For example, the US Food and Drug Administration Good Manufacturing Practices regulations (1969) and many of the codes of hygiene practice developed by the FAO/WHO Codex Alimentarius Commission's Committee on Food Hygiene embody the traditional approach to microbiological control through inspection. This mentality is also reflected in the extreme by the continuous inspection approach employed by some countries, including the USA, for slaughtering and meat processing. The problem with the inspectional approach, however, is that the food laws are often vague, general, and leave much to the interpretation of the inspector. The American Medical Association's Council on Scientific Affairs (1993) has also discussed the limitations of the inspectional approach. They point out that inspectors cannot see *Salmonella*, *Campylobacter* or *Listeria* and that protecting the consumer from these organisms must come from other means, such as sanitary handling, refrigeration and proper cooking. The position of the American Medical Association is that inspection should be viewed as only one component of an overall program for improving food safety. The ICMSF (1974) was also among the first to point out that microbiological methods are inadequate to appraise the microbiological quality of food without a satisfactory sampling plan. The ICMSF asserts that the cost of sampling and testing, and inadequate awareness of the principles involved nearly always preclude the use of an ideal sampling plan. This approach ultimately leads to a false sense of security because of an unrecognized, elevated level of risk.

Finally, even if microbiological monitoring is instituted based on a sound sampling plan, traditional microbiological methods generally are too slow and laborious to meet today's food manufacturing needs, including just-in-time delivery, real-time process monitoring and sanitation control.

9.3 HACCP approach to microbiological control

The concept of the Hazard Analysis and Critical Control Point approach to food safety management was first publicly introduced at the 1971 Conference on Food Protection (US Department of Health, Education and Welfare, 1972). HACCP is a systematic approach to food safety management, encompassing the entire process literally from farm to table. It identifies those parts of the system which are critical to safety and emphasizes control of microbiological hazards, as well as chemical and physical hazards. The HACCP approach was pioneered by the Pillsbury Corporation as part of the National Aeronautics and Space Agency (NASA) space food program (Bauman, 1992). NASA wanted 100% assurance that the food products Pillsbury was producing for space use would not be contaminated with bacterial or viral pathogens, toxins, chemicals or physical hazards that could cause illness or injury. NASA lacked confidence in traditional approaches at the time, which primarily depended on inspection and end-product testing to assure safety.

HACCP is intended to be a preventive program which minimizes finished product testing. The sole focus of HACCP is on product safety. Its focus should not be confounded with non-safety issues, such as product quality or economic adulteration.

Since 1971, HACCP has continued to gain impetus as the most rational approach to food safety management. HACCP was formally endorsed in 1985 by an expert committee of the National Research Council/National Academy of Sciences as the best means for controlling the microbiological safety of the US food supply. Acting on this endorsement, the United States government created the National Advisory Committee on Microbiological Criteria for Foods (NACMCF) in 1987. The NACMCF has since re-affirmed HACCP as the most logical approach to food safety. HACCP has also been endorsed by the ICMSF (1988), the Codex Alimentarius Commission (1993), as well as other national and local regulatory bodies, worldwide. The NACMCF (1990) has outlined seven principles, shown in Table 9.1, for implementing HACCP. While HACCP obviates the need for extensive end-product testing, it does not eliminate microbiological testing. Within the context of the seven HACCP principles outlined in Table 9.1, direct microbiological testing still plays an important role. Typically, in the HACCP framework, there is a shift from end-product microbiological testing to in-process and environmental monitoring. For example, in many processes, microbiological testing will be required for assessing the hazards throughout the production chain and for determining specific Critical

Table 9.1 The HACCP principles as defined by the National Advisory Committee on Microbiological Criteria for Food (1990).

Principle no. 1	Assess the hazard associated with growing, harvesting, raw materials and ingredients, processing, manufacturing, distribution, marketing, preparation and consumption of the food.
Principle no. 2	Determine Critical Control Points (CCPs) required to control the identified hazards.
Principle no. 3	Establish the critical limits that must be met at each identified CCP.
Principle no. 4	Establish procedures to monitor CCPs.
Principle no. 5	Establish corrective action to be taken when there is a deviation identified by monitoring a CCP.
Principle no. 6	Establish effective record-keeping systems that document the HACCP plan.
Principle no. 7	Establish procedures for verification that the HACCP system is working correctly.

Control Points (CCPs) required to control each identified hazard (Principles 1 and 2). Microbiological analyses may be required to monitor certain CCPs (Principle 4). They may also be required for validating and verifying the effectiveness of the HACCP plan (Principle 7). Examples may include pre-operation checks of cleaning and sanitizing, screening of sensitive raw ingredients, validation of the HACCP plan via microbiological challenge testing, and monitoring of critical sites for microbiological build-up during processing.

Quite often, during the development of a HACCP plan for a food process, there is confusion and debate surrounding what constitutes a true CCP. There is a tendency to confuse true CCPs with simple control points (quality or non-safety related) and therefore to have redundant CCPs. Several diagnostic tools have been developed to aid in discerning true CCPs. The NACMCF (1990) has developed a risk assessment model for microbiological hazards which has been recently expanded to include chemical and physical hazards (Corlett and Stier, 1991). Another very useful tool is the CCP decision tree show in Table 9.2, developed by the Codex Alimentarius Committee on Food Hygiene (1993) and recently adopted by the NACMCF HACCP subcommittee (1992). Use of the CCP decision tree will help assure that the number of CCPs are kept to the minimum needed to assure product safety. Minimizing the number of CCPs will

Table 9.2 Critical control point decision tree

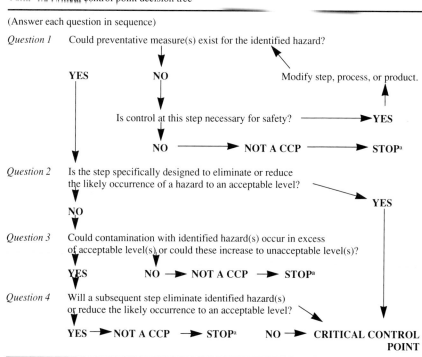

(Answer each question in sequence)

Question 1 Could preventative measure(s) exist for the identified hazard?

YES   NO   Modify step, process, or product.

Is control at this step necessary for safety? ──────► YES

NO ──────► NOT A CCP ──────► STOP[a]

Question 2 Is the step specifically designed to eliminate or reduce the likely occurrence of a hazard to an acceptable level?

YES

NO

Question 3 Could contamination with identified hazard(s) occur in excess of acceptable level(s) or could these increase to unacceptable level(s)?

YES   NO ─► NOT A CCP ─► STOP[a]

Question 4 Will a subsequent step eliminate identified hazard(s) or reduce the likely occurrence to an acceptable level?

YES ─► NOT A CCP ─► STOP[a]   NO ─► CRITICAL CONTROL POINT

[a] Proceed to next step in the described process.

help the HACCP plan remain 'user friendly' and avoid the serious pitfall of being too cumbersome to function effectively. There has been a number of practical guides to the implementation of HACCP published in recent years (ICMSF, 1988; Pierson and Corlett, 1992; NFPA, 1993a, b). These and others can aid the food manufacturer in tailoring a plan to his specific process.

9.4 Regulatory perspectives

As alluded to earlier, regulatory bodies worldwide are embracing the HACCP approach as the best way to manage food safety. In the USA, several HACCP initiatives have been undertaken by the federal government. In 1989, the US Department of Agriculture announced their intent to apply HACCP to meat and poultry inspection as part of the inspection modernization initiative (Adams *et al.*, 1992). The intent of this initiative was to systematically develop a HACCP-based inspection program. As part of this initiative, the USDA embarked upon an in-depth study which included:

- solicitation of input from employees, employee organizations, consumer representatives, and industry during five public hearings
- workshops with industry experts facilitated by the USDA to determine generic HACCP plans for select categories of products
- in-plant evaluation of the generic HACCP plans
- application of predetermined evaluation criteria.

This USDA initiative found HACCP to be a viable approach, with support from Agency employees and industry. This project is still active and recommendations have been developed for the use of HACCP in meat and poultry inspection.

Another initiative undertaken in the United States has been the proposed mandatory requirement that the seafood industry apply HACCP plans to their process. This is an outcome of the National Marine Fisheries Service's Model Seafood Surveillance Project (1987), initiated in response to the US Congress directive to design a program of certification and surveillance for seafood consistent with the HACCP system. The NMFS and USFDA have concluded that HACCP is a viable approach for ensuring the safety of seafood, supported by the vast majority of the seafood industry.

In addition to these US initiatives, the United Nation's Codex Alimentarius Commission, as well as the European Community, are in the process of adopting HACCP as the international standard for producing safe food. Under proposed rulemaking, beginning in 1995, all seafood exported to the EC will have to be produced under standards certified by the exporting country and accepted by the EC as complying to their HACCP standards (Taylor, 1993). Another initiative in which HACCP plays a role is the European International Standards Organization (ISO) 9000 standards. Prepared by the Technical Committee

ISO/TC 176 on Quality Assurance, the ISO 9000 series (ISO 9000–9004) is a set of documents devised to harmonize national and international quality standards (Lamprecht, 1991). The ISO 9000 initiative encompasses all the requirements of a good quality assurance system and provides a template for establishing a due diligence system. The ISO 9000 initiative is currently a voluntary process that entails certification by third-party assessment. The scope of the assessment is agreed prior to review. While ISO 9000 is not specific to the food industry, those food companies choosing to pursue certification should include a well-documented HACCP plan as part of the process. Because of the increasing trend in international trade, US companies are increasingly interested in achieving ISO 9000 certification (Saunders, 1991).

Microbiological specifications and criteria can serve several purposes. They can serve as a determinant of the acceptability of an ingredient, finished product or process. In this regard, they may encompass acceptability based on microbiological safety or microbiological quality. In practice, microbiological specifications typically are used both as an internal tool by the manufacturer to judge acceptability against pre-determined standards and as an external measure against customer or governmental standards. In the context of HACCP, microbiological specifications and criteria play a role in the monitoring of critical control points in food processing and distribution. It is crucial to identify the means needed to control hazards at each CCP. All of these must be clearly documented, either as clear statements or as specifications including tolerances, where appropriate.

Establishment of microbiological specifications is dependent upon the product, process, customer and regulatory requirements. A specification dealing with microbiological quality (e.g. mould count for spices used in salad dressings) is usually determined by its impact on consumer or customer acceptability. However, a specification dealing with public health (e.g. *Salmonella* specification for infant formula) is normally developed to comply with regulatory standards.

In the US, the Food and Drug Administration (1969) has issued a number of regulations covering Good Manufacturing Practices (GMPs). The so-called 'umbrella' GMPs provide only general guidelines for the food processor and regulatory inspectors. No specific details were included that could be used for specific CCps in a HACCP framework. Subsequently, however, the US FDA (1973) issued detailed GMP regulations and guidelines for specific food commodities. One example of a detailed GMP regulation that can be directly used in HACCP plans is the regulation on Thermally Processed Low Acid Foods Packaged in Hermetically Sealed Containers. These regulations specify the precise conditions and methods of monitoring CCPs in the production of thermally processed canned foods. The US FDA (1993) has recently issued an updated food code that incorporates the HACCP approach for the retail segment of the food industry. The purpose of the 1993 Food Code is to assist food control jurisdictions at all levels of government by providing them with a scientifically sound technical and legal basis for regulating the retail segment of the food industry.

In Europe, in addition to those specified by individual countries, microbiological criteria are contained within several of the product specific hygiene directives of the EC (1992a,b). These EC directives outline specific microbiological criteria and specifications required for intra-Community trade for specific commodities such as meat and milk.

Table 9.3 EC microbiological standards for minced meat products. Production plants and independent production units must ensure that, in accordance with Chapter VI of Annex I and with the methods of interpretation set out, minced meat and meat in pieces of less than 100 g intended to be marketed without further processing or as an ingredient in meat preparations, comply with specified standards.

	M^a	m^b
Aerobic mesophilic bacteria $n^c = 5$ $c^d = 2$	$5 \times 10^6/g$	$5 \times 10^5/g$
Escherichia coli $n = 5$ $c = 2$	5×10^{2g}	$50/g$
Sulphite-reducing anaerobes $n = 5$ $c = 1$	$10^2/g$	$10/g$
Staphylococci $n = 5$ $c = 1$	$5 \times 10^2/g$	$50/g$
Salmonella $n = 5$ $c = 0$	absence in 25 g	

[a] M = acceptability threshold, above which results are no longer considered satisfactory, where M equals 10 m when the count is made in a solid medium and M equals 30 m when the count is made in a liquid medium.
[b] m = threshold below which all results are considered satisfactory.
[c] Number of units making up the sample.
[d] Number of units in the sample giving values between m and M.

The establishment of microbiological specifications related to the presence of foodborne pathogens or spoilage organisms should always include the type of sampling plan. The ICMSF (1974) provides an excellent treatise on this subject, as discussed earlier. The ICMSF approach states that the stringency of sampling plans should be based on the hazard to the consumer, which is a function of the types and numbers of organisms present. The ICMSF approach utilizes attribute data where the decision is based on the number of sample units above or below a specified level. In a two class attribute plan, the decision is made based on whether or not the measure is above some critical measure designated m. The stringency of the sampling plan depends upon the number of samples tested per lot (n) and the number of samples (c) allowed to exceed the critical limit. For example, the recommended sampling plan for Salmonella in ice cream is: $n=10$, $c=0$, $m=0$, meaning 10 samples per lot are tested, and that no sample should test positive for Salmonella (zero tolerance). In many cases, however, it is not necessary to have a zero tolerance specification. For these cases, the ICMSF has incorporated the three-class attribute scheme. In this scheme, the acceptability of the product in terms of microbiological criteria is divided into three classes, based on two levels of counts (m and M). A count above M for any sample unit is unacceptable. Counts between m and M are not ideal, but some can be tolerated. In the ice cream example, while *Salmonella* cannot be tolerated, some level of *Staphylococcus aureus* is acceptable without posing a health risk. In this case, the recommended sampling plan for *S. aureus* in ice cream is: $n=5$, $c=1$, $m=10$, $M=100$. In this case, five samples per lot are tested. The lot is rejected if any sample has an *S. aureus* count greater than 100/g. One sample of the five may have a count between 10 and 100/g. If more than one sample has a county between 10 and 100/g, the lot is rejected.

Table 9.3 shows the EC (1988) microbiological standards for minced meat products which incorporate both the ICMSF two class and three class attribute plans. The attractive feature of attribute sampling plans is that probabilities are practically independent of distribution within the lot. This is an important practical advantage, because the distribution of microorganisms in any given lot is usually unknown.

9.5 Future directions

Over the past 15 years, the US Centers for Disease Control and Prevention have reported that the group of foodborne disease-producing microorganisms of concern increased from 5 to 13 (Wolf, 1992). Microbiological safety continues to be a high priority with regulatory authorities worldwide. This has been recently underscored by the *Escherichia coli* O157:H7 and *Listeria monocytogenes* outbreaks in the USA (Mermelstein, 1993) and France (Anon., 1992), respectively.

As already discussed, there is a definite worldwide shift to the application of the HACCP approach to managing food safety. In a report by the research committee of the Institute of Food Technologists, several needs were emphasized (Cliver, 1993). Among these is the need for better detection methods for foodborne pathogens and toxins. There remains a need for detection methods that are quicker, more sensitive and less costly. Much energy is being focused by researchers worldwide to develop innovative alternatives to the traditional cultural methods. Table 9.4 summarizes some of these approaches. While many of these methods have shown vast improvement in sensitivity and specificity, most have not been adopted for routine monitoring of CCPs. Ease of use and cost are primary considerations for lack of implementation. Another reason is that many of these methods have not received official recognition by regulatory or other sanctioning bodies. In the USA, for example, many microbiological test methods are evaluated and validated through a rigorous collaborative study process sanctioned by the Association of Official Analytical Chemists (AOAC) (Andrews, 1987). In the AOAC process, a method is adopted as an official

Table 9.4 Summary of selected innovative methods for rapid detection of foodborne pathogens, toxins and spoilage organisms.

Target	ELISA	Impedance/ conductance	ATP bioluminescence-	DNA probes	Genetic amplification	Latex agglutination	Enrichment serology	Immunomagnetic capture
Salmonella	X	X	–	X	X	X	X	X
Listeria monocytogenes	X	X	–	X	X	–	–	X
Escherichia coli	X	X	–	X	X	X	–	X
Staphylococcal enterotoxin	X	–	–	–	–	–	–	–
Campylobacter	–	X	–	X	X	X	–	–
Yersinia	X	X	–	X	–	–	–	–
Clostridium botulinum toxin	X	–	–	–	–	–	–	–
Lactobacilli	–	X	X	X	X	–	–	–
Yeast	X	X	X	X	X	–	–	–

method after successful completion of an interlaboratory collaborative study, and review, publication, and acceptance by the Official Methods Board. Use of an approved method ensures that it has been peer-reviewed and validated for its intended purpose. It also has the advantage of being recognized as an acceptable method by the regulatory authorities and in cases where litigation may be involved.

Listeria monocytogenes has received much attention in the USA over the past decade, and more recently in Europe. Much research has been conducted indicating that this organism does not merit a zero tolerance designation across all food types (Cliver, 1993; Jay, 1992). Given the ubiquity of this organism, it is being consumed at low levels in our food supply every day. Research is needed to determine why outbreaks of listeriosis are rare given this fact. Methods are needed that will distinguish between virulent and avirulent strains, and will accurately quantitate their presence in foods. It is also important to determine which foods will support the growth of *Listeria monocytogenes* to high levels. This will be valuable in helping the regulatory authorities establish acceptable levels in foods. *Escherichia coli* O157:H7 detection is a unique challenge because it is a pathogen that is not detected by traditional *E. coli* methods. Much progress has been made in the detection of this *E. coli* O157:H7, although improvements are still needed. The detection of this strain in mixed populations of *E. coli* is still difficult. As HACCP becomes more entrenched as the method for controlling food safety, methods are required that will yield results, ideally in a matter of minutes (or at least same day) as opposed to days. In a streamlined food production process, it is economically desirable to be able to react to a deviation from a microbiological CCP on a real-time or near real-time basis. As previously discussed, there is also an economic need to detect spoilage microorganisms or their metabolites on a real-time or near real-time basis, as well.

Despite the focus given to development of rapid methods over the past decade, most of the commercially-available rapid methods for both pathogens and spoilage organisms only provide retrospective data. The detection of pathogens, in a real-time, at-line or off-line fashion, is still a longer term proposition. The best promise for achieving this goal includes a combination of simple, rapid and automated purification/concentration techniques, combined with some type of amplification and/or detection system. One such approach is the combination of immunomagnetic separation and genetic detection techniques. Immunomagnetic techniques allow for the rapid separation and concentration of the target organism(s) (Skjerve *et al.*, 1990). When combined with genetic amplification techniques such as the polymerase chain reaction (PCR), this system could potentially provide real-time or near real-time results.

Another novel method that shows promise for concentrating target organisms of importance is the technique of dielectrophoresis (Patel, 1993). In this technique, microorganisms such as *L. monocytogenes* can be concentrated in minutes by high frequency voltage application to the suspending medium. This technique exploits the polarization of charges on cell surfaces. Manipulation of frequency and voltage settings can result in a differential concentration of microbial species. This technique, combined with advances in the development and application of biological sensors, also offers promise for real-time monitoring.

Other novel methods that have been reported for detecting specific foodborne organisms include the lux and ice nucleation gene technologies. The lux gene approach exploits the use of specific phages encoding the luminescent genes from *Vibrio fischeri* (Stewart and Williams, 1992). When combing with the target organism, the bioluminescent gene encoded by the phage causes the target organism to luminesce. Another gene technique is the Bacterial Ice Nucleation Diagnostic assay (BIND), which employs the use of a phage encoding for an ice nucleation gene from species of *Pseudomonas* or *Erwinia* (Wolber and Gruen, 1990). Following combination of the phage with the target organism, ice nucleation proteins are expressed. Detection is then achieved through freezing of the test samples in the presence of an indicator dye. Positive samples freeze and do not take up the dye. Negative samples do not freeze and show green fluorescence from uptake of the dye.

Combinations of physicochemical techniques and these novel amplification and detection systems offer the best opportunity, to date, of realizing real-time or near-real time monitoring methods. A Delphi survey of experts conducted in 1980–1981 predicted that traditional methods for enumeration and detection of microorganisms in foods will be automated by the end of the century (Vasavada, 1993). Much work, however, still needs to be done to bring these techniques to a state that fulfills that Delphi prophecy.

In conclusion, concerted efforts must be focused by academia, industry and government on the following needs, as outlined by the Research Committee of the IFT (Cliver, 1993).

1. Determine the true incidence, causes and financial/personal impact of foodborne diseases.
2. Develop more effective methods of testing foods (routine monitoring/epidemiological tools).
3. Devise better methods of controlling microbiological risks associated with foods.
4. Develop better means of hazard assessment, including degree of exposure, pathogenicity/toxicity for humans and differential risks among special population groups such as the immunocompromised.

Focusing on the above areas will provide a holistic approach for managing food safety in this era of streamlined processing. Incorporation of HACCP into programs such as ISO-9000 and Total Quality Management (TQM) provides a comprehensive system for process optimization and continuous improvement. Carefully integrating HACCP into a TQM system will integrate food safety, quality and productivity into a total systems approach to manufacturing and distribution (NFPA, 1992). This must be managed in such a way as not to dilute the emphasis on safety, but rather emphasize its focus in a comprehensive management framework. As this trend takes us into the 21st century, the focus on improved rapid methods is a paramount need.

References

Adams, C.E., Garrett, E.S., Hudak-Roos, M., *et al.* (1992) HACCP system in regulatory inspection programs: case studies of the USDA, USDC, and DOD, in *HACCP Principles and Applications* (eds. Pierson, M.D. and Corlett, D.A.) Von Nostrand Reinhold, New York.

American Medical Association, Council on Scientific Affairs (1993) Food safety, federal inspection programs. *Arch. Fam. Med.*, **2**, 210–214.

Andrews, W.H. (1987) Recommendations for preparing test samples for AOAC collaborative studies of microbiological procedures for foods. *J.A.O.A.C.*, **70 (6)** 931–936.

Anon. (1992) Listeriosis in France. *CDR Weekly*, 2, 29.

Banwart, G.J. (1989) *Basic Food Microbiology*, Reinhold, Van Nostrand, New York.

Bauman, H.E. (1992) Introduction to HACCP, in *HACCP Principles and Applications*, (eds Pierson, M.D. and Corlett, D.A.) Van Nostrand Reinhold, New York, pp. 1–5.

Cliver, D.O. (1993) America's food research needs: into the 21st century, in *Food Technol.*, **47 (3)**, 105–135.

Codex Alimentarius Commission – Codex Committee on Food Hygiene (1993) Guidelines for the application of hazard analysis critical point (HACCP system). Alinorm 93/13A, Appendix B. Food and Agriculture Organization/World Health Organization, Rome.

Corlett, D.A. and Stier, R.F. (1991) Risk assessment within the HACCP system. *Food Control*, **44 (7)** 71–72.

EC Council Directive 88/657 (1988) Minced meat directive. *Off. J. Eur. Communities.* No. L382/13–382/14.

EC Council Directive 92/5 (1992a) Amending and updating directive 77/99/EEC on health problems affecting intra-community trade in meat products and amending directive 64/433/EEC. *Off. J. Eur. Communities.* No. L57/1–57/26.

EC Council Directive 92/46 (1992b) Laying down the health rules for production and placing on the market of raw milk, heat-treated milk and milk-based products. *Off. J. Eur. Communities.* No. L268/1–268/34.

Grossklaus, D., Gerik, K., Kolb, H. and Zastrow, K.D. (1991) Zar welweiten zunahme von enteritis infectiosa-fallen-kritische anmerkungen. *Arch. Lebensmittlelhyg.*, **42** 136–140.

International Commission on Microbiological Specifications for Foods (1974) *Microorganisms in Foods 2 – Sampling for microbiological analysis: principles and specific applications.* University of Toronto Press.

International Commission on Microbiological Specifications for Foods (1988) *Microorganisms in Foods 4 – Application of the hazard analysis critical control point (HACCP) system to ensure microbiological safety and quality.* Blackwell Scientific, Oxford.

Jay, J.M. (1992) Microbiological food safety. *Crit. Rev. in Food Sci. Nut.* **31 (3)** 177, 190.

Lamprecht, J.L. (1991) ISO 9000 implementation strategies. *Quality*, **November**, 14–17.

Lechowich, R.V. (1992) Current concerns in food safety, in *Food Safety Assessment*, (eds Finley, J.W., Robinson, S.F. and Armstrong, D.J.). ACS Symp. Ser. 484:232–242, American Chemical Society, Washington, DC.

Mermelstein, N.H. (1993) Controlling *E. coli* O157:H7 in meat. *Food Technol.* **47 (4)** 90–91.

National Advisory Committee on Microbiological Criteria for Foods (1990) HACCP principles for food production. USDA–FSIS Information Office, Washington, DC.

National Advisory Committee on Microbiological Criteria for Foods (1992) Hazard analysis and critical control point system. USDA–FSIS Information Office, Washington, DC.

National Food Processors Association, Microbiology and Food Safety Committee (1992) HACCP and Total Quality Management – winning concepts for the 90's: a Review. *J. Food Prot.*, **55**, 459–462.

National Food Processors Association, Microbiology and Food Safety Committee (1993a) Implementation of HACCP in a food processing plant. *J. Food Prot.*, **56**, 548–553.

National Food Processors Association, Microbiology and Food Safety Committee (1993b) HACCP implementation: a generic model for chilled foods. *J. Food Prot.*, **56**, 1077–1084.

National Marine Fisheries Service (1987) Plan of operation – model seafood surveillance project. Office of Trade and Industry Services, National Seafood Inspection Laboratory, Pascagoula, Mississippi.

National Research Council/National Academy of Sciences (1985) An evaluation of the role of microbiological criteria for foods and food ingredients. National Academy Press, Washington, DC.

Patel, P.D., (1993) private communication.

Pierson, M.D. and Corlett, D.A. (1992) *HACCP Principles and Applications.* Van Nostrand Reinhold, New York.

Saunders, M. (1991) EC testing and certification procedures: how will they work? *Business America*, **112 (4)** 27–28.

Skjerve, E., Rorvik, L.M. and Olsvik, O. (1990) Detection of *Listeria monocytogenes* in foods by immunomagnetic separation. *Appl. Envir. Microbiol.* **56** 3478–3481.

Stewart, G.S.A.B., and Williams, P. (1992) Lux genes and the application of bacterial bioluminescence. *J. Gen. Microbiol.*, **138**, 1289–1300.

Taylor, M.R. (1993) FDA's plans for food safety and HACCP – institutionalizing a philosophy of prevention. Presented at *Symposium o n Foodborne Microbiological Pathogens*, International Life Sciences Institute in Conjunction with the International Association of Milk, Food and Environmental Sanitarians National Meeting, Atlanta.

Todd, E.C.D. (1992) Foodborne disease in Canada – a 10–year summary from 1975–1984. *J. Food Prot.*, **55** 123–132.

US Department of Health, Education and Welfare (1972) Proceedings of the 1971 conference on Food Protection. US Government Printing Office, Washington, DC.

US Food and Drug Administration (1969) Good manufacture practice in manufacturing, processing, packing or holding human food. Section 37C.

US Food and Drug Administration (1973) Thermally processed low acid foods packed in hermetically sealed containers GMP. *Federal Register*, **38 (16)** 2398–2410.

US Food and Drug Administration/Public Health Service (1993) *Food Code.* US Dept. of Commerce Technology Administration, Springfield, Virginia.

Vasavada, P.C. (1993) Rapid methods and automation in food microbiology: beyond Delphi forecast., *J. Rapid Meth. Automat. Microbiol.* **2 (1)** 1–7.

Wolber, P.K. and Gruen, R. L. (1990) Detection of bacteria by transduction of ice nucleation genes. *Trends Biotechnol.* **8**, 276–279.

Wolf, I.D. (1992) Critical issues in food safety, 1991–2000. *Food Technol.* **46 (1)** 64–70.

10 Detection and identification of foodborne microbial pathogens by the polymerase chain reaction: food safety applications

W.E. HILL and Ø. OLSVIK

10.1 Introduction

Traditional microbiological techniques for the isolation and identification of bacteria from foods have depended on obtaining pure cultures. Enrichment and selection steps are often time-consuming and the biochemical identification of a particular species may add several days to the procedure. Such methods are insufficient when an outbreak of foodborne disease is occurring or food products with short shelf-lives must be tested. DNA hybridisation methods first developed for research use in the molecular biology laboratory, such as DNA gene probes used for colony hybridisation (Grünstein and Hogness, 1975), and the polymerase chain reaction (PCR, Saiki et al., 1985) have shortened the time required for analysis, by obviating the need for pure cultures. Very rapid PCR methods have been developed that allow the analysis of some food samples within 3–4 h. The basic mechanism of the PCR and its variations will be discussed. Examples of how PCR has been applied to the detection of particular foodborne pathogens will be presented.

The development of DNA nucleotide sequencing methods, the storage of this information in a computer-searchable database, and the invention of automated oligonucleotide synthesis methods have all contributed to the rapid implementation and widespread use that PCR now enjoys. Another important step for PCR use was the discovery that thermostable DNA polymerases were suitable (Saiki et al., 1988); thus, thermocycling could be automated, dramatically increasing its attractiveness. This has resulted in the publication of more than 18 000 papers referring to PCR; currently, papers reporting PCR uses are appearing in print at a rate greater than 500 per month.

The use of PCR to detect foodborne pathogens is closely related to the clinical applications of this test for the detection of disease-causing microorganisms. The significant differences arise when sample matrices and the level of target cells are considered. In a clinical situation, there may be a half dozen or so major sample types such as blood, stool, urine and other body fluids, while there are a considerable number and wide variation in the types of foods that might have to be examined. Because the human body can be a good enrichment medium for pathogenic microbes, clinical samples often contain high numbers of the strain responsible for the disease and samples from healthy individuals may be sterile. Foods, on the other hand, often contain high levels of indigenous microflora, especially in fresh commodities. A major challenge for the food microbiologist has been to devise enrichment and selection schema to increase the number of the sought-after species to levels sufficient for their detection by conventional cultural techniques.

Finally, the nature of food safety microbiology has changed over the past decade or two. The last generation of microbiologists was usually content with the taxonomic identification of a particular isolate. Now that much more is known about the mechanisms of microbial pathogenicity and strain variation within species, isolates must be tested for their ability to cause disease, which is sometimes more important than assigning to them the proper genus and species names. As our knowledge of the genetics of pathogenic determinants has increased, it has been possible to focus our efforts to detect disease-causing bacteria by testing for the presence of particular genes that confer pathogenicity. This was true for the gene probe tests and continues to be the case for PCR-based assays. However, as more nucleotide sequences of ribosomal RNA genes become available, genus and species specific PCR tests are being developed. Thus, PCR methods can be designed to be general enough to detect particular taxonomic groups, as well as specific enough to detect pathogenic strains.

10.2 Foodborne diseasees

For most of the foodborne outbreaks in the USA reported to the Centers for Disease Control, the aetiologic agent is never confirmed (Bean *et al.*, 1990); the important microbes have been reviewed (Archer and Young, 1988). The number of reported outbreaks undoubtedly underestimates the true incidence of foodborne disease by a large factor (Synder, 1992), but the projected amounts of morbidity and mortality are still considerable (Roberts, 1989; Todd, 1989). Data from the outbreaks in which the aetiologic agent is confirmed show that *Salmonella* species are responsible for more than half of the outbreaks and cases. Within the past decade the incidence of *Listeria monocytogenes, Campylobacter jejuni, Vibrio cholerae, Salmonella enteritidis* and, just recently, *Escherichia coli* O157:H7 has increased dramatically. Because of the difficulties in detecting foodborne viruses and parasites, these agents could be the most underreported contributors to foodborne illness. Increases in outbreaks caused by bacteria could be due, in part, to surveillance artifacts, increased public awareness of the hazards associated with foodborne microorganisms, and, perhaps, better detection methods. Nevertheless, the ability to recognise, control and prevent outbreaks depends to a large extent on having rapid, sensitive and cost-effective techniques for the detection and identification of a wide range of foodborne microbes.

10.3 PCR fundamentals

There are several reviews that discuss the mechanism of PCR, including a growing number of variations and enhancements (Arnheim and Levinson, 1990;

Innis *et al.*, 1990; Erlich *et al.*, 1991; Wolcott, 1992). The PCR is an *in vitro* method for the selective, repeated duplication of a specific segment of DNA. The region of DNA to be amplified (the target DNA) is defined by a pair of oligonucleotide primers, the 3' ends of which serve as initiation points for DNA replication. In the first phase, the DNA strands are separated (denatured) and then cooled in the presence of a billion-fold excess of primers and a thermally stable DNA polymerase (Figure 10.1). When the strands are cooled, many target DNA–primer hybrid complexes will be formed; DNA replication will be initiated and the target region will consist of one original strand and a new strand that has a defined 5' end. The temperature of this annealing step is critical in controlling the accuracy (stringency) of the amplification process. Too low an annealing temperature will result in non-specific amplification and an array of PCR products. When the annealing temperature is too high, primers will not be stably associated with the target DNA and no amplification will occur. When a thermally stable DNA polymerase is used, there is no need to add additional polymerase after each denaturing step, so the entire thermocycling procedure can be automated by using a programmable heating block. The temperature is then increased again to separate the double strands and cooled to allow primer binding and DNA replication. The template strands synthesised in the first round of replication have one defined end and when they themselves serve as a template, the resulting new strand will have two defined ends and, therefore, a uniform length. This length can be predicted; it is the distance between the 5' ends of each primer. The thermocycling is repeated, usually for 30–40 cycles, and million-fold increases in DNA are common.

There are several ways in which PCR products can be characterised. A common method is to determine the size of the amplified fragments by using gel electrophoresis. However, other than the observed size of the fragment matching the predicted size, there is no other evidence that the product is, in fact, the expected region. Several techniques are available to provide more information. The amplified fragment can be sequenced, as this is laborious and not well-suited for the routine analysis of many samples, but it does provide the strongest confirmatory data. Because the sequence of the amplified region is usually known, conveniently located internal restriction endonuclease cleavage sites can often be located and the size of the digestion products known. This can be confirmed by electrophoresis of the cleavage products and, once again, by comparing the size of the observed fragments with the predicted size. An internal hybridisation probe can also be used on a Southern blot of a gel or on a dot blot of the completed reaction prior to electrophoresis. A final method is to use an internal oligonucleotide as a primer and dilute the amplified product several 1000-fold, add the internal primer and the appropriate member of the original primer pair, and conduct another amplification. The size of this second amplification product can be compared with the predicted size or used in a dot blot with the other original primer serving as a negative hybridisation control and the 'third' primer as a positive control.

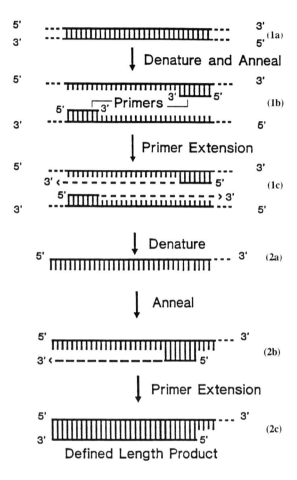

Figure 10.1 The basic PCR mechanism. (1a) A region of DNA is chosen for amplification. Usually, the nucleotide sequence of the region is known. The dashed lines at the ends of the molecules represent strands that do not have fixed ends. (1b) The strands are separated (denatured) by heating and then cooled in a million- to billion-fold excess of the primers. When the reaction is cooled to the annealing temperature, the primers hybridise specifically to the template DNA. (1c) The temperature is raised to the optimum for a thermostable DNA polymerase and the primers are extended using the sequence of the template strands (dashed lines with arrows indicating direction of elongation). This completes the first cycle and there are now two copies of the region defined by the primers, although the newly synthesised strands have one defined end (the primer end) and one variable end. (2a) For illustrative purposes, only one of the newly made strands is shown after a second round of heating. (2b) When cooled, primers bind to the newly made strand which serves as a template. At the enzyme optimum, the primers are extended only to the defined 5' end of the template strand. (2c) The newly made strand has two defined ends, each of which is determined by the location of the primers. This moiety will increase exponentially and will become the dominant PCR product. Because of the defined length, these molecules will appear as single bands after separation by gel electrophoresis.

10.4 PCR: some practical considerations

10.4.1 Samples

The application of PCR tests to 'real world' problems (instead of basic research procedures in the molecular biology laboratory) requires that considerable attention be paid to the preparation of samples. While PCR usually works well with highly purified DNAs, it is difficult to easily prepare such pristine templates from food samples. Some investigators have chosen to attempt purification of bacterial DNA directly from homogenised foods, but, in the majority of cases, a growth step is included prior to template preparation. This has several advantages. The number of copies of target DNA is increased, thereby increasing test sensitivity. This growth increase means that a greater dilution can be added to the reaction, further reducing the concentration of PCR inhibitors which might be present in the food. Finally, dead bacterial cells may contain DNA that can be amplified, so that a positive result may be observed when no viable cells are present. However, such an outcome might signal a potential public health problem.

10.4.2 Primers

There are a number of important factors to be considered during the course of PCR primer set development. The regions selected as primer sites should be unique and some assurance of this can be obtained by searching for similar regions in computerised nucleotide sequence databases such as GenBank or the European Molecular Biology Laboratory (EMBL). Another factor is the thermal stability of the primer when annealed to target sequences. For best results, the two primers should have similar dissociation temperatures (T_d). The 3' end of the primer must be complementary to the template strand or amplification will not occur. (This requirement forms the basis of the ligase chain reaction.) Several computer programs have been developed to search for predicted thermal stabilities, for primer self- and intrastrand-complementarity (hairpin loop formation), and interprimer complementarity to reduce the chance of primer–dimer formation.

10.4.3 Reaction conditions

There are a number of factors to be considered when seeking optimal reaction conditions for DNa amplification (Saiki, 1989; Innis et al., 1990). Mullis (1989) gives some practical advice for starting parameters and makes suggestions on what to vary. When preparing templates from contaminated food samples it is important to have a reproducible preparation with respect to contaminants, because they can have a considerable effect on the amplification efficiency. Other parameters to consider include divalent cation concentration (especially Mg^{2+}), pH, primer sequence and concentration, template concentration, hybridisation stringency, number of cycles, length of elongation step, the size of the amplified fragments and the rate of the temperature change between cycles.

10.5 PCR format variations and applications

10.5.1 Multiplex PCR

It is possible, under some sets of circumstances, to have two or more sets of primers function simultaneously within the same reaction tube. Therefore, several gene targets can be amplified to detect the presence of several genes or multiple microorganisms at the same time (Bej *et al.*, 1991). Primer sets should be selected so as to have similar thermal stabilities. It may be necessary to alter reaction conditions because there may be some interaction between the primers.

10.5.2 Nested PCR

Once a region has been amplified, confirmatory information can be obtained by diluting the product several thousand fold, and conducting a subsequent PCR using primers that are targeted to sequences on the 3' side of each of the original primers. This new product will be shorter than the original, but its size can be predicted. Because there can be problems in trying to increase amplification reaction sensitivity or productivity by simply increasing the number of cycles, the nested PCR may provide an alternative.

10.5.3 Reverse transcription PCR

Some foodborne viruses such as hepatitis A virus contain RNA as their genetic material. To detect specific RNA nucleotide sequences by using PCR, it first must be copied, using reverse transcriptase, into a strand of cDNA. This is accomplished by using a 'downstream primer' and then the cDNA product is used as a template for a conventional PCR.

10.5.4 Ligase chain reaction (LCR)

Often, methods are needed to distinguish between two templates that differ only in a single nucleotide. To carry out the LCR, two oligonucleotides are constructed as complementary to directly adjacent regions on the target DNA. The locations of the two oligonucleotides are chosen so that the 3' end of one of the primers is at the site that varies. Then DNA ligase is used to covalently join the two oligonucleotides. If the 3' end is not complementary to the target strand, ligation will not occur and the two oligonucleotides will remain separated.

10.5.5 Detection of immobilised amplified nucleic acid (DIANA)

A recent development in PCR is the use of a set of nested primers and magnetic separation (MS) of the PCR-generated fragments (MS-PCR), followed by different signal transducing systems (Uhlen, 1989; Lundberg *et al.*, 1990; Uhlén *et al.*, in press). The MS-PCR test principle has been designated Detection of Immobilised Amplified Nucleic Acid (DIANA) (Uhlén *et al.*, 1990). The inner primers are labelled, one with $\gamma^{32}P$ or with a tail of a partial sequence of the lac operator (*lac*Op) gene, and the other with biotin. An alternative labelling method is incorporation of digoxigenin-11-dUTP in the PCR reaction. The biotin- and the labelled amplicons are separated from the solution by streptavidin-coated magnetic beads. Using parts of the *lac*Op gene as a label for one of the primers, this can be detected by the *lac* repressor protein (*lac*I) fused with the enzyme β-galactosidase (Lundeberg *et al.*, 1990).

The PCR amplicon from the DIANA can be used as the template in robot-assisted DNA sequencing. The biotin-labelled amplicons from the target genes are bound to streptavidin-coated magnetic beads and then made single-stranded by denaturing with NaOH or heat. The bead-bound and the free dissolved strands are separated by the magnet; both strands can be recovered for sequencing (Uhlén, 1989; Olsvik *et al.*, 1993). The amplicons generated by DIANA can be sequenced with a standard sequencing primer corresponding to the *lac*Op sequence tailing the PCR product. Commercial primers, such as those for M13 sequences, can also be used if a tail of corresponding M13 sequences is used in the primer not labelled with biotin (Olsvik *et al.*, 1993). Using the described methodology, an advanced microbiology laboratory equipped with a PCR thermocycler, robotic station and an automatic DNA sequencer could sequence 300–600 bases from 20 to 30 strains within 6–12 h. DNA sequences from many strains should be invaluable in addressing the distribution and evolution of genes and bacterial strains.

10.6 Immunomagnetic separation and PCR

Small (2–5 μm), super-paramagnetic particles or beads coated with antibodies against surface antigens of microorganism, have been shown to be efficient for isolation of such cells from different biological and environmental samples (Ugelstad *et al.*, 1992; Olsvik *et al.*, in press; see chapter 4). Immunomagnetic separation (IMS), or immunomagnetic enrichment, is assisted by the fact that bacteria immunologically bound to magnetic beads usually remain viable and can continue to multiply if nutritional requirements are provided. Both polyclonal and monoclonal antibodies have been employed in IMS (Lund *et al.*, 1988; Skjerve and Olsvik, 1991). These antibodies can be linked to the beads either directly or indirectly using beads precoated with anti-mouse or anti-rabbit antibodies. The target bacteria are separated from the environment and are concentrated from a large volume to a volume suitable for the detection method.

Growth inhibitors present in the sample are removed by this technique, also enhancing success for positive cultivation of bead-bound bacteria. Selective growth media are usually less efficient for isolation of pathogenic variants if the sample also contains large numbers of non-pathogenic variants of the same species (Olsvik *et al.*, 1991a,b). Traditional cultivation of food samples for selective enrichment of enterotoxigenic *Escherichia coli* in broths favours strains of environmental origin over those of human origin (Hill and Carlisle, 1981). The pathogenic strain of human origin might therefore not be detected in a food sample. However, IMS provides the potential for selective isolation of strains possessing specific surface epitopes, such as fimbriae, associated with the ability to induce disease (Lund *et al.*, 1988, 1991; Olsvik *et al.*, 1991a,b).

PCR detection of foodborne pathogens can have certain disadvantages and, thus, limit the technique for diagnostic use. The sample volumes traditionally used in PCR range from less than 1 μl to 20 μl, and this volume is not sufficient when, for example, testing for *Salmonella* in foods. Such sample volumes restrict the test sensitivity to a theoretical minimum of 5 000–100 000 organisms per ml (Olsvik *et al.*, 1991, in press). An additional factor hindering the diagnostic use of PCR directly on samples is the sensitivity of the *Taq* polymerase to inhibitor elements in samples (De Francis *et al.*, 1988; Beutler *et al.*, 1990; Hornes *et al.*, 1991).

Use of IMS as a pre-PCR step appears to solve several of these problems. The bacteria in the sample are concentrated from 1–10 ml to a suitable volume of 10–100 μl, and specific *Taq* polymerase inhibitors are simultaneously removed. Material that can prevent the organisms from lysing and making the nucleic acid available for PCR are also removed. After IMS, boiling the extracted fraction for 10–30 min is generally sufficient for lysing most Gram-negative bacteria (Hornes *et al.*, 1991; Olsvik *et al.*, 1991; Islam and Lindberg, 1992; Kapperud *et al.*, in press). Freeze–thawing of the IMS sample can also be used to open the bacterial cells and make the nucleic acid available for PCR. In this case, DNA is not denatured, which can be an advantage if the sample contains little DNA; otherwise single-stranded DNA can attach to the plastic in the lysing tubes and, therefore, not be amplified in the PCR.

Adding the magnetic bead fraction to a growth medium for precultivation can in some instances increase the number of target organisms for the PCR. Samples that have been frozen often contain non-viable cells, but these cells can still be extracted with IMS and identified by PCR (Hornes *et al.*, 1991). This method can also be important for identifying the origin of strains involved in foodborne outbreaks if only non-viable bacteria remain in the implicated food sample.

10.7 PCR detection and identification of foodborne microorganisms: Bacteria

New PCR-based methods for the detection of pathogenic microorganisms in foods are being reported continuously. The species listed below are some that play an important role as agents of foodborne disease, although some of the methods referenced are based on clinical or environmental samples.

10.7.1 Campylobacter

During the last few years, *Campylobacter jejuni* has been recognised as a major agent of sporadic diarrhoeal disease in the USA and may have an incidence similar to that of *Salmonella* species. *Campylobacter coli* appears to be the only other member of this species that is of public health significance, but this may be a reflection of inadequate isolation methods, as microaerophilic conditions are required to culture *Campylobacter* in the laboratory. (Recently, *C. pylori* was reclassified as *Helicobacter pylori*.) The bacterium can be acquired from mammals and from birds, but a major vehicle of infection appears to be raw or undercooked meat. Poultry appears to be a primary source, as more than half of the carcasses are contaminated (Hood *et al.*, 1988). A PCR-based assay for *Campylobacter* species was developed by Giesendorf *et al.* (1992), by targeting the variable regions of the 16SrRNA genes of *C. jejuni*, *C. coli* and *C. lari*. While about 12 cells could be detected in pure cultures, the sensitivity was much lower with naturally infected samples of poultry parts. By using a conserved region in the 5' end of the *C.coli* flagellar protein gene *(flaA)*, Oyofo *et al.* (1992) developed an assay specific only for species *C.coli* and *C.jejuni*.

10.7.2 Clostridium botulinum

Because of the life threatening consequences of botulinal neurotoxin, it is important to identify *Clostridium botulinum* rapidly. Animal tests and other bioassays are too time-consuming, so PCR-based methods are attractive. While most foodborne pathogens cause disease by an active infection within the host, *C.botulinum* (and *Staphylococcus aureus*) cause disease when a pre-formed toxin is ingested and viable cells may not be present at the time the food is consumed. Because PCR can be used to amplify the DNA of non-culturable or even nonviable bacteria, this method can be quite useful for determining if a product had been contaminated with *C.botulinum*. Because the conformation of DNA molecules is quite thermally stable when compared with proteins, PCR is used more successfully for testing heat processed foods than are immunologically-based techniques.

The nucleotide sequences of several neurotoxin genes have been determined and they are popular PCR targets for the detection and identification of *C.botulinum*. Szabo *et al.* (1992) amplified a region of the type B neurotoxin gene and obtained a sensitivity of 10 pg of *C.botulinum* DNA (representing about 3000 cells), when the PCR products were analysed by gel electrophoresis. Sensitivity was increased about 100-fold when a biotinylated PCR product was used as a gene probe and hybridised to Southern blots. Ferreira *et al.* (1992)

targeted the type A neurotoxin and was able to detect 100 cells per gram in some of the seeded commercially canned products. The PCR products were detected by using a digoxigenin-labelled internal gene probe. This data suggests that unsporulated *C. botulinum* cannot be detected by this method, ostensibly because the lysing procedure used to release DNA from vegetative cells was ineffective when applied to spores. Gurtler *et al.* (1991) were able to distinguish between *C. perfringens* and *C. botulinum* by amplifying ribosomal RNA genes with primers based on conserved regions of the *Escherichia coli* 16S rRNA genes and generating restriction endonuclease fragment patterns by digesting the products.

10.7.3. Escherichia coli *and* Shigella *spp.*

Once thought to be a normal, indigenous and harmless inhabitant of the human gut, some strains of *E.coli* are now recognised as potentially fatal foodborne pathogens. Infection with enterotoxigenic strains of *E. coli* (ETEC) is the leading cause of morbidity and mortality among children in underdeveloped countries. The genes encoding these enterotoxins have been used as PCR amplification targets (Olive *et al.*, 1988; Olive, 1989). ETEC were detected by PCR in surface water and in soft cheese without the need for a growth step (Furrer *et al.*, 1990; Candrian *et al.*, 1991; Meyer *et al.*, 1991). Strains of ETEC seeded into minced meat were enriched by two 24 h incubations and then detected by PCR amplification (Wernars *et al.*, 1991). The limit of detection after 2 days of growth was 3 cells per 25 grams of contaminated meat.

Bej *et al.* (1990) developed a number of primer sets for the detection of groups of bacteria important in the assessment of water quality. The *lacZ* gene was used as an *E. coli*-specific target. *Escherichia coli, Salmonella typhimurium* and *Shigella flexneri* were detected by specific PCR amplification of the *lamB* gene (which encodes the bacteriophage lambda receptor). The *uidR* encoding the regulatory gene for β-D-glucuronidase in *E. coli* and *Shigella sonnei* was used as a PCR target for the detection of these species.

Some of the nucleotide sequences of the invasion factor encoding genes of *E. coli* and *S. flexneri* are similar. A region of a large plasmid found in invasive strains of *E. coli* and *S. flexneri* was used to detect the former species seeded into unpasteurized milk (Keasler and Hill, 1992) and the latter in seeded lettuce (Lampel *et al.*, 1990). The methods used to prepare template DNA in these two reports were quite different. *Shigella flexneri* plasmid DNA was extracted using a method that favours the recovery of plasmid DNA, while total DNA methods were used on the milk samples.

Strains of enterohaemorrhagic *E. coli* (EHEC) elaborate toxins resembling the one produced by some strains of *Shigella* (the Shiga-like toxins SLTI and SLTII) and can cause haemolytic-uraemic syndrome and haemorrhagic colitis. The organism appears to be carried by asymptomatic cattle and is a potential health hazard in insufficiently cooked beef and raw milk (Doyle, 1992). One especially well-known serotype, O157:H7, caused several recent foodborne outbreaks and resulted in several deaths. The SLT genes have been targets for PCR-based assays (Karch and Meyer, 1989; Brian *et al.*, 1992; Gannon *et al.*, 1992; Hill *et al.*, 1993). Jackson (1991) labelled SLT gene PCR products with digoxigenin and detected them with a nonisotopic chromogenic substrate coupled with anti-digoxigenin antibody.

10.7.4. Listeria monocytogenes

Not recognised as a major foodborne pathogen until just a few years ago, *Listeria monocytogenes* has received considerable attention (Farber and Peterkin, 1991; Schuchat *et al.*, 1991). In the USA, attention was focused on *L. monocytogenes* after discovering that it was the agent of a large outbreak that occurred in 1985 (Linnan *et al.*, 1988). A favourite PCR target is the gene listeri-olysin O (*hlyA, lisA*), which has been identified as a virulence factor (Menguad *et al.*, 1988; Cossart *et al.*, 1989). The gene has been amplified by Bessesen *et al.* (1990), Border *et al.* (1990), Deneer *et al.* (1991), Fitter *et al.* (1991), Furrer *et al.* (1991), Golsteyn Thomas *et al.* (1991) and Candrian *et al.* (1992). Another target is the invasion-associated protein (*iap*) that was amplified by Rossen *et al.* (1991), Bubert *et al.* (1992), Herman and De Ridder (1992) and Jaton *et al.* (1992). The delayed hypersensitivity gene (*Dth-18*) was targeted by Wernars *et al.* (1991), the *L. monocytogenes* antigen gene (*lmaA*) by Johnson *et al.* (1992) and the ribosomal RNA genes by Border *et al.* (1990), Jensen *et al.* (1993) and Wang *et al.* (1993). A variety of foodstuffs has been examined: milk (Bessesen *et al.*, 1990; Furrer *et al.*, 1991; Golsteyn Thomas *et al.*, 1991), cheese (Fitter *et al.*, 1991; Rossen *et al.*, 1991; Wernars *et al.*, 1991), beef (Golsteyn Thomas *et al.*, 1991), and poultry (Fitter *et al.*, 1991; Wang *et al.*, 1993). A ligase chain reaction (LCR) was also developed to detect *L. monocytogenes*-specific 16S rDNA (Wiedmann *et al.*, 1992).

10.7.5. Salmonella

About half of confirmed outbreaks and cases of foodborne disease are attributed to *Salmonella* (Bean *et al.*, 1990). In 1991 alone, over 48 000 cases were reported (Centers for Disease Control, 1992). Even though this organism is responsible for a considerable fraction of foodborne disease, not many reports of PCR-based assays have been made. This might be due in part to the difficulty in developing primer sets that can inclusively detect the large number of serogroups within this genus. By targeting an evolutionarily conserved region of the chromosomal origin of replication, Widjojoatmodjo *et al.* (1991) developed a *Salmonella*-specific PCR assay which they coupled with an immunomagnetic separation based on a monoclonal antibody. The *lamB* gene was targeted by Bej *et al.* (1990) to develop a PCR test to detect *E. coli, Salmonella typhimurium* and *Shigella flexneri*. The selectivity of this assay could be increased to give positive results with only strains of *E. coli* and *S. flexneri* by using primers to amplify a region of the *uidR* gene. The strain positive for *lamB* but negative for *uidR* would be *Salmonella*.

10.7.6 Staphylococcus aureus

As an important agent of foodborne illness, members of the species *Staphylococcus aureus* can elaborate a group of related enterotoxins designated types A to E. the manner of disease is similar to *C. botulinum*, in that illness is incurred by the ingestion of a preformed toxin; the staphylococcal enterotoxins can induce diarrhoea and/or vomiting within a few minutes to several hours. Of public health concern is the fact that these toxins are relatively heat-stable; active toxin can exist in heat processed products even though viable organisms may no longer be present. Therefor, identification of enterotoxin-encoding genes and toxin type can provide important epidemiological information because live cultures might not be recovered from a suspect food. Johnson *et al.* (1991) developed PCR primer sets for the five types of staphylococcal entero-toxin genes. Some strains of *S. aureus* produce an extracellular thermostable nuclease (TNase which is encoded by the *nuc* gene), to which a specific primer set was developed (Johnson *et al.*, 1991). Strains of *S. aureus* that elaborate enterotoxin type C can contaminate the milk of cows afflicted with bovine mastitis. The entire *entCl* gene was amplified by Wilson *et al.* (1991) and test sensitivity was increased by using nested primers to amplify an internal segment. The detection sensitivity in these assays was between 10^3 and 10^4 cells per reaction, which is considerably lower than in some tests described for other foodborne microorganisms. These results suggest that PCR-inhibiting substances may be present which could reduce the efficiency of amplification and, therefor, lower the amount of product formed. Also, the efficiency of template DNA preparation from Gram-positive bacteria might be lower and this would contribute to the apparent low sensitivity.

10.7.7 Vibrio spp.

Even though cholera is endemic in parts of Asia, this disease was observed only sporadically in the Western Hemisphere until the outbreak in Latin America (Tauxe and Blake, 1992). In addition to *Vibrio cholerae*, *V. parahaemolyticus* and *V. vulnificus* are important pathogens. Vibrios are naturally occurring marine microorganisms, making it important to have rapid methods specific for these pathogens because traditional indicators of faecal pollution do not always correlate well with their presence. Dorsch *et al.* (1992) used 16S ribosomal RNA sequences to propose a revised phylogeny for this genus. The *ctx* gene which encodes the classical cholera enterotoxin (and shares significant nucleotide sequence homology with the heat-labile enterotoxin gene of *E. coli*) was amplified by Shirai *et al.* (1991) who tested clinical samples. A *V. cholerae* El

Tor Inaba strain isolated from a patient during the Latin American outbreak was used to seed oysters, shrimp and lettuce (Koch *et al.*, 1993). The A and B cistrons of the region encoding cholerae enterotoxin were amplified by using a boiled template from alkaline peptone water enrichments of homogenates of the seeded foods. Positive results for samples of oysters and lettuce seeded with 1 cell of *V. cholerae* per gram could be detected when sampled after 6 hours of enrichment. Fields *et al.* (1992) used a PCR targeted to the *ctxA* region and were able to identify this gene in strains within four hours of strain receipt.

During the warmer months, *V. vulnificus* can be recovered from some marine molluscan shellfish harvested from USA coastal waters. Apparently, most healthy individuals are not at risk from this bacterium, but for persons with iron metabolism abnormalities or immune system defects the mortality rate approaches 50% (Tackett *et al.*, 1984). A *V. vulnificus*-specific cytolysin has been cloned and sequenced (Yamamoto *et al.*, 1990) and used as a PCR amplification target (Hill *et al.*, 1991). Homogenised oyster meat inhibits the PCR, so several template preparation schemes were evaluated by using purified *V. vulnificus* DNA as a template, but adding increasing amounts of a *V. vulnificus*-free homogenate extract. *V. vulnificus* can enter into a viable but non-culturable phase (Oliver *et al.*, 1991) and the DNA from these cells was inefficiently amplified (Brauns *et al.*, 1991).

While most strains of *V. parahaemolyticus* isolated from nature lack the expression of haemolysin, the vast majority of patient isolates exhibit this virulence factor (Miyamoto *et al.*, 1969). However, development of PCR tests has been complicated by the fact that there exists a family of related haemolysins in this species (Nishibuchi and Kaper, 1990), although some assays have been established (Tada *et al.*, 1992).

These three important pathogenic species of the genus *Vibrio* may share an ecological niche, so there could be situations when it would be advantageous to be able to test for the three disease-causing species simultaneously. A group of three primer sets was developed by Trost *et al.* (1993) for the *ctx* gene of *V. cholerae*, the *tdh* (thermostable direct haemolysin) gene of *V. parahaemolyticus*, and the cytolysin gene of *V. vulnificus*. These primers all function under the same PCR conditions, although not in the same tube, as in a true multiplex PCR. However, the primers will amplify their respective targets during the same thermocycling run and produce fragments of sufficiently different lengths so as to be easily distinguishable by gel electrophoresis.

10.7.8 Yersinia enterocolitica and Y. pseudotuberculosis

Of the many *Yersinia enterocolitica* serotypes found in nature, only a few have been linked to diseases in humans. While virulent strains harbour a 60–70 kilobase pair plasmid, this extrachromosomal DNA can be lost during culture or storage. Therefore, plasmid-borne genes for invasion are not good candidates as sole PCR targets. Chromosomal virulence factors include *inv* and *ail,* which encode factors for the invasion of eukaryotic cells, and *yst*, a heat-stable entero-toxin (Miller *et al.*, 1989; Pierson *et al.*, 1990). The chromosomal target, *ail*, and the plasmid-borne gene, *virF*, were used by Feng *et al.* (1992) to detect virulent strains in blood by using PCR. Ibrahim *et al.* (1992) used primers for *yst* to differentiate serogroups more common to North America from those more prevalent in Europe; PCR products of different sizes were observed. Kwaga *et al.* (1992) also targeted the *ail* gene to differentiate between virulent strains of *Y. enterocolitica* and other species of *yersinia*. Primer sets for the *inv* gene of *Y. pseudotuberculosis*, the *ail* gene of *Y. enterocolitica*, and the plasmid target, *virF*, were used by Wren and Tabaqchali (1990) to develop a PCR assay specific for virulence plasmids found in both species.

10.8 PCR detection and identification of foodborne microorganisms: Viruses

About 4.5 % of the confirmed foodborne outbreaks in the USA are attributed to viruses (Bean *et al.*, 1990). Because of insufficient methods, this incidence is underestimated probably to a greater extent than is the incidence of bacterial foodborne disease (Archer and Kvenberg, 1985). Infectious doses are thought to be quite low; sensitive methods are required and many foodborne viruses cannot be propagated in laboratory tissue culture cells. While PCR offers the promise of both rapid and sensitive detection, foodborne viruses such as hepatitis A virus and Norwalk virus contain RNA as their genetic material. Therefore, reverse transcriptase coupled PCR (RT-PCR) methods must be developed.

10.8.1 Norwalk virus

Because it cannot be grown in cell culture, identification of Norwalk virus is difficult to accomplish by traditional methods. Improved methods are needed because this agent is probably the leading cause of non-bacterial gastroenteritis among adults in the USA (Kaplan *et al.*, 1982). The RT-PCR has been used to detect Norwalk virus in stools by precipitating virions with polyethylene glycol and extracting with cetyltrimethylammonium bromide (Jiang *et al.*, 1992). This same method was applied to the detection of rotavirus and hepatitis A virus RNA from shellfish (Zhou *et al.*, 1991). Sensitivity was judged to be between 10^2 and 10^4 virus particles. Atmar *et al.* (1993) extracted viral RNA from oysters and used an RT-PCR targeted to the viral polymerase gene. Sensitivity was claimed to be about 50–500 virus particles in seeded oysters and 1.5–15 particles in stool samples.

10.8.2 Rotavirus

RT-PCR was used to amplify the major outer capsid glycoprotein gene (*vp7*) to detect rotavirus in faecal samples from humans (Gouvea *et al.*, 1990). Primer pairs were targeted to type-specific regions within the *vp7* gene, whereby each viral serotype produced a characteristic length fragment. Sensitivity was about 10^8 viral genomes, but when rotaviral RNA was bound to a CF11 cellulose column, apparent inhibitors were removed and the sensitivity was increased more than a thousand fold (Wilde *et al.*, 1990). De Leon *et al.* (1992) removed inhibitors by Sephadex G-200 gel chromatography and used primers to amplify the RNa polymerase gene and a putative immunogenic protein. About 10^6 virions could be detected when the PCR product was hybridised by a digoxigenin-labelled internal oligonucleotide.

10.8.3 Hepatitis A virus (HAV)

Shellfish, if grown in contaminated waters and consumed raw, can easily spread HAV to humans *(Desenclos* et al., 1991). HAV is readily passed from human to human because it is easily transmitted by the faecal–oral route; it can survive for extended periods in artificially contaminated environments, and the communicable period precedes the onset of clinical symptoms. Viral particles were captured by an anti-HAV monoclonal antibody and the viral RNA amplified by a RT-PCR (Jansen *et al.*, 1990). Sensitivity was estimated to be 3–30 viral particles. HAV was not detectable after being seeded into whole oysters, but was detected after being spiked into oyster extracts (Atmar *et al.*, 1993). It was determined that HAV particles were removed during the flocculation step, although this method was acceptable for the detection of Norwalk virus. To test oyster homogenates and environmental samples from a HAV outbreak, Desenclos *et al.* (1991) used antigen capture and RT-PCR. All ELISA positive samples were also positive with the RT-PCR assay, but two ELISA negative samples were RT-PCR positive, suggesting that the latter method may be more sensitive.

10.8.4 Enteroviruses

The RNA-containing enterovirus group is diverse and its members are responsible for a number of childhood infections. The availability of nucleotide sequence data for several serotypes has allowed the development of RT-PCR methods (Rotbart, 1990). Conserved regions of the 5' end were targeted and sensitivity is estimated to be of the order of 10^6 molecules. However, it is important to note that the time required to identify these viruses has been shortened from 4–8 days to less than 2 days by using RT-PCR.

10.9 PCR detection and identification of foodborne microorganisms: Parasites

In developing countries, protozoal parasites are a major cause of disease and the rapid detection of these organisms is needed to minimise the public health consequences. Even though several DNA hybridisation probes have been developed, PCR-based methods promise to be more sensitive (Baker, 1990). While most of the disease is caused by consumption of contaminated water, it is possible that most foodborne parasitic diseases may be overlooked at present because of insensitive methods.

10.9.1 Giardia

In the USA, *Giardia duodenalis* and *G. lamblia* are the most frequent causes of waterborne gastroenteritis. By using giardin (a cytoskeletal protein of the ventral disc) as a target for amplification, Mahbubani *et al.* (1991) developed a PCR method. They also developed an RT-PCR test for giardin mRNA as an amplification template, which increased the yield of PCR product. A different region of the giardin gene was used as a *G. duodenalis*-specific amplification target (Mahbubani *et al.*, 1992) and sensitivity was as low as 1 cyst per 100 ml of spiked water, but false-negative results were obtained with 400 l sample volumes.

10.9.2 Entamoeba

About half a billion persons are infected with the enteric protozoan *Entamoeba histolytica* and it may cause as many as 70 000 deaths per year. Tannich and Burchard (1991) discovered that restriction endonuclease cleavage of PCR amplified products could distinguish between pathogenic and nonpathogenic strains with a sensitivity of fewer than 10 cells.

10.10 Summary

PCR is being used to detect a wide range of microorganisms in an array of matrices. While the basic mechanism of PCR is relatively straightforward, the use of this technique with hostile samples requires some ingenuity to create efficient template preparation schemes. The applications of PCR will continue to increase as new primer sets are developed and new techniques are used for the preparation of inhibitor-free DNAs to allow the efficient amplification of specific genetic segments. While PCR currently enjoys use in the research laboratory, it is not in routine use for food testing. As the PCR technology is transferred to more testing laboratories, we will see its use increase sharply, first as a screening tool and then, perhaps, as a determinative method.

Acknowledgments

The authors thank Peter Feng, Marleen Wekell, Karen Jinneman, Ronald

Manger, Charles Kaysner and Ann Adams of FDA and Kaye Wachsmuth of CDC for their helpful comments and criticisms.

References

Archer, D.L. and Kvenberg, J.E. (1985) Incidence and cost of foodborne diarrhoeal disease in the United States. *J. Food Prot.*, **48**, 887–894.

Archer, D.L. and Young, F.E. (1988) Contemporary issues: diseases with a food vector. *Clin. Microbiol. Rev.*, **1**, 377–398.

Arnheim, N. and Levenson, C.H. (1990) Polymerase chain reaction. *Chem. Engng News*, 1 oct. 36–47.

Atmar, R.L., Metcalf, T.G., Neill, F.H. and Estes, M.K. (1993) Detection of enteric viruses in oysters by using the polymerase chain reaction. *Appl. Environ. Microbiol.*, **59**, 631–635.

Baker, R.B., Jr. (1990) DNA probe diagnosis of parasitic infections. *Exp. Parasitol.*, **70**, 494–499.

Bean, N.H., Griffin, P.M., Golding, J.S. and Ivey, C.B. (1990) Foodborne disease outbreaks, 5-year summary, 1983–1987. *Morb. Mortal. Weekly Rep.*, **39**, 15–57.Bej, A.K., Steffan, R.J., DiCesare, J., *et al.* (1990) Detection of coliform bacteria in water by polymerase chain reaction and gene probes. *Appl. Environ. Microbiol.*, **56**, 307–314.

Bej, A.K., McCarty, S.C. and Atlas, R.M. (1991) Detection of coliform bacteria and *Escherichia coli* by multiplex polymerase chain reaction: comparison with defined substrate and plating methods for water quality monitoring. *Appl. Environ. Microbiol.*, **57**, 2429–2432.

Beutler, E., Gelbart, T. and Kuhl, W. (1990) Interference of heparin with the polymerase chain reaction. *BioTechniques*, **9**, 166.

Border, P.M., Howard, J.J., Plastow, G.S. and Siggens, K.W. (1990) Detection of *Listeria* species and *Listeria monocytogenes* using polymerase chain reaction. *Lett. Appl. Microbiol.*, **11**, 158–162.

Brian, M.J., Frosolono, M., Murray, B.E. *et al.*, (1992) Polymerase chain reaction for diagnosis of enterohaemorrhagic *Escherichia coli* infection and hemolytic-uremic syndrome. *J. clin. Microbiol.*, **30**, 1801–1806.

Bubert, A., Kohler, A. and Goebel, W. (1992) The homologous and heterologous regions within the *iap* gene allow genus- and species-specific identification of *Listeria* spp. by polymerase chain reaction. *Appl. Environ. Microbiol.*, **581**, 2625–2632.

Buffone, G.J., Demmler, G.J., Shimbor, C.M. and Greer, J. (1991) Improved amplification of cytomegalovirus DNA from urine after purification of DNA with glass beads. *Clin. Chem.*, **37**, 1945–1949.

Candrian, U., Furrer, B., Höfelein, C. *et al.* (1991) Detection of *Escherichia coli* and identification of enterotoxigenic strains by primer-directed enzymatic amplification of specific DNA sequences. *Int. J. Food Microbiol.*, **12**, 339–351.

Candrian, U., Hofelein, C. and Luthy, J. (1992) Polymerase chain reaction with additional primers allows identification of amplified DNA and recognition of specific alleles. *Molec. Cell. Probes*, **6**, 13–19.

Centers for Disease control (1992) Summary of notifiable diseases, United States, 1991. *Morb. Mortal. Weekly Rep.*, **40**, 3.

Cossart, P., Vincente, M.F., Mengaud, J., *et al.* (1989) Listeriolysin O is essential for virulence of *Listeria monocytogenes*: direct evidence obtained by gene complementation. *Infect. Immunol.*, **57**, 3629–3936.

Cudjoe, K.S., Patel, P.D., Olsen, E., *et al.* (1993) Application of immunomagnetic separation techniques to detect pathogenic bacteria in foods, in *Application of New Techniques in Food and Beverage Microbiology* (eds Kroll, R.G., Gilmour, A. and Sussman, M.). Academic Press, London, pp. 17–31.

De Francis, R., Cross, N.C.P., Foulkes, N.S. and Cox. T.M. (1988) A potent inhibitor of *Taq* polymerase co-purifies with human genomic DNA. *Nucleic Acids Res.*, **16**, 10355.

De Leon, R., Matsui, S.M., Baric, R.S., *et al.* (1992) Detection of Norwalk virus in stool specimens by reverse transcriptase-polymerase chain reaction and nonradioactive oligoprobes. *J. Clin. Microbiol.*, **30**, 3151–3157.

Deneer, H.G. and Boychuk, I. (1991) Species-specific detection of *Listeria monocytogenes* by DNA amplification. *Appl. Environ. Microbiol.*, **57**, 606–609.

Desenclos, J.-C.A., Klontz, K.C., Wilder, M.H., *et al.* (1991) A multistate outbreak of hepatitis A caused by the consumption of raw oysters. *Am. J. Public Health*, **81**, 1268–1272.

Dorsch, M., Lane, D. and Stackebrandt, E. (1992) Towards a phylogeny of the genus *Vibrio* based on 16S rRNA sequences. *Int. J. Syst. Bacteriol.*, **42**, 58–63.

Doyle, M.P. (1991) *E.coli* O157:H7 and its significance in foods. *Int. J. Food Microbiol*, **12**, 289–301.

Erlich, H.A., Gelfand, D. and Sninsky, J.J. (1991) Recent advances in the polymerase chain reaction. *Science*, **252**, 1643–1651.

Farber, J. and Peterkin, P. (1991) *Listeria monocytogenes*, a food-borne pathogen. *Microbiol. Rev.*, **55**, 476–511.

Feng, P., Keasler, S. and Hill, W.E. (1992) Direct identification of *Yersinia enterocolitica* in blood by using polymerase chain reaction amplification. *Transfusion* **32**, 850–854.

Ferreira, J.L., Hamdy, M.K., McCay, S.G. and Baumstark, B.R. (1992) An improved assay for identification of type A *Clostridium botulinum* using the polymerase chain reaction. *J.Rapid Methods Automat. Microbiol.*, **1**, 29–39.

Fields, P.I., Popovic, T., Wachsmuth, K. and Olsvik, O (1992) Use of polymerase chain reaction for detection of toxigenic *Vibrio cholerae* O1 strains from the Latin American cholera epidemic. *J. Clin. Microbiol*, **30**, 2118–2121.

Fitter, S., Heuzenroeder, M. and Thomas, C.J. (1991) A combined PCR and selective enrichment method for rapid detection of *Listeria monocytogenes. J. Appl. Bacteriol.*, **73**, 53–59.

Furrer, B., Candrian, U. and Lüthy, J. (1990) Detection and identification of *E.coli* producing heat-labile enterotoxin type 1 by enzymatic amplification of a specific DNA fragment. *Lett. Appl. Microbiol.*, **10**, 31–34.

Furrer, B., Candrian, U., Höfelein, C. and Lüthy, J. (1991) Detection and identification of *Listeria monocytogenes* in cooked sausage products and in milk by *in vitro* amplification of haemolysin gene fragments. *J. Appl. Bacteriol.*, **70**, 372–379.

Frankel, G., Giron, J.A. Valmasoi, J. and Schoolnik, G.K. (1989) Multi-gene amplification: simultaneous detection of three virulence genes in diarrhoeal stool. *Molec. Microbiol.*, **3**, 1729–1734.

Fratamico, P.M., Schultz, F.J. and Buchanan., R.L. (1992) Rapid isolation of *Escherichia coli* O157:H7 from enrichment cultures of foods using an immunomagnetic separation method. *Food Microbiol.*, **9**, 105–113.

Gannon, V.P.J., King, R.K., Kim, J.Y. and Golsteyn Thomas, E.J. (1992) Rapid and sensitive method for detection of Shiga-like toxin-producing *Escherichia coli* in ground beef using the polymerase chain reaction. *Appl. Environ. Microbiol.*, **58**, 3809–3815.

Giesendorf, B.A.J., Quint, W.G.V., Henkens, M.H.C., *et al.* (1992) Rapid and sensitive detection of *Campylobacter* spp. in chicken products by using the polymerase chain reaction. *Appl. Environ. Microbiol.*, **58**, 3804–3808.

Golsteyn Thomas, E.J., King, R.K., Burchak, J. and Gannon, V.P.J. (1991) Sensitive and specific detection of *Listeria monocytogenes* in milk and ground beef with the polymerase chain reaction. *Appl. Environ. Microbiol.*, **57**, 2576–2580.

Gouvea, V., Glass, R.I., Woods, P., *et al.* (1990) Polymerase chain reaction amplification and typing of rotavirus nucleic acid from stool specimens. *J. Clin. Microbiol.*, **28**, 276–282.

Grünstein, M. and Hogness, D.S. (1975) Colony hybridisation: method for the isolation of cloned DNAs that contain a specific gene. *Proc. Nat. Acad. Sci. USA*, **72**, 3961–3965.

Gurtler, V., Wilson, V.A. and Mayall, B.C. (1991) Classification of medically important clostridia using restriction endonuclease site differences of PCR-amplified 16S rRNA. *J.Gen. Microbiol.*, **137**, 2673–2679.

herman, L. and De Ridder, H. (1992) DNA hybridization and amplification techniques to accelerate the identification of *Listeria monocytogenes. Milchwissenschaft*, **47**, 354–357.

Hill, W.E. and Carlisle, C.L. (1981) Loss of plasmids during enrichment for *Escherichia coli Appl. Environ. Microbiol.*, **41**, 1046–1048.

Hill, W.E., Keasler, S.P., Trucksess, M.W., *et al.* (1991) Polymerase chain reaction identification of *Vibrio vulnificus* in artificially contaminated oysters. *Appl. Environ. Microbiol.*, **57**, 707–711.

Hill, W.E., Jinneman, K.C., Trost, P.A., *et al.* (1993) Multiplex polymerase chain reaction detection of Shiga-like toxin genes in *Escherichia coli. FDA Lab. Inf. Bull.*, Submitted.

Hood, A.M., Pearson, A.D. and Shahamat, M. (1988) The extent of surface contamination of retailed chickens with *Campylobacter jejuni* serogroups. *Epidemiol. Infect.*, **100**, 17–25.

Hornes, E., Wasteson, Y. and Olsvik, O. (1991) Detection of *Escherichia coli* heat-stable enterotoxin genes in pig stool specimens by an immobilized, colorimetric nested polymerase chain reaction. *J. Clin. Microbiol.*, **29**, 2375–2379.

Ibrahim, A., Liesack, W., Pike, S. and Stakebrandt, E. (1992) the polymerase chain reaction: an epidemiological tool to differentiate between two clusters of *yersinia enterocolitica* strains. *FEMS Microbiol. Lett.*, **97**, 63–66.

Innis, M.A., Gelfand, D.H., Sninsky, J.J. and White, T.J. (1990) *PCR Protocols, A Guide to Methods and Applications.* Academic Press, New York.

Islam, D. and Lindberg, A.A. (1992) Detection of *Shigella dysenteriae* type 1 and *Shigella flexneri* in feces by immunomagnetic isolation and polymerase chain reaction. *J. Clin. Microbiol.*, **30**, 2801–2806.

Islam, D., Tzipori, S., Islam, M. and Lindberg, A.A> (1993) Rapid isolation of *Shigella dysenteriae* and *Shigella flexneri* in faeces by O-antigen specific monoclonal antibody coated immunomagnetic beads. *Eur. J. Clin. Microbiol. Infect. Diseases*, **12**, 25–32.

Jackson, M.P. (1991) Detection of Shiga toxin-producing *Shigella dysenteriae* type 1 and *Escherichia coli* by using polymerase chain reaction with incorporation of digoxigenin-11–dUTP. *J. Clin. Microbiol.*, **29**, 1910–1914.

Jansen, R.W., Siegl, G. and Lemon, S.M. (1990) Molecular epidemiology of human hepatitis A virus defined by an antigen-capture polymerase chain reaction method. *Proc. Nat. Acad. Sci. USA*, **87**, 2867–2871.

Jaton, K., Sahli, R. and Bille, J. (1992) Development of polymerase chain reaction assays for detection of *Listeria monocytogenes* in clinical cerebrospinal fluid samples. *J. Clin. Microbiol.*, **30**, 1931–1936.

Jiang, X., Estes, M.K. and Metcalf, T.G. (1987) Detection of hepatitis A virus by hybridization with single-stranded RNA probes. *Appl. Environ. Microbiol.*, **53**, 2487–2495.

Johnson, W.M., Tyler, S.D., Ewan, E.P., *et al.* (1991) Detection of genes for enterotoxins, exfoliative toxins, and toxic shock syndrome toxin 1 in *Staphylococcus aureus* by the polymerase chain reaction. *J. Clin. Microbiol.*, **29**, 426–430.

Kaplan, J.E., Feldman, R., Campbell, D.S., *et al.* (1982) The frequency of a Norwalk-like pattern of illness in outbreaks of acute gastroenteritis. *Am. J. Public Health*, **72**, 1329–1332.

Kapperud, G., Vardun, T., Skjerve, E., *et al.* (1994) Detection of pathogenic *Yersinia enterocolitica* in food and water by immunomagnetic separation, nested polymerase chain reactions, and colorimetric detection of amplified DNA. *Appl. Environ. Microbiol.* (submitted).

Karch, H., and Meyer, T. (1989) Single primer pair for amplifying segments of distinct Shiga-like toxin genes by polymerase chain reaction. *J. Clin. Microbiol.*, **27**, 2751–2757.

Keasler, S.P. and Hill, W.E. (1992) Polymerase chain reaction identification of enteroinvasive *Escherichia coli* seeded into raw milk. *J. Food Prot.*, **55**, 382–384.

Koch, W.H., Paynes, W.L., Wentz, B.A. and Cebula, T.A. (1993) Rapid polymerase chain reaction method for detection of *Vibrio cholerae* in foods. *Appl. Environ. Microbiol.*, **59**, 556–560.

Kwaga, J., Iversen, J.O. and Misra, V. (1992) Detection of pathogenic *Yersinia enterocolitica* by polymerase chain reaction and digoxigenin-labelled polynucleotide probes. *J. Clin. Microbiol.*, **30**, 2668–2673.

Lampel, K.A., Jagow, J.A., Trucksess, M. and Hill, W.E. (1990) Polymerase chain reaction for detection of invasive *Shigella flexneri* in food. *Appl. Environ. Microbiol.*, **56**, 1536–1540.

Linnan, J.J., Mascola, L., Lou, X.D., *et al.* (1988) Epidemic listeriosis associated with Mexican-style cheese. *New Eng. J. Med.*, **319**, 823–828.

Luk, J.M.C. and Lindberg, A.A. (1991) Rapid and sensitive detection of *Salmonella* (O:6,7) by immunomagnetic monoclonal antibody-based assays. *J. Immunol. Meth.*, **137**, 1–8.

Lund, A., Hellemann, A.L. and vartdal, F. (1988) Rapid isolation of K88+ *Escherichia coli* by using immunomagnetic particles. *j. Clin. Microbiol.*, **26**, 2572–2575.

Lund, A., Wasteson, Y. and Olsvik, O. (1991) Immunomagnetic separation and DNA hybridization for detection of enterotoxigenic *Escherichia coli* in a piglet model. *J. Clin. Microbiol.*, **29**, 2259–2262.

Lundeberg, J., Wahlberg, J.M., Holmberg *et al.* (1990) Rapid colorimetric detection of *in vitro* amplified DNA sequences. *DNA Cell. Biol.*, **9**, 287–292.

Mahbubani, M.H., Bej, A.K., Perlin, M.H. *et al.* (1991) Detection of *Giardia* cysts by using the polymerase chain reaction and distinguishing live from dead cysts. *Appl. Environ. Microbiol.*, **57**, 3456–3461.

Mahbubani, M.H., Bej, A.K., Perlin, M.H. *et al.* (1992) Differentiation of *Giardia duodenalis* from other *Giardia* spp. by using polymerase chain reaction and gene probes. *J. Clin. Microbiol.*, **30**, 74–78.

Mengaud, J., Vincente, M.F., Chenevert, J., *et al.* (1988) expression in *Escherichia coli* and sequence analysis of the listeriolysin O determinant of *Listeria monocytogenes*. *Infect. Immunol.*, **56**, 766–772.

Meyer, R., Luthy, J. and Candrian, U. (1991) Direct detection by polymerase chain reaction (PCR) of *Escherichia coli* in water and soft cheese and identification of enterotoxigenic strains. *Lett. Appl. Microbiol.*, **13**, 268–271.

Miller, V.L., Farmer, J.J., III, Hill, W.E. and Falkow, S. (1989) The *ail* locus is found uniquely in *Yersinia enterocolitica* serotypes commonly associated with disease. *Infect. Immun.*, **57**: 121–131.

Miyamoto, Y., Kato, T., Obara, Y., *et al.* (1969) *In vitro* hemolytic characteristic of *Vibrio parahaemolyticus*: its close correlation with human pathogenicity. *J. Bacteriol.*, **100**, 1147–1149.

Muir, P., Nicholson, F., Jhetam, M., *et al.* (1993) Rapid diagnosis of enterovirus infection by magnetic bead extraction and polymerase chain reaction detection of enterovirus RNA in clinical specimens. *J. Clin. Microbiol.*, **31**, 31–38.

Mullis, K.B. (1989) The polymerase chain reaction: why it works, in *Polymerase Chain Reaction* (eds. Erlich, H.A., Gibbs, R. and Kazazian, H.H., Jr.) Cold Spring Harbor Laboratory Press, Cold Spring Harbor, NY, pp.237–243.

Nishibuchi, M. and Kaper, J.B. (1990) Duplication and variation of the thermostable direct haemolysin *(tdh)* gene in *Vibrio parahaemolyticus*. *Molec. Microbiol.*, **4**, 87–99.

Nustad, K., Paus, E. and Bömer, O.P. (1991) Magnetic particles as solid phase in immunoassays, in *Magnetic Separation Techniques Applied to Cellular and Molecular Biology*, (ed. Kemshead, J.T.). Wordsmiths Conference Publications, Somerset, pp. 39–46.

Okrend, A.J.G., Rose, B.E. and Lattuada, C.P. (1992) Isolation of *Escherichia coli* O157:H7 using O157 specific antibody coated magnetic beads. *J. Food Prot.*, **55**, 214–217.

Olive, D.M. (1989) Detection of enteropathogenic *Escherichia coli* after polymerase chain reaction amplification with a thermostable DNA polymerase. *J. Clin. Microbiol.*, **27**, 261–265.

Olive, D.M., Atta, A.I. and Sethi, S.K. (1988) Detection of toxigenic *Escherichia coli* using biotin-labelled DNA probes following enzymatic amplification of heat labile toxin gene. *Molec. Cell. Probes.*, **2**, 47–57.

Oliver, J.D., Nilsson, L. and Kjellberg, S. (1991) formation of nonculturable *Vibrio vulnificus* cells and its relationship to the starvation state. *Appl. Environ. Microbiol.*, **57**, 2640–2644.

Olsvik, Ø., Hornes, E., Wasteson, Y. and Lund, A. (1991) detection of virulence determinants in enteric *Escherichia coli* using nucleic acid probes and polymerase chain reaction, in *Molecular Pathogenesis of Gastrointestinal Infections* (eds. Wadström, T., Mäkelä, P.H., Svennerholm, A.-M and Wolf-Watz, H.). Plenum Press, New York, pp. 267–272.

Olsvik, Ø., Popovic, T. and Fields, P.I. (1993) Polymerase chain reaction for detection of toxin genes in strains of *Vibrio cholerae*, in *Diagnostic Molecular Microbiology: Principles and Applications* (eds. Persing, D.H., Tenover, F.C., Smith, T.F. and White, T.J.) American Society for Microbiology, Washington, DC, pp. 266–270.

Olsvik, Ø., Popovic, T., Skjerve, E., *et al.* Magnetic separation techniques in clinical microbiology. *Clin. Microbiol. Rev.* (in press).

Olsvik, Ø., Rimstad, E., Hornes, E. *et al.* (1991) A nested PCR followed by magnetic separation of amplified fragments for detection of *Escherichia coli* Shiga-like toxin genes. *Molec. Cell. Probes*, **5**, 429–435.

Olsvik, Ø. and Strockbine, N.A. (1993) Polymerase chain reaction for detection of heat-stable, heat-labile and Shiga-like toxin genes in *Escherichia coli*, in *Diagnostic Molecular Microbiology: Principles and Applications*. (eds Persing, D.H., Tenover, F.C., Smith, T.F. and White, T.J.). American Society for Microbiology, Washington, DC, pp. 271–276.

Olsvik, Ø., Wahlberg, J., Petterson, B., *et al.* (1993) Use of automated sequencing of PCR-generated amplicons to identify three types of cholera toxin subunit B on *Vibrio cholerae* O1 strains. *J. Clin. Microbiol.* **31**, 22–25.

Olsvik, Ø., Wasteson, Y., Lund, A. and Hornes, E. (1991). Pathogenic *Escherichia coli* found in food. *Int. J. Food Microbiol.*, **12**, 103–114.

Oyofo, B.A., Thornton, S.A., Burr, D.H., *et al.* (1992) Specific detection of *Campylobacter jejuni* and *Campylobacter coli* by using polymerase chain reaction. *J. Clin. Microbiol.*, **30**, 2613–2619.

Pierson, D.E. and Falkow, S. (1990) Nonpathogenic isolates of *yersinia enterocolitica* do not contain functional *inv*-homologous sequences. *Infect. Immunol.* **58**, 1059–1064.

Popovic, T., Olsvik, Ø., Blake, P.A. and Wachsmuth, I.K. (1993) Cholera in the Americas, food borne aspects. *Int. J. Food Prot.*, **56**, 811–821.

Roberts, T. (1989) Human illness costs of foodborne bacteria. *Am. J. Agric. Economics*, **71**, 468–474.

Rossen, L., Holmström, K., Olsen, J.E. and Rasmussen, O.F. (1991) A rapid polymerase chain reaction (PCR)-based assay for the identification of *Listeria monocytogenes* in food samples. *Int. J. Food Microbiol*, **14**, 145–152.

Rotbart, H.A. (1990) Enzymatic RNA amplification of the enteroviruses. *J. Clin. Microbiol.*, **28**, 438–442.

Saiki, R.K. (1989) The design and optimization of the PCR, in *PCR Technology, Principles and Applications for DNA Amplification* (ed. Erlich, H.A.). Stockton Press, New York, pp. 7–16.

Saiki, R.K., Scarf, S., Faloona, F.A., *et al.* Arnheim, N. (1985) Enzymatic amplification of B-globin sequences and restriction site analysis for diagnosis of sickle cell anemia. *Science*, **230**, 1350–1354.

Saiki, R.K., Gelfand, D.H., Stoffel, *et al.* (1988) Primer-directed enzymatic amplification of DNA with a thermostable DNA polymerase. *Science*, **239**, 487–494.

Schuchat, A., Swaminathan, B. and Broome, C.V. (1991) Epidemiology of human listeriosis. *Clin. Microbiol. Rev.*, **4**, 169–183.

Shirai, J., Nishibuchi, M., Ramamurthy, T., *et al.* (1991) Polymerase chain reaction for detection of the cholera enterotoxin operon of *Vibrio cholerae*. *J.Clin. Microbiol.*, **29**, 2517–2521.

Skjerve, E. and Olsvik, Ø. (1991) Immunomagnetic separation of *Salmonella* from foods. *Int. J. Food Microbiol.*, **14**, 11–18.

Skjerve, E., Rørvik, L.M. and Olsvik, Ø. (1990) Detection of *Listeria monocytogenes* in food by immunomagnetic separation. *Appl. Environl. Microbiol.*, **56**, 3478–3481.

Skjerve, E., Hornes, E., Cudjoe, K.S. *et al.* (1994) Detection of *Salmonella* in milk powder using immunomagnetic separation followed by colorimetric detection of a polymerase chain amplified virulence associated gene. *Appl. Environ. Microbiol.* (in press).

Synder, O.P. (1992) HACCP – an industry food safety self-control program. Part IV. *Dairy Food Environ.*, 230–232.

Szabo, E.A., Pemberton, J.M, and Desmarchelier, P.M. (1992) Specific detection of *Clostridium botulinum* Type B by using the polymerase chain reaction. *Appl. Environ. Microbiol.*, **58**, 418–420.

Tackett, C.O., Brenner, F. and Blake, P.A. (1984) Clinical features and an epidemiological study of *Vibrio vulnificus* infections. *J. Infect. Diseases*, **149**, 558–561.

Tada, J., Ohashi, T., Nishimura, N., *et al.* (1992) Detection of the thermostable direct hemolysin gene (tdh) and the thermostable direct hemolysin-related hemolysin gene (*trh*) of *Vibrio parahaemolyticus* by polymerase chain reaction. *Molec. Cell. Probes*, **6**, 477–487.

Tannich, E. and Burchard, G.D. (1991) Differentiation of pathogenic from nonpathogenic *Entamoeba histolytica* by restriction fragment analysis of a single gene amplified *in vitro*. *J. Clin. Microbiol.*, **29**, 250–255.

Tauxe, R.V. and Blake, P.A. (1992) Epidemic cholera in Latin America. *J. am. Med. Assoc.*, **267**, 1388–1390.

Todd, E.C.D. (1989) Preliminary estimates of costs of foodborne disease in the United States. *J. Food Prot.*, **52**, 595–601.

Trost, P.A., Hill, W.E., Kaysner, C.A. and Wekell, M.M. (1993) Detection of three pathogenic *Vibrio* species by using the polymerase chain reaction. *FDA Lab. Info. Bull.*, No. 3733, January.

Ugelstad, J., Berge, A., Ellingsen, T., *et al.* (1992) Preparation and application of new monosized polymer particles. *Prog. Polym. Sci.*, **17**, 87–161.

Uhlén, M. (1989) Magnetic separation of DNA. *Nature*, **340**, 733–734.

Uhlén, M., Lundberg, J. and Wahlberg, J. (1990) DNA diagnosis using the polymerase chain reaction, in *Application of Molecular Biology in Diagnosis of Infectious Diseases*, (ed. Olsvik, Ø. and Bukholm, G.). Norwegian College of Veterinary Medicine, Oslo, pp. 86–90.

Uhlén, M., Olsvik, Ø. and Hornes, E. (1993) Affinity separation of nucleic acids on monosized magnetic beads, in *Molecular Interactions in Bioseparation*, (ed. Ngo, T.T.) Plenum Press, New York (in press)

Wernars, K., Heuvelman, C.J., Chakraborty, T. and Notermans, S.H.W. (1991) Use of the polymerase chain reaction for direct detection of *Listeria monocytogenes* in soft cheese. *J. Appl. Bacteriol.*, **70**, 121–126.

Widjojoatmodjo, M.N., Fluit, A.C., Torensma, R., *et al.* (1991) Evaluation of the magnetic immuno

PCR assay for rapid detection of *Salmonella*. *Eur. J. Clin. Microbiol.*, **10**, 935–938.

Wiedmann, M., Czajka, J., Barany, F. and Batt, C.A. (1992) Discrimination of *Listeria monocytogenes* from other *Listeria* species by ligase chain reaction. *Appl. Environ. Microbiol.*, **58**, 3443–3447.

Wilde, J., Eiden, J. and Yoken, R. (1990) Removal of inhibitory substances from human fecal specimens for detection of group A rotaviruses by reverse transcriptase and polymerase chain reactions. *J. Clin. Microbiol.*, **28**, 1300–1307.

Wilson, I.G., Cooper, J.E. and Gilmour, A. (1991) Detection of enterotoxigenic *Staphylococcus aureus* in dried skimmed milk: use of the polymerase chain reaction for amplification and detection of staphylococcal enterotoxin genes *entB* and *entC1* and the thermonuclease gene *Nuc*. *Appl. Environ. Microbiol.*, **57**, 1793–1798.

Wolcott, M.J. (1992) Advances in nucleic acid-based detection methods. *Clin. Microbiol. Rev.*, **5**, 370–386.

Wren, B.W. and Tabaqchali, S. (1990) Detection of pathogenic *Yersinia enterocolitica* by the polymerase chain reaction. *Lancet*, **336**, 693.

Yamamoto, K., Wright, A.C., Kaper, J.B. and Morris, J.G., Jr. (1990) The cytolysin gene of *Vibrio vulnificus*: sequence and relationship to the *Vibrio cholerae* El Tor hemolysin gene. *Infect. Immunol.*, **58**, 2706–2709.

Zhou, Y.-J., Estes, M.K., Jiang, X. and Metcalf, T.G. (1991) Concentration and detection of hepatitis A virus and rotavirus from shellfish by hybridization tests. *Appl. Environ. Microbiol.*, **57**, 2963–2968.

Index